RESEARCH DATA MANAGEMENT IN THE ECOLOGICAL SCIENCES

THE BELLE W. BARUCH LIBRARY IN MARINE SCIENCE NUMBER 16

Research Data Management in the Ecological Sciences

Edited by
William K. Michener

Published for the Belle W. Baruch Institute for Marine Biology

and Coastal Research by the

UNIVERSITY OF SOUTH CAROLINA PRESS

Published in Columbia, South Carolina, by the
University of South Carolina Press

First Edition

Manufactured in the United States of America

Library of Congress Cataloging-in-Publication Data
Main entry under title:

Research data management in the ecological sciences.

   (The Belle W. Baruch library in marine science ;
no. 16)
   Papers from a symposium held Nov. 4-6, 1984 at
Hobcaw Barony, Georgetown, S.C.
   Includes index.
   1. Ecology--Data processing--Congresses.
I. Michener, William K. II. Belle W. Baruch Institute
for Marine Biology and Coastal Research. III. Series.
QH541.15.E45R47 1986   574.5'072   85-26557
ISBN 0-87249-476-4

LIST OF CONTRIBUTORS

P.B. ALABACK, Department of Forest Science, College of Forestry, Oregon State University, Corvallis, OR 97331-5704.

MATTHEW P. ANDERSON, Department of Psychology, Computing Research Laboratory, New Mexico State University, Las Cruces, NM 88003

J.L. BELL, Mathematics Department, Colorado School of Mines, Golden, CO 80401.

JACKSON O. BLANTON, Skidaway Institute of Oceanography, P.O. Box 13687, Savannah, GA 31416.

CARL J. BOWSER, Department of Geology and Geophysics, and Center for Limnology, University of Wisconsin, Madison, WI 53706.

ELGENE O. BOX, Geography Department, University of Georgia, Athens, GA 30602.

CRAIG BRANDT, Science Applications International Corporation, Oak Ridge, TN 37830.

FRANK M. BROOKFIELD, Illinois Natural History Survey, Champaign, IL 61820.

CHARLES COMISKEY, Science Applications International Corporation, Oak Ridge, TN 37830.

WALT CONLEY, Department of Biology, Computing Research Laboratory, New Mexico State University, Las Cruces, NM 88003

G.B. CUNNINGHAM, Coweeta Hydrologic Laboratory, Otto, NC 28763.

vi

MELVIN I. DYER, Environmental Sciences Division, Oak Ridge National Laboratory, Oak Ridge, TN 37831.

TERRELL FARMER, Science Applications International Corporation, Oak Ridge, TN 37830.

ROBERT R. FREEMAN, Deputy Director, Assessment and Information Services Center, National Environmental Satellite, Data, and Information Service, National Oceanic and Atmospheric Administration, Washington, DC 20235.

MARTIN E. GURTZ, Division of Biology, Ackert Hall, Kansas State University, Manhattan, KS 66506.

DONALD A. HOLZWORTH, Program Resources, Inc., 703 Giddings Avenue, Suite M4, Annapolis, MD 21401.

PAUL KANCIRUK, Environmental Sciences Division, Oak Ridge National Laboratory, Oak Ridge, TN 37831.

THOMAS B. KIRCHNER, Natural Resource Ecology Laboratory, Colorado State University, Fort Collins, CO 80523.

V. KLEMAS, Center for Remote Sensing, College of Marine Studies, University of Delaware, Newark, DE 19716.

M.W. KLOPSCH, Department of Forest Science, College of Forestry, Oregon State University, Corvallis, OR 97331-5704.

V. KOMARKOVA, Institute of Arctic and Alpine Research and Department of Environmental, Organismic and Population Biology, University of Colorado, Boulder, CO 80309.

M.F. MAROZAS, Belle W. Baruch Institute for Marine Biology and Coastal Research, University of South Carolina, Georgetown, SC 29442.

G. RICHARD MARZOLF, Division of Biology, Kansas State University, Manhattan, KS 66506.

R.A. MCCORD, Science Applications International Corporation, Oak Ridge, TN 37830.

WILLIAM MICHENER, Belle W. Baruch Institute for Marine Biology and Coastal Research, University of South Carolina, Columbia, SC 29208.

RICHARD J. OLSON, Environmental Sciences Division, Oak Ridge National Laboratory, Oak Ridge, TN 37831.

PAUL G. RISSER, Illinois Natural History Survey, Champaign, IL 61820.

CHANNING H. RUSSELL, Vice-President, BBN Software Products Corporation, Cambridge, MA 02138.

FRANKLIN B. SCHWING, Skidaway Institute of Oceanography, P.O. Box 13687, Savannah, GA 31416.

STEVEN K. SEILKOP, Program Resources, Inc., 703 Giddings Avenue, Suite M4, Annapolis, MD 21401; and Program Resources, Inc., P.O. Box 12794, Research Triangle Park, NC 27709.

RICHARD A. SITZE, Department of Computer Science, Computing Research Laboratory, New Mexico State University, Las Cruces, NM 88003

R.L. SLAGLE, Department of Forest Science, College of Forestry, Oregon State University, Corvallis, OR 97331-5704.

BRIAN M. SLATOR, Department of Computer Science, Computing Research Laboratory, New Mexico State University, Las Cruces, NM 88003

S.G. STAFFORD, Department of Forest Science, College of Forestry, Oregon State University, Corvallis, OR 97331-5704.

L.W. SWIFT, JR., Coweeta Hydrologic Laboratory, Otto, NC 28763.

COLIN G. TREWORGY, Illinois Natural History Survey, Champaign, IL 61820.

K.L. WADDELL, Department of Forest Science, College of Forestry, Oregon State University, Corvallis, OR 97331-5704.

K.C. ZINNEL, Ecology and Behavioral Biology Department, University of Minnesota, Minneapolis, MN 55455.

PREFACE

Collecting, managing, and analyzing research data requires a large expenditure of time and money. Processing ecological data in a timely and efficient manner for informed decision-making requires that ecologists utilize the most suitable state-of-the-art techniques. However, it is often difficult for ecologists to maintain awareness of technological advances in the data management field. Refereed journal page limitations preclude lengthy discussion of data management and analysis as well as the general paucity of outlets for this type of information contribute to the dilemma.

Ecologists, data managers, statisticians, and computer scientists were brought together for a symposium held Nov. 4-6, 1984 at Hobcaw Barony, Georgetown, South Carolina in an attempt to describe standard data management techniques, the latest technological developments, and the future direction of data management in ecology.

To maintain the high quality of the Belle W. Baruch Library in Marine Science series, all papers have been externally peer reviewed. Special thanks are extended to the numerous scientists who served in this capacity. Timely publication of this volume would not have been possible without the considerable effort expended by Ms. Mary K. Hall and the members of the Editorial Board. The encouragement and support provided by Drs. Dennis Allen, Elizabeth Blood, and F. John Vernberg are especially appreciated.

The Long-Term Ecological Research Program Coordinating Committee (through a grant provided by the National Science Foundation) and The Belle W. Baruch Institute for Marine Biology and Coastal Research, University of South Carolina are gratefully acknowledged for co-sponsoring the symposium. The mention of trade names or products does not constitute endorsement or recommendation for use by the co-sponsoring institutions.

William K. Michener

CONTENTS

DATA MANAGEMENT AND LONG-TERM
ECOLOGICAL RESEARCH

William Michener

INTRODUCTION

Ecology is the study of the relationship between organisms and their environments. Ecology began as a descriptive science; patterns were observed and attempts were made to formulate "rules" which described those patterns. Scientists quickly moved beyond the descriptive phase and explored methods for quantifying observed relationships. Ecological focus has evolved in an orderly fashion from studies of populations, to communities and ecosystems, and most recently to ecosystem interactions, "landscape ecology" (Naveh & Lieberman, 1984).

Data collected by ecologists exhibit many characteristics including variety, volume, and scale. Data sets may contain physical and chemical parameters (e.g., water quality, meteorological conditions), quantitative descriptors of natural systems (marsh, forest, etc.), and population data (e.g., abundance and life history information for zooplankton species). Also, descriptors of community dynamics (e.g., predator-prey interactions), organism response to changing environmental conditions, and data reflecting other complex relationships are routinely collected. Ecological data sets span the range from small (e.g., a data set containing limited observations of species abundance over a defined time period) to large (e.g., a data set containing meteorological parameters collected every minute for an indefinite period). The scale of ecological data is quite broad: temporal scales range from seconds to centuries and millenia; spatial scales vary from microhabitat to global.

Understanding relationships exhibited in diverse ecological data sets requires that a wide variety of analyses
be available. It is essential also that variables can be
integrated across data sets and between different temporal
and spatial scales.

As ecology increases in scope and sophistication, it is
important that tools be made available to ecologists which
will help them answer questions in a timely and efficient
manner.

RESEARCH DATA MANAGEMENT

"Research data management" is a task whereby various
tools (computers, statistics, etc.) are used to organize,
store and retrieve, integrate, and analyze research data.
Data entry, quality assurance, documentation, and statistical and graphical analysis are all tasks that fall under
the broad definition of data management (Table 1).

Table 1.   Research data management activities.

TASK	EXAMPLES
Experimental Design	Randomized Complete Block Design Systematic Survey Sampling
Data Entry	Forms Design Program Design Manual/Electronic Data Entry
Quality Assurance	Entry Error Checking Range Checking Statistical/Graphical Outliers
Documentation	Database Catalog Program Documentation
Storage and Retrieval	Database Additions Device Dependent Programs
Security	Password Protection Multiple Storage Locations
File Manipulation	Unit Conversion Sort Merge
Analysis	Statistics Graphics Modeling

The level of management a data set receives depends upon several factors, including investigator initiative, interest, and expertise; questions being addressed; envisioned project longevity; data complexity and analyses required; degree of secondary data utilization; requirements set forth by granting agencies; and funding levels. Many factors are inherently determined by the research project focus. Most funded ecological research projects are short-term (2-3 years maximum), limited in scope (addressing a single specific hypothesis), and involve one or few investigators. Data management characteristics for such projects vary considerably. For example, a single investigator may accept or reject hypotheses by analyzing relatively few data points collected over a limited period. Data management activities may simply involve manual data tabulation and statistical analysis with an electronic calculator. Quality assurance may consist of entering the data twice into the calculator and comparing results. The investigator may then append a few notes to his data sheets and store them in a filing cabinet for future reference. Under certain circumstances, this level of data managment may be entirely adequate.

In contrast, sophisticated data management may be required for the successful completion of a larger project. Such projects may employ many investigators from diverse disciplines. Complex studies generate large volumes of data that must be integrated across numerous spatial and temporal scales to answer the questions posed. Particular attention must be paid to data structure so that integration across scales can be performed. Rigorous quality assurance should be inherent in the data management scheme. For example, data may be entered twice for validation purposes, range checks may be routinely performed on data to locate outliers, and the data may be graphed routinely to evaluate sampling adequacy or locate errors. Quality assurance may also entail routine laboratory instrument calibration to ensure precision and accuracy in the data. In addition, a database catalog may be generated which contains detailed documentation regarding the experimental design, techniques used, and information about data structure and storage mechanisms employed. A well-conceived database catalog is usually necessary for researchers to effectively utilize a database years or decades after a project has terminated.

LONG-TERM ECOLOGICAL RESEARCH

Effective data management techniques are necessary to accommodate expansion of ecological foci. Long-term Ecological Research (LTER), a pilot program funded by the National Science Foundation, marks a new direction in ecological research. Long-term studies requiring many investigators from various discplines are currently being supported and it is likely that this trend will continue (Callahan, 1984).

The LTER program was established with several objectives, including:

1. initiating intensive baseline studies of site-specific phenomena
2. measuring and interpreting yearly variablity in the system
3. coordinating research between sites for comparative analyses
4. investigating ecological phenomenon that occur on time scales of decades and centuries (Halfpenny & Ingram, 1983).

Eleven sites in the United States have been funded under LTER (Fig. 1). Although numerous ecosystem types are represented, each site is addressing questions and hypotheses in five core research areas: "(1) pattern and control of primary production; (2) spatial and temporal distribution of populations selected to represent trophic structure; (3) pattern and control of organic matter accumulation in surface layers and sediments; (4) patterns of inorganic input and movements through soils, groundwater, and surface waters; and (5) patterns and frequency of disturbance to the research site" (Callahan, 1984).

Data sets generated at LTER sites will be a valuable resource not only to site-specific and affiliated researchers, but scientists from other research institutions as well. In response to the concerns regarding data protection, accessibility, and other factors, the National Science Foundation encouraged LTER sites to develop data management systems and reference collections (Callahan, 1984).

Recognition of the value of institutional data management systems resulted in a series of annual LTER meetings and workshops. Individuals designated as data managers by their respective institutions were brought together to develop guidelines for data management system development and maintenance. The workshops had two goals: to exchange data management ideas and to encourage convergence of data management techniques that would tend toward compatibility among sites without compromising the utility of data management for site-specific needs. Means were sought to enhance data utility and accessibility for both current and future investigators.

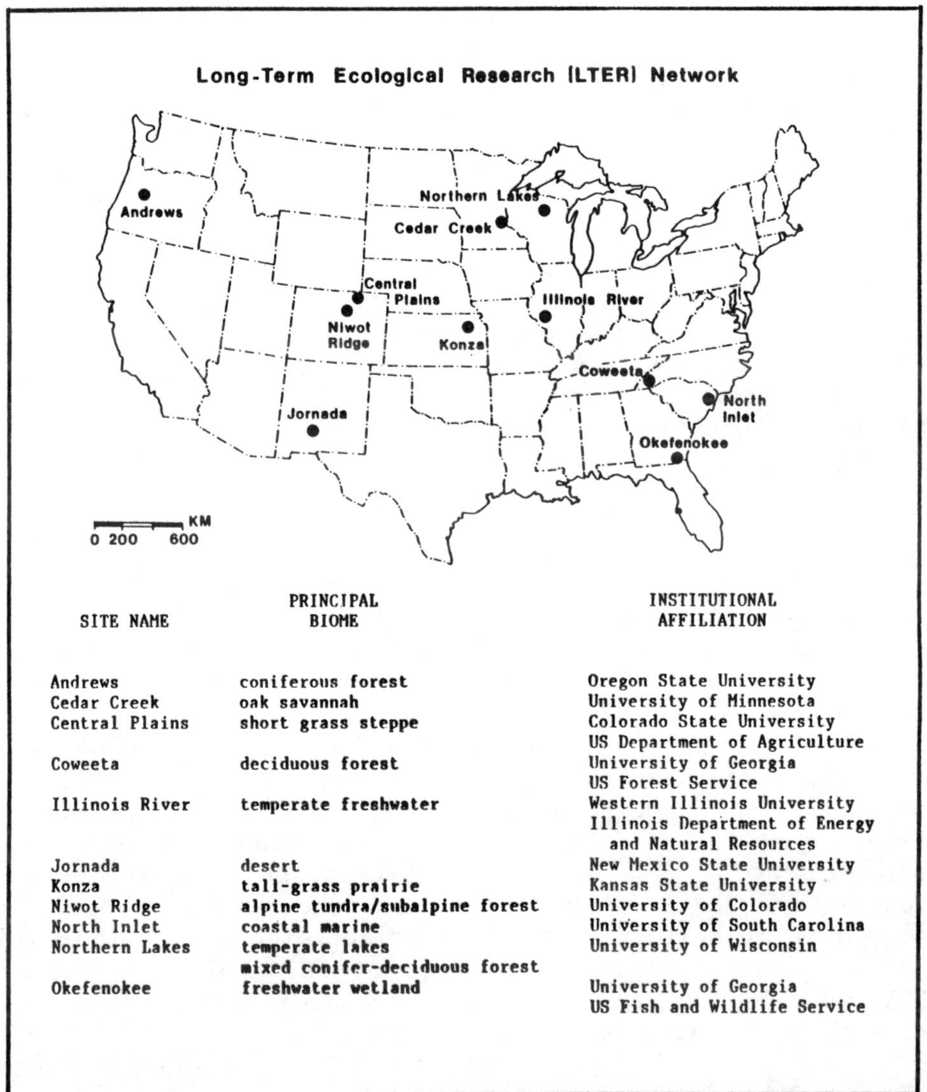

SITE NAME	PRINCIPAL BIOME	INSTITUTIONAL AFFILIATION
Andrews	coniferous forest	Oregon State University
Cedar Creek	oak savannah	University of Minnesota
Central Plains	short grass steppe	Colorado State University US Department of Agriculture
Coweeta	deciduous forest	University of Georgia US Forest Service
Illinois River	temperate freshwater	Western Illinois University Illinois Department of Energy and Natural Resources
Jornada	desert	New Mexico State University
Konza	tall-grass prairie	Kansas State University
Niwot Ridge	alpine tundra/subalpine forest	University of Colorado
North Inlet	coastal marine	University of South Carolina
Northern Lakes	temperate lakes mixed conifer-deciduous forest	University of Wisconsin
Okefenokee	freshwater wetland	University of Georgia US Fish and Wildlife Service

Fig. 1.   Sites for Long-Term Ecological Research projects (modified after Halfpenny and Ingraham, 1983, and Callahan, 1984).

The first workshop, held at the W.K. Kellogg Biological Station of Michigan State University on May 17-20, 1982, included 45 individuals from various biological field stations and Federal agencies. Major topics included views of data management, databases, computer software systems, data administration, and information exchange (Lauff, undated). This meeting set the stage for the LTER workshops which followed.

Data managers from LTER sites and other interested agencies convened November 22-23, 1982, at the University of Illinois in Champaign-Urbana, Illinois. Discussion focused on data management protocols established at each site, database documentation, and intersite data exchange (Sinclair, 1983). One prime accomplishment was the formulation of essential and optional descriptors to be included in LTER database documentation. The procedures developed for data documentation have now been implemented at most LTER sites and many other institutions as well.

The workshop held at Oregon State University in Corvallis, Oregon on November 4-6, 1983, included representatives from LTER sites and Federal agencies. Major discussion topics included cataloging and documentation, intersite communication, utilization of microcomputers, and data retrieval. Protocols for intersite communication were established.

A symposium entitled "Research Data Management in the Ecological Sciences" and workshop entitled "Data Management in the Long-Term Ecological Research Program" were held November 4-7, 1984, at Hobcaw Barony in Georgetown, South Carolina. The meetings were attended by 61 participants, 27 directly affiliated with the Long-Term Ecological Research program. The other participants represented various educational institutions, Federal agencies (NOAA, National Park Service, etc.), state and municipal agencies, and private consulting firms. Proceedings from the symposium are included in this volume. Workshop discussion topics included software currently under development for ecologists, database documentation, intersite communication, and future directions for data management in LTER.

INTRODUCTION TO THE SYMPOSIUM

Several factors led to the conception of the symposium:
1. Scientists are becoming increasingly aware of the need for collective stewardship of data and

unpublished baseline information related to ecological research sites. Ecosystems are complex,
and data collection is expensive and requires the
talents and skills of several disciplines. Fast
and reliable communication of diverse information
requires that we apply the most careful and complete management techniques to handling and storing data. Recent advances in data handling technology have provided the means for effectively
achieving this goal.

2. Considerable confusion exists between traditional
   business data management approaches and evolving
   data management techniques for application in the
   ecological sciences.
3. There are few recognized methods for managing research data in the ecological sciences.
4. Numerous institutions are involved in implementing
   or enhancing research data management at their
   sites. Information published about recent trends
   and practices in data management could save many
   sites from having to "reinvent the wheel."

Specific symposium objectives were to serve as a reference manual for general concepts and methodologies within
research data management; aid ecologists and administrators
in implementing data management schemes at their sites; synthesize past and current developments within the field, and
outline future directions and goals.

The symposium and proceedings were divided into two
sections. The first section included general concepts and
research data management methods (e.g., data entry, scientific database characteristics, quality assurance, and data
management system implementation). The second section included specific examples of data management applications and
described developing and currently available technology
(e.g., ecological modeling aids, real-time data collection,
management of satellite imagery data, geographical information systems).

The symposium was intended to increase awareness of the
value of effective data management and the tools available
to research sites and individuals for data management.

REFERENCES

Callahan, J.T. 1984. Long-term ecological research. Bio-
    Science 34(6): 363-367.

Halfpenny, J.C. and K.P. Ingraham (eds.). 1983. Long-Term
    Ecological Research in the United States: A Network of
    Research Sites. (3rd ed., revised). Forest Sciences
    Laboratory, 3200 Jefferson Way, Corvallis, OR 97331.
    28 pp.
Lauff, G.L. (ed.). (undated). Data Management at Biologi-
    cal Field Stations. W.K. Kellogg Biological Station,
    Michigan State University. 46 pp.
Naveh, Z. and A.S. Lieberman. 1984. Landscape Ecology.
    Springer-Verlag, Inc., New York. 356 pp.
Sinclair, R.A. (ed.). 1983. Long-term Ecological Research
    Data Management Workshop held at Urbana-Champaign,
    Illinois, November 22-23, 1982. Illinois State Water
    Survey. Miscellaneous Publication 72, Champaign, IL
    24 pp.

OVERVIEW OF ECOLOGICAL RESEARCH DATA MANAGEMENT

Paul G. Risser and Colin G. Treworgy

ABSTRACT

Management of research data in the ecological sciences
has developed into a fascinating challenge. This challenge,
exacerbated by the rate of hardware and software advances,
has been created by the need to simultaneously integrate
administrative approaches that motivate rigorus science;
data manipulation procedures that stimulate ideas and tests
of ideas; and methods for cataloging and saving data and
information which will support the scientific enterprise now
and in the long-term future.

This paper discusses scientific, administrative, and
practical issues concerning data management in the ecological sciences. Several examples are suggested as models of
using data management to stimulate ecological ideas.
Finally, several recommendations are made about the conceptual design and operation of data management systems that
will encourage synthesis of ecological ideas and analyses in
both the basic and applied context.

INTRODUCTION

An old adage says that much ecological research can be
accomplished with a notebook, a pencil, and a keen observing
eye. Over the past 50 years, many new techniques have been
developed to assist that observing eye. Field pH kits have
been replaced with electronic digital pH meters, careful
stalking has been augmented with radio telemetry, electro-
phoresis now assists in distinguishing biochemical characteristics that are invisible to the naked eye, and time-
motion photography records images that later can be observed

in slow motion or even backward and forward. Our progress with the pencil and notebook also has been notable and equally dramatic. Data, which were recorded in notebooks and later graphed on strip charts, are now recorded automatically in digital form. Similarly, data from telemetry radios, hygrothermographs, and inductively coupled argon plasma spectrophotometers can appear almost instantaneously in computerized data files.

The engineering, business, and communication sectors provided the impetus for the electronic management and display of information; from these efforts came mathematical computing packages and database management systems. It became apparent to ecologists that these computational and data management systems were potentially powerful tools, that is, better notebooks and pencils. Database management systems now are used routinely in ecological research for such activities as storing, compiling, comparing, analyzing, and displaying data. With respect to these activities, some guidelines for research data management in the ecological sciences will be briefly discussed.

In addition to providing better notebooks and pencils, ecological research data management has the potential for enhancing the keen eye-to-notebook interaction. Specifically, research data management systems should facilitate both the generation and the testing of ecological ideas by increasing the researcher's power to observe, and to capture and manipulate data.

THE NEED FOR ECOLOGICAL RESEARCH DATA MANAGEMENT

Implementing research data management systems costs money and time, and frequently disrupts conventional and comfortable ways of conducting research. The decision to invest in an ecological research data management system must be based on the need for such a system (i.e., the types of experiments being conducted and the intentions of the scientists at the research site). The following objectives might assist researchers' making decisions about the implementation of an ecological research data management system for a project or a research site. The objectives are general and proposed systems need not address all of these objectives.

    1.   To maintain a research memory by providing an organized means for capturing the results of a research program. This is especially important when studying long-term ecological phenomena,

    which may persist and be studied over periods longer than the normal duration of a single investigator's tenure.

2.   To encourage the transfer of valuable data from storage in file cabinets to an active working data file. In many instances, unreduced data sets are simply overwhelming without summarization, and their massive size precludes convenient manual summarization.

3.   To motivate the preparation of more rigorously examined and compatible data sets. Frequently, the mere prospect of contributing to a database prompts a more thorough evaluation of the data and encourages an exploration of the possibility of constructing the database in a manner that will be compatible with other data sets.

4.   To permit the convergence of multiple data sets that can be brought to bear on a specific topic or research management question of interest. This convergence is more likely if some forethought has accompanied the construction of the database system.

5.   To provide easy access to data sets that are commonly used by more than one investigator. For example, basic information about the climate or soils of a research site can be made readily available to both resident and visiting investigators. High-speed data transmission networks now make it possible for scientists in different cities, and even different countries, to have access to common data sets.

6.   To reduce the time spent by scientists in locating, compiling, and synthesizing data, thereby increasing the time available to analyze and test hypotheses. Researchers need time-saving tools to cope with today's explosion of information and to tackle site-specific ecological issues as well as global problems, such as desertification and deforestation.

7.   To increase the insight into and comprehensiveness of the research questions being asked or to expand the possible approaches to addressing resource management problems. A research data system may permit comparisons to be made or ideas to be tested which would simply be too expensive without a data management system.

8.   To present persuasive arguments in the decision-making arena. By their nature, ecological issues are often complex and interactive, and furthermore, frequently these ecological processes of interest occur over long periods. Displaying such information in an easily comprehensible manner may be facilitated by, or even require, electronic data management systems.

9.   To be able to incorporate ecological data from rapidly changing field observations into real-time modeling programs. Outbreaks of pests, forest fires, and oil spills are examples of fast breaking events, the damage from which could be minimized by timely information from ecologists.

A decision to implement an ecological data management system will be based on meeting some or all of the objectives. Once a decision is made to implement a data management system, it is then necessary to consider a number of details about the specific data sets to be included and how these data sets are to be managed (National Science Foundation, 1984; Treworgy, 1984).

TYPES OF DATA TO BE INCLUDED AND SYSTEM DESIGN ISSUES

After establishing the objectives of the system, it is necessary to make decisions about the appropriate types of data to be incorporated into the ecological database. The first step is to inventory the candidate data sets, which might include some or all of the following: results from specific projects, data from long-term monitoring studies, species lists, label information from samples or collections, bibliographies, current and past project descriptions or summaries, place names or gazetteers, lists of software or programs, common sample or grid locations, and lists of experts in various topical areas. Once the data sets have been identified, it is necessary to consider the size of each data set, the rate at which it might expand in the future, and access considerations, such a frequency of demand and protection from unauthorized use. Different types of data sets have different characteristics with respect to management. For example, generalized data sets, such as routine climate data, will have a large number of potential users, will need to be easily accessible, and will be updated frequently. Similarly, site descriptive data, such as soils information or the location of a sampling grid, may need to be easily accessed, yet may be updated only infrequently. On the other hand, data generated from an ongoing project may be used only by the investigator, may contain erroneous data that have not been corrected yet, and may be used very frequently but not in a form that is easily interpreted by anyone else.

Data to be included in the database need not be confined to conventional alphanumeric forms (e.g., tables and maps can be stored in a database). The new video disk technology makes it feasible to store data as pictures. Real estate agencies are using this technology to store pictures of buildings and lots. The agent enters a customer's requirements for price, number of bedrooms, and so forth, and the computer retrieves and displays images of the houses that match (Parker, 1984). One can easily imagine applications to ecological data management, such as pictures of specimens, sample localities, or unique habitats.

In designing a system or improving an existing one, a convenient analytical technique may consist of constructing a matrix listing the data sets along the vertical axis, perhaps according to such general categories as site descriptive information and project data. The columns of the matrix can list the size of each data set, its projected increase in size through time intervals, the types of data manipulation likely to be used with the data, the frequency of access, user requirements, and any unusual characteristics of the data set. If costs for data storage on the system are available, as well as costs of various manipulations, it will be possible to obtain an estimate of the system's operational costs. Also, if a hardware system is to be purchased, this information can be used to determine the system configuration and size.

An important aspect in the actual database design process is to search for data elements that are shared between two or more data sets. By standardizing the classification codes and format of these shared elements, several important design goals can be achieved.

    1.   Storage of redundant information or inconsistencies between data sets that contain the same information is eliminated. For example, sample location descriptions may be shared by several data sets. When redundant locations are kept in separate files, there is often a problem of keeping all corrections to the locations consistent.

2.   Links are established between related data sets. For example, data from files on fishes and stream water quality may be linked if both sets use the same codes. These linkages between data sets are a powerful feature of electronic ecological databases.

3.   Documentation is simplified if standard codes and formats are used throughout the system. Researchers adjust quickly to handling new data sets if there is a structural and nomenclatural commonality between files.

4.   Programs and lookup tables can be shared. Even though two files contain unrelated data, if they use the same format for common items (such as locations or taxonomic codes), a program written to manipulate one file may be easily modified to be used with the other.

Whenever possible, a database coordinator experienced in database design should review the proposed designs of individual data sets. Often, small changes in format can make significant improvements in consistency and compatibility between data sets, reduce storage requirements, and improve operating efficiencies.

In ecological data management systems, several specific issues are likely to arise in designing the database. For example, a decision must be made whether to include both long-term background data sets or just short-term active data sets. Both types are likely to be useful for addressing ecological questions; however, it may be desirable to store only frequently used data sets on-line. Also, there is the question of whether to include only raw data, data summaries, or some combination. A third general issue is whether to incorporate data sets that are available elsewhere, but might be included for the convenience of the scientists using the system. The answers to these issues will be determined by the objectives of the data management system and the resources available for purchasing and operating the system.

ISSUES OF SYSTEM OPERATION

Any system developed for access by multiple users must have a uniform format for documentation. This documentation will aid in tracking archived data, in identifying changes in methods of data acquisition (such as sample locations or identifying codes), and will assist in describing changes in software or analytical procedures.

Interactive data entry capability and automated data verification are important for most systems that will include ecological data. Whether data entry is on-line into the main database or is handled by a local processor (such as a microcomputer) will depend on 1) the complexity of the file, 2) the need for immediate data verification as opposed to a later check by batch process, 3) the system load, and 4) the transmission speed. It may be best to enter data from remote field stations on a microcomputer at the field station. The microcomputer can be programmed to display input forms and do simple field checking. Periodically, the file of new data would be transferred to the main computer and merged into the database. Transmission might be by high-speed leased telephone lines, unconditioned dial-up lines, or even by sending floppy disks through the mail. Consideration should also be given to interfaces with automatic data recording and collection devices.

If a data management system is to be accessed by a large number of users, particularly visiting scientists, the index system to the data files is of paramount importance. The index allows the user to identify the data sets that are available and how to relate the data sets to each other. There are several considerations in designing the most appropriate index system. For example, it might be desirable to use a controlled vocabulary (dictionary) or thesaurus. Such a standardized approach provides uniform access to the information included in the system, but in large data management systems the vocabularies may become long, cumbersome, and exhausting unless simplified versions are also available. In addition, there are fundamental decisions to be made as to whether such a system should emphasize topical material, taxonomic entities, or both. The index system might be designed to relate the on-site or project data to published information and perhaps to databases located elsewhere.

In the near future, world databases and global information networks will be established. Data catalogs may be one of the researcher's most important tools for keeping track of the massive amounts of scientific data that are being collected. Consideration should be given to including in the index references to all ecological data regardless of its format, (e.g., digital form or in a glass jar). For data catalogs to be successful, scientists must get in the habit of documenting and reporting their data collections.

Resistance to this reporting may come from fears about the Freedom of Information Act, unreasonable possessiveness, or (most likely) the apathy of many of us about documenting our basic data.

Table 1.   Optimum capabilities for a research data management system.

---

1. The system would include the capacity to search creatively and sort various data files.

2. The system should contain the capability to subset, merge, aggregate, statistically manipulate, and summarize data sets with minimum restrictions imposed by the software or hardware.

3. Data management capabilities should provide for multiple entry so that the investigator can enter individual data sets by several pathways.

4. File management strategies should be convenient, logical, and consistent throughout the system, and they should facilitate the archiving of data and the migration of rarely used files from disk storage to tape and vice versa.

5. The system should provide a reasonable response time so that the investigator will be likely to pursue questions or test alternative approaches to ecological ideas.

6. Provision should be made for updating files, and the updating procedure should provide a historical record of the updatings.

7. The system should incorporate various types of quality checks in the data input scheme as well as within the data summarization manipulations.

8. The security system should permit various levels of access to users with different levels of security authorization.

9. Graphical and tabular data should be easily integrated to facilitate manipulations between these two data types.

10. Multiple display modes should be available so that the output can be tailored to the audience.

11. The hardware and software interfaces should include those with statistical packages, graphic systems, image processors, photographic equipment, word processors, and typesetting capability.

12. Data file structure should be such that data sets can be easily transferred to other sites which may have different hardware and software.

13. The system should include multiple hardware paths for transporting data to and from other sites. These paths might include tape drives, floppy disk drives, dial-up modems, remote job entry links, local area networks, and satellite transmission stations.

Of particular interest in ecological databases is the location and identification of field sampling points. This represents another method by which data can be indexed. To make geographic locations useful in an electronic data management system, ecologists must adopt more consistent and precise forms of describing geographic locations. Location descriptions such as "3 miles northeast of Champaign" are not useful for many computer applications such as plotting sample locations or retrieving all points within a given area. Furthermore, although dot maps of sample locations can be digitized, this method is not precise unless large scale maps are used. Digitizing from a 1:500,000 scale map, (a common scale for plotting statewide sample locations) may result in location errors of a mile or more, a serious enough error that fish sampling points might be shifted to the wrong stream or even fall in an upland area. Whenever possible, location descriptions should be recorded in a standard cartographic system, such as latitude-longitude, UTM, or state plane coordinates.

DATA MANAGEMENT ISSUES

Once the data sets to be included in the management system have been identified and operational characteristics have been considered, the issues of data management itself must be addressed. There are a number of system characteristics that ideally should be incorporated if the data management system is to be most useful in expediting the scientific endeavor.

A research data management system incorporating all of the optimum capabilities listed in Table 1 will be expensive, and a staff will be required to manage such a system. If the expense for such a system is too great, then the researchers should select system capabilities of the highest priority and configure hardware and software accordingly.

ADMINISTRATION OF THE DATA MANAGEMENT SYSTEM

The administrator of an ecological research data management system has a number of responsibilities. The administrator, with the research and data management staff, must set goals, priorities, and policies as they relate to the database and the users of the database. Decisions must be made about access rights to the information and consequent security procedures. Also the administrator must determine budget and staffing needs of the system.

A research database has a large human component. An administrator must ensure that a positive and synergistic relationship exists between data managers and researchers. It may be important to provide reference personnel in related areas (e.g., statistics or electronics). A mechanism should be established for obtaining feedback from various users so that the performance of the data management system can be evaluated. This information must then be used conspicuously in making adjustments in the operation of the system. It is the administrator's responsibility to ensure that the research data management system develops in such a manner that the highest quality research effort is encouraged.

An administrator of an ecological research data management system should have sufficient data management technical skill to understand the general hardware and software capabilities. In addition, ecological concepts and field and laboratory sampling protocols should be familiar to the administrator. Most importantly, the adroit administrator must learn how to use the data management system to enhance the quality of ecological science.

RESEARCH DATA MANAGEMENT SYSTEMS AND ECOLOGICAL SCIENCE

Today, we have just begun to consider ways in which research systems can be employed to enhance ideas about ecological systems; it is useful to consider attributes of data management systems which facilitate innovative analyses. A discussion of the capabilities that the ideal system might have follows.

The system would be easily usable by researchers with little knowledge of file structures, programming syntax, operating system commands, or network links. The system should be "intelligent" enough to accept natural-language commands, locate the files needed, retrieve specific records by using the most efficient search technique, reformat the data if necessary, and execute the appropriate programs. Procedures would be available to permit data to be shared among investigators, particularly during a discussion. Furthermore, such data sets should have variable amounts of instantaneous documentation. A person unfamiliar with the data could obtain information about how the data were collected, but the person already familiar with the data would not have to see this information.

Draft text would be available on a terminal in front of each participant in a discussion so that changes could be made by anyone and could be agreed upon by everyone. Standard information useful for discussing ecological ideas would be easily called to the discussion electronically. Such data sets or descriptors might include site descriptions of soil characteristics or records of climatic conditions. With well developed geographic information systems, sites with similar ecological conditions could be identified and located. Electronically searchable bibliographic files, with abstracts, would be available easily and would become a routine part of any discussion about ecological issues.

Spatial and tabular data sets could be easily integrated and manipulated to test alternative hypotheses about the spatiotemporal behavior of ecological systems or the consequences of translating ecological information from one hierarchical scale to another.

A whole set of statistical and analytical procedures could be applied to the data, again in an interactive mode which would be responsive to discussion. For example, one could determine the type of curve which best fits a data set, select a multiple regression which explains the most variability, or identify the most similar populations or sites as defined by alternative multivariate techniques. Readily useable and easily manipulated simulation models would be available so that investigators could test various parameters, coefficients, and formulations.

Implementation of such a utopian system would require several ingredients. First, the amount and quality of ecological research must increase to provide the requisite data. Second, data managers and scientists would need to enhance the everyday working relationship to augment existing procedures. Third, algorithms and hardware will have to be refined and perhaps programming techniques should be more process-oriented rather than storage-oriented. Fourth, hardware costs must decrease so that interconnected data management systems can be used from inexpensive input-output devices. Fifth, artificial intelligence concepts, such as natural-language interfaces and expert systems, need to be implemented so that scientists can spend their time working with data rather than contending with machines and software.

These constraints are real and undoubtedly affect the ways in which ecologists interact with data management systems. However, if ecologists push the systems to their maximum potential, future data management developments will respond to these pressures. To return to the original point, ecologists must investigate ways to improve the notebook and pencil, the keen observing eye, and the interaction between the notebook and the eye. By the latter, we mean to increase the quality of ecological ideas and testing of these ideas. Until data management systems become so simple to operate that little effort is required, the ecological scientist and the data manager must cooperate closely. In some organizations, this process has become standard practice. However, in organizations of every size, ecologists need to attempt more creative ways of using research database management systems. These first attempts will require careful preparation and perhaps even some choreography to assist reluctant or unsophisticated users. Undoubtedly, many first attempts may be clumsy and contrived, and some will be failures. However, a creative failure usually is a learning experience, which forms the constructive base for the next experiment. The following discussion proposes some ideas to stimulate experimentation.

If the research team has an operational database, simply placing a terminal in the conference room may prompt creative uses of the system. The next time a group is to consider a document (e.g., site description, proposal, manuscript), the word processor screen could be used as the basis for the discussion. This could be done using one screen or multiple screens driven by one processor, or each person could have the document on a floppy disk with an individual or shared microcomputer. The research team could consider a specific data set and its ecological explanation. If the data manager were able to perform manipulations (at least statistical analyses) during the discussion, the team discussion might profit from instantaneous feedback. The conventional alternative is to send the data manager off to conduct a series of prescribed analyses which would then be discussed at some later time. However, this standard approach will not capture the spontaneity of idea generation at the time of the initial discussion.

Simulation models might be presented in explicit formats. In addition to the computational steps, the program upon demand could display the formula used to calculate a particular process and the initial conditions and coefficients used in the particular model run. Then the researchers could pose alternative initial conditions, parameters, and formulations of individual processes or even the model structure. The model could be run again and the output could be evaluated in an iterative process with turn-around times consistent with the discussion format.

Geographic information systems that combine tabular and mapped data will be particularly powerful tools (Marble et al., 1984) in several areas of ecology. Currently, many of these systems are used at the level of regional planning and thus do not include data appropriate to the examination of ecological questions, for example, the natural history of particular organisms or specific biological processes. Yet numerous ecological questions can only be answered by processing information at the landscape scale (Allen et al., 1984; Risser et al., 1984). Therefore, as these databases are being designed, ecologists need to pose germane ecological questions to ensure that the data system is adequate for these analyses. Once operational, the geographic information systems should be used in the same manner as simulation models. For example, the system could be an integral part of discussions about the habitat requirements of a particular species, similarity of specific habitats, or the relative productivity of different ecosystems.

Both simulation models and geographic information systems should be useful in presenting ecological issues in ways that will be valuable to decision makers. For example, a siting question being discussed with regulatory authorities could use a geographic information system to depict the soil types around the proposed site, the physical and chemical characteristics of these soils, a model to predict migration of chemicals through these soils, graphical descriptions of the surrounding vegetation, and a simulation model to predict the loss of primary and secondary productivity caused by project-related influences.

Ecologists should use research data management systems to address questions that combine ecological processes and public opinion. For example, overlays of ecological information can be used to discuss the suitability of a county for certain types of land use. In evaluating the conditions, a model can be employed to predict suitability; however, the relative importance of the included attributes is a value judgment. Thus, ecologists should be able to place a computer terminal on the conference table and assist the decision-maker to explore the consequences of giving the ecological attributes alternative weights of importance.

These suggestions are designed to prompt the mode of thinking which makes ecological research data management an inherent part of investigative processes. Ecologists and data managers have an enormously exciting opportunity to make giant strides in assisting the individual researcher and in facilitating imaginative ecological investigations at spatial and temporal scales not heretofore possible.

LITERATURE CITED

Allen, T.F.H., R.V. O'Neill, and T.W. Hoekstra. 1984. Interlevel Relations in Ecological Research and Management: Some Working Principles from Hierarchy Theory. USDA Forest Service General Technical Report RM-110. Rocky Mountain Forest and Range Experiment Station, Fort Collins, CO. 11 pp.

Marble, D.F., H.W. Calkins, and D.J. Peuquet. 1984. Basic Readings in Geographic Information Systems. SPAD Systems, Ltd., Williamsville, NY. 362 pp.

National Science Foundation. 1984. Data Management at Biological Field Stations. Report of a Workshop, May 17-20, 1982. W.K. Kellogg Biological Station, Michigan State University, Hickory Corners, MI. 43 pp.

Parker, W. 1984. Interactive video: Calling the shots. PE World 2(11): 98-108.

Risser, P.G., J.R. Karr, and R.T.T. Forman. 1984. Landscape Ecology: Directions and Approaches. Illinois Natural History Survey Special Publication 2. Champaign, IL. 18 pp.

Treworgy, C.G. 1984. Design and Development of a Geographic Information System for Illinois. Spatial Information Technologies for Remote Sensing Today and Tomorrow, Pecora IX Proceedings. IEEE Computer Society Press, Silver Spring, MD. 4 pp.

DEVELOPMENT OF A RESEARCH
DATA MANAGEMENT SYSTEM

Martin E. Gurtz

ABSTRACT

Ecological research sites share some common goals with respect to research data management (RDM). Minimum requirements, regardless of the size of the database or other site-specific factors, include creation of a data set inventory, organization and documentation of data sets, and development of procedures for data entry, quality assurance, archival, maintenance, and administration. The RDM budget must balance expenses among personnel, hardware and software, storage, and data exchange. Characteristics of the research program that influence these choices include the purpose of the data collection, the form of the data, size and complexity of the database, type of analyses planned, dynamics of the database, number and expertise of investigators, and the need for network capabilities. Experiences in RDM at Konza Prairie support the idea that the data manager plays a key role and should be trained in the field of research. The RDM system should be institutionalized so that it can continue after the original staff is gone.

INTRODUCTION

Consider this scenario: You have been hired by a team of scientific researchers at a major university. The team has collected data at a nearby site for a decade, but many of the original investigators are no longer working on the project, and some are no longer with the university. The early data sets will become extremely valuable in interpreting future data collections, as the present team is initiating experiments and observations on ecological phenomena that will extend beyond their careers. Your task is to salvage past data sets and organize present and future data collections. The data must be usable by individuals other than the current investigators, in some cases using analytical techniques that are not anticipated (or perhaps do not even exist) at present. What can you do?

This situation is similar to the one that faces data managers at ecological research sites, including those of the National Science Foundation's Long-Term Ecological Research (LTER) program. Similar problems face researchers at biological field stations, state or federal environmental quality agencies, environmental consulting firms, or private nonprofit conservation organizations. Although tasks and approaches to research data management (RDM) differ in each of these situations, there are certain common principles involved in developing an RDM program.

Scientific investigators typically have taken a pragmatic approach to managing their data. For example, data collected for a specific short-term purpose might be stored only in the form most convenient for currently planned analyses, with little effort directed toward possible later use. Documentation and internal organization of each data file under these conditions are often interpretable only by the original investigator. Sometimes published papers are the only records for future use. This lack of recognition by the investigator that the data may be useful for future purposes is a disservice to the scientific community. Indeed, the data may have their greatest value in future examinations different from those of the original investigator, for example, to test new hypotheses or use the data to address societal problems (e.g., Lide, 1981).

Data management principles apply to ecological databases, regardless of their size, complexity, or the budgetary limits of the research project (Gorentz et al., 1983; Sinclair, 1983). The objective of this paper is to present elements of RDM that can be implemented by any individual or group, and to discuss economic considerations as well as characteristics of the research program that can influence decisions regarding RDM.

BASE-LEVEL RDM

Minimum requirements of any data management system include the ability to locate, interpret, and use data by all interested researchers. If data are accessible and usable only by the original investigator, these requirements have not been met. This base level of RDM does not require computers, but involves several organizational steps, each of which is described briefly below and examined more thoroughly in other papers in this volume.

Data Set Inventory

The starting point in establishing an RDM program is to identify the existing data sets. This may include historical data sets from previous research at the site as well as ongoing and planned data collections. This step includes a qualitative evaluation to determine which data sets should be incorporated into the data management system, or at least setting priorities among them. For example, ongoing and planned research may have the highest priority in order that immediate goals of data analysis can be met. Historical data sets, student research projects, or other independent research at the site may be added to the database later, but in some cases waiting too long may result in permanent loss of the data.

Organization of Data Sets

Indexing and classifying data collections will facilitate their storage and retrieval. This may include assignment of unique codes to data sets, using a logical structure whereby relationships among data sets can be most clearly seen. Establishment of procedures for encoding commonly encountered data items (e.g., sites, taxonomic groups, and collection dates) is desirable. Investigators using different codes for the same parameters will cause considerable confusion for themselves or others who may wish to compare the results.

Data Set Documentation

Documentation means a careful recording of the origin of the data (e.g., site, time, and methods), the variable formats, the mode and location of data storage, and records of updates and usage of the data. Standardized forms that require all pertinent information on each data set make documentation fast and simple, and also make subsequent searching more efficient. A methods manual for a project can provide supplemental details.

Data Entry and Quality Assurance

Transfer of data from raw data sheets to a computer storage medium can be made more rapidly using sophisticated data entry programs, which also may reduce errors through range and format validations. Laboratory instruments that are interfaced directly with computers and automated data recording devices for field stations can eliminate this step for some data. Even data not stored in a computerized form require a scheme of standardized labeling prior to storage; a coded system should include the physical location, the type of material, and other relevant information.

Quality assurance includes procedures that reduce the frequency of occurrence of errors in a data set. This may be facilitated in several ways, beginning with improved design of raw laboratory or field data forms that reduce or eliminate the need to transcribe data prior to storage. Double-entry verification (entering data twice and comparing for inconsistencies) and error checking (e.g., examining data for values outside of a specified range) will improve quality of the data. If several versions of a data set are maintained at a given time, there should be a mechanism for distinguishing between raw data, verified data, and summarized or derived data. Data are ready for the archival step only after they have been verified by the investigator responsible for their collection (i.e., their accuracy had been confirmed by comparing them against the original data sheets).

Data Archival and Maintenance

Archival may involve transfer of the data to another storage medium (e.g., magnetic tape), with an appropriate level of backup copies (including multiple media, i.e., printed copies in addition to backup tapes). Procedures should also be developed for file maintenance, including updating and periodic examination of both data and documentation.

Data Administration

Administrative structure and procedures must be developed so that responsibilities are clearly defined and assigned to appropriate personnel; for example, which tasks are to be taken care of by the data manager, and which ones are to be left to the investigators.

Beyond the base level, RDM development must weigh the merits and costs associated with personnel, hardware/software, storage and maintenance, and data exchange. Decisions must take into consideration the characteristics and requirements of the particular research program.

ECONOMIC CONSIDERATIONS

Data management costs are a small fraction of original research costs (Stockmayer, 1978). Costs of collecting ecological data are high and many studies of natural phenomena cannot be replicated (time is unidirectional!), so skimping on data management makes little sense.

Funds available for RDM come from a variety of sources. Individual research grants generally provide RDM funds only for the short-term goals of data analysis. Provisions for salaries of RDM personnel, long-term storage and maintenance of data, or major computer purchases may require pooling of resources among several investigators or through an umbrella research institution.

Personnel

In a team-research environment, the need for trained data management personnel has often been overlooked or understated. In contrast, it is not uncommon for personnel costs in business data management to consume 50 percent of the total RDM budget (Kanter, 1981). In ecological research, computer facilites may be provided, but it is often left to individual researchers to determine how to best utilize those facilities.

Assigning the task of overseeing data management operations to one person is the most important first step in establishing an RDM system. This person need not be a full-time data manager, but could be a staff researcher with organization skills. This person can facilitate individual research efforts by freeing the investigators from data maintenance responsibilities. An organized and standard framework for documentation, storage, analysis, and maintenance of the data removes the need to develop a separate system for each project.

Beyond the individual level, the need for a data manager is even more clear. Interdisciplinary research, traditionally involving a direct exchange of data between two or more investigators, can be accomplished much more efficiently (particularly within a site research team) with a centralized data management system using common formats and procedures. A strong data management staff (whether there is a single data manager or a team involving an administrator, programmers, data entry personnel, and others) can also reduce costs associated with training users of the RDM system. The more decentralized the system, the greater the burden placed on individual researchers, and therefore the greater the costs of training (or learning by trial-and-error).

The role of the data manager in the ecological sciences is changing. Data managers trained in the field of research become a member of the team, not merely a provider of data handling services. With an ecologist as data manager, there is also greater understanding by the data manager of the need for data integrity, as well as improved communication with other researchers. Further, there is greater potential for integration of data into a synthetic framework. The data manager may have a better overview of site-related research, as well as availability of data from other sites, than any other team member, and may be more facile with integrative analytical tools.

Hardware/Software

An evaluation of hardware and software needs begins with a careful examination of existing facilities. Are microcomputers available to the researchers? Is a mainframe system located conveniently to the project, easily accessed when needed, and supported by a staff willing and able to help with applications? Does existing hardware have the speed, memory, and compatibility required to execute desired software using the amount of data expected? Can existing software meet the current needs? Are sufficient peripherals such as printers and plotters available? These questions need to be addressed often and regularly.

If hardware or software is needed, an additional cost factor (which is sometimes forgotten) is the personnel costs and development time for installing it and getting it operative. Using or upgrading the current system may be the best option, especially if the study will be of relatively short duration. Resultant savings can be used to increase the personnel staff to help offset the loss of convenience.

An important caution is not to be "hardware-bound." Considering the rapidly changing technologies, a micro- or minicomputer or associated peripherals may become obsolete (and/or replacement parts difficult to obtain) within the lifetime of the research project. Data or software stored on one type of medium may not be interpretable by later "improved" devices. If the RDM system has two or more identical smaller systems rather than a single larger system, it can continue to function while one computer is being repaired.

Many sites have several microcomputers of different brands and models for use by researchers; such variety makes available a broad spectrum of software, but multiple incompatible systems require a tremendous amount of effort to learn (or teach others), to maintain, and to transfer files or programs between systems for various applications.

Data stored on micro- or mini-based systems can be made more secure if an alternative backup storage medium (e.g., a 9-track tape system as backup for floppy or hard disk) is part of the system or if a mainframe is used for backup storage. If microcomputer hardware becomes obsolete, then data can be accessed from the backup medium. Otherwise, data sets may have to be reconstructed from original hard copies (data sheets, field notes, etc.), at great cost in personnel time. Data integrity would decrease since there is an increased chance for errors during reentry (especially if the original investigators are no longer involved). If the RDM system is integrated with an institutional mainframe, then transitions are necessary as mainfames change; data storage and program formats can be modified as the new system is brought on-line.

Commercially available software is almost always preferable to in-house developed software. Development costs of custom software can easily exceed expectations and may be hidden in salaries of personnel spending major amounts of time developing, documenting, and modifying programs. Also, in-house developed software may decline in utility when the original developer leaves. Although some commercially available software is poorly documented and may lack flexibility, selection of software with good vendor support can minimize these problems.

Database management software (DBMS) can help assure data integrity through data-checking procedures, reduce redundancy of data storage, facilitate data retrieval, and provide automatic backups and automatic recovery of data after a system "crash." Once it is installed and applications are developed, a good DBMS can reduce or eliminate the need for a programmer interface between the scientific user and the database. However, many DBMS's require large amounts of programming time to set up the database, write internal quality assurance programs, and increase user friendliness. Ecological research teams may benefit from integrated software packages that combine features of DBMS, graphics, and statistical packages. These are becoming more available at the level of mini- and microcomputers.

Hardware repairs can be expensive and should be anticipated. Annual service contracts allow hardware maintenance costs to be more predictably budgeted. Maintenance costs for in-house software include updating programs and their documentation as projects change. Commercial software may carry with it an annual maintenance or leasing fee.

Storage

Costs associated with storage of data, software, and documentation include the media themselves, physical storage requirements, and the preparation and periodic examination of backup copies. Choices for computer storage of data include paper hard copy (i.e., as backup storage), microcomputer floppy disks, hard disks, magnetic tape, permanent mainframe disk space, and others. Decisions regarding duration of storage of raw data (e.g., original field and laboratory data sheets, notebooks, hard copies of the archived data sets), may be based on the amount of physical space required or the costs of cabinets and environmental control needed to protect the media. Adequate protection from fire, floods, or other catastrophes can exceed the cost of the media themselves, but should be provided unless alternative procedures are used, such as routine transporting of backup copies to a place remote from their primary locations. Institutional mainframes may provide reasonable protection for storage of electronic media. The Kansas State University Computing Center, for example, maintains backup copies of some magnetic tapes in the deep salt mines in Rice County, KS. Memory requirements also need to be considered before purchasing large computer programs, whether it means upgrading micro- or minicomputer peripherals or paying a continual mainframe storage cost.

Data Exchange

Traditional means of data exchange include transporting data in forms such as printed (or handwritten) hardcopy, punched cards, magnetic tapes, or floppy disks. The added convenience and timeliness of direct transfer of data via telephone lines, microwave signals, or other means may be worth the cost of the equipment and installation. On a smaller scale, local hard-wired networks may increase the accessibility of the data; this is especially advantageous for groups of scientists working together. In multi-user networks, the use of a common database improves data integrity by making it possible to make updates and error corrections concurrently for all users, rather than providing new copies or updates to each user. Ultimately, networks of ecological research data may improve our abilities to accomplish cross-site comparisons or achieve a new level of scientific synthesis.

DECISION CRITERIA

Ecological databases have certain management requirements in common, but features of individual research programs must be taken into consideration when establishing a new RDM system.

Purpose of the Data Collection

The purpose for collecting ecological data may vary greatly in terms of broadness of scope, spatial scale, and temporal framework of the study. A research team may be involved in many projects that differ greatly in these characteristics. For example, one goal of biological field stations is collection of data for general research use (e.g., species lists, site and ecosystem descriptions, land-use history, meteorological data, long-term monitoring of ecosystems). Basic ecological research at the same site may be oriented toward short-term phenomena (e.g., life histories of ephemeral taxa, experimental manipulations to examine immediate responses) or long-term phenomena (e.g., climatic changes, long-term effects of temporal disturbance patterns, successional trends). All data have value that extends beyond the initial publication of results. Short-term experiments help in the interpretation of long-term trends; such studies should be archived as part of the database, even if the primary purpose is the evaluation of long-term phenomena.

The expected longevity of the RDM system will influence decisions concerning major hardware or software purchases or the need for long-term data storage. Ecological consulting firms, for example, may have a much more short-term perspective than a biological field station; however, consulting firms may place an even greater emphasis on quality control since their data may be very closely scrutinized (in a legal sense) in government hearings or courtroom situations.

Form of the Data

RDM should consider data in its broadest sense; computerized data are not the only kinds of information to be managed. Published papers or reports, theses, dissertations, and computer programs should all be maintained in a way that facilitates their retrieval. This is not meant to imply that a computer-searchable cross-referenced index of published materials is the goal; a simple listing and organized storage may be all that is needed. Other forms of data include maps, photographs, taxonomic reference collections, and samples to be stored as standards against which to compare future technologies. Just as loss of irreplaceable historic data can be tragic, so can the loss of an insect reference collection.

Size and Complexity of the Database
and Types of Analyses Planned

The hardware and software must be capable of storing the database (or an appropriate part of it) and performing the analyses desired, including interactive graphics if needed. Larger databases generally require a greater administrative support structure to manage them.

Data must be used to have value, even if that use occurs 50 years after collection of the data. Proper data management should increase ease of data access. In multidisciplinary research teams, it becomes especially important to be able to conduct exploratory analyses easily, and to be able to interactively examine the data for trends and compare parameters. An organizational scheme that allows comparisons across data sets and interactive graphics capabilities is valuable for this purpose. Enhanced ability to look at data in new ways may be necessary to fully interpret integrative studies of ecosystems.

Dynamics of the Database

Most scientific databases are dynamic or open-ended, that is, new parameters are added, old ones are dropped, and new ways of examining data are expected. This contrasts with traditional types of databases in the business world in which no (or few) changes in measurements or analyses are expected. Long-term databases require a flexible RDM system to allow such changes. This is especially important to consider in selection of software for a DBMS.

Some databases may be entirely static (that is, the data are the result of previous studies, and no new data are being added) or at least have static (historic) components. Such databases nevertheless have maintenance costs, which include correcting errors that might be discovered long after the termination of a study, periodic examination of data to assure their integrity, periodic backup procedures which help prevent problems of deterioriation of storage media, alertness to changes in technology, and retrieval of data upon request.

Number and Expertise of Investigators

A solitary investigator can apply the principles of base-level RDM to an individual project. A large team, however, requires an RDM system with components that are compatible among users, flexibility to accommodate varied needs (and expandable to include new ones), and an administrative structure.

Previous experience of the team members with computers should be considered. Although computer-literate scientists may successfully interact with a modern computer system, the system should still be user friendly to others. It should not be necessary for a programmer to be part of each analysis. If investigators feel that their access to needed data is limited, then the RDM system can decrease the overall productivity of the research team. The system should actually encourage participation by those unfamiliar with computers. A particularly effective tool is an easy-to-read user's manual for the system, which introduces the entire team, not just the data processing personnel, to the hardware and software that are available. Such a manual, coupled with regular training sessions provided by the data management staff, will increase the independence of the researchers with respect to data analysis, and will greatly reduce the time required to address each individual's needs as they arise.

Need For Network Capabilities

Cooperation among research sites will become more common as ecologists focus their efforts on broader scale questions involving intersite and regional comparisons (regardless if they have a common source of funding such as the National Science Foundation-funded LTER program). Sharing of information through computer networking is more easily attained if comparable data management procedures are followed at all sites and compatible equipment is available. With sound data management practices followed at all sites, future coordination efforts will be facilitated and there will be a greater trust in another site's data if it has been demonstrated that they follow accepted methods for quality assurance.

PERSPECTIVES FROM KONZA PRAIRIE

Development of a centralized RDM system at Konza Prairie Research Natural Area benefited greatly from data management workshops for biological field stations and LTER sites. We were able to learn from the experiences of other sites and could choose facets of existing programs that seemed to meet our needs the best. These workshops, this symposium, and the resulting volume provide an excellent means for communication and serve to promote the goal of intersite exchange of data.

During the development of the data set catalog, detailed instructions for completing the documentation forms proved to be useful. General information concerning hardware and software available to scientists and data entry personnel were prepared to lessen the data manager's task of training users. A methods manual containing details of procedures and site locations for each of the data sets provided further support for the summaries included in the documentation forms.

Data entry software suitable for our needs was not available for our existing hardware. The decision to proceed with development of in-house software contradicts current RDM recommendations. Data entry programs were individually modified to the formats of each data set, with appropriate automatic features such as decimal points, skipped spaces, repeated variable values, and insertion of species abbreviations upon entry of numeric codes. These programs continue to be useful for LTER data collections since data are entered each year using the same formats. They were tailored to our needs and allowed us to postpone purchase of new hardware. However, the costs of development (including extensive testing and debugging) were hidden in salaries of personnel involved. Further, substantial effort was required to prepare documentation of the programs with sufficient detail so that a new data manager could make modifications as needed. Our experience makes us even more cautious about developing in-house software in the future. Hardware purchases should be made with software needs in mind. A thorough search for commercially available software that has the desired features and flexibility and is

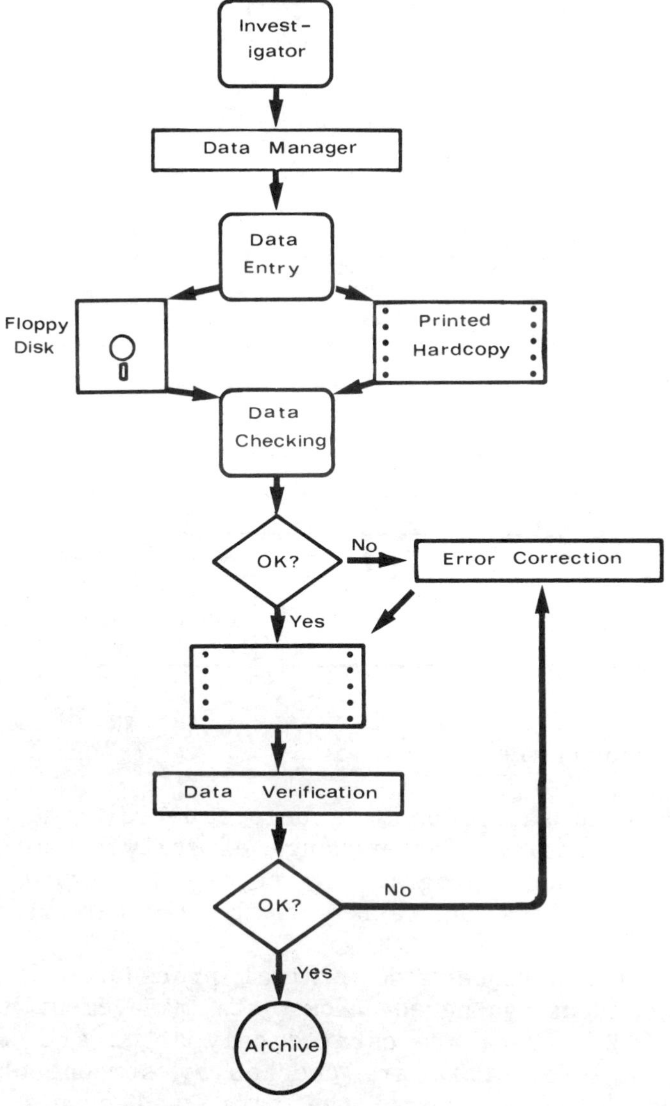

Fig. 1.   Data flow chart for Konza Prairie RDM: from investigator to the archival step.

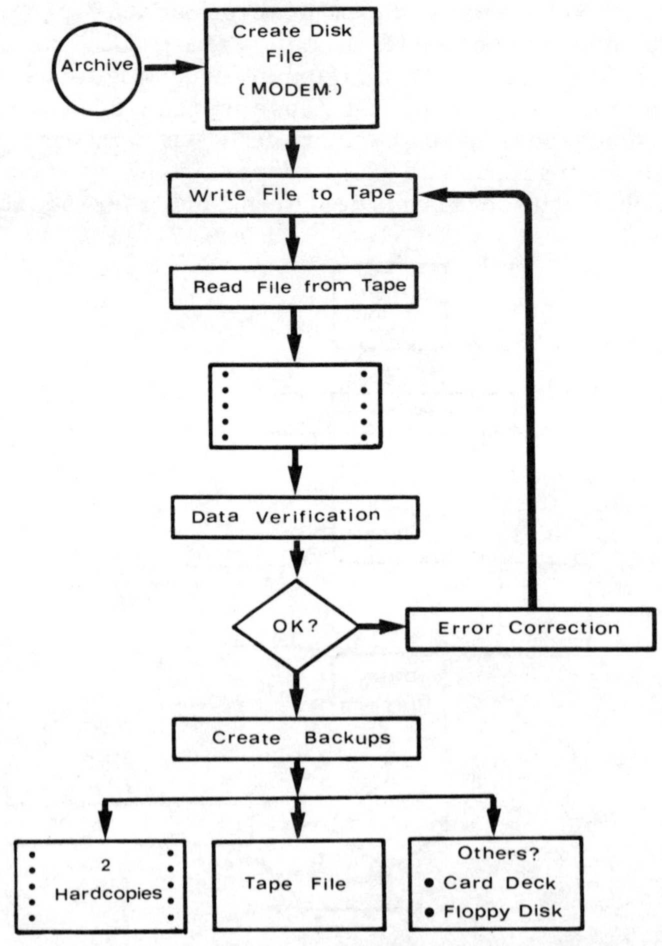

Fig. 2.   Data flow chart for Konza Prairie RDM: archival procedures.

well-documented should be made before deciding to write new programs. Mechanisms for exchange of software among scientists will reduce programming needs, but such software should nevertheless be tested to be certain it can meet one's needs.

Quality assurance and archival procedures were developed from ideas gathered from data management workshops (Figs. 1 & 2). Data are entered only after all documentation is complete. Data are checked by someone other than the individual who entered the data, and errors found at that time are corrected prior to the data being returned to the investigator, who carefully rechecks the data and verifies that they are correct. Archival involves transmission via modem to the mainframe and writing the data file onto magnetic tape. When the data have been successfully read back from the tape, the file is considered to be archived, and hardcopy and tape backups are made.

Throughout the process of development and implementation of the Konza Prairie RDM system, interaction between the data manager and the team of scientists has been essential. From designing field or laboratory data sheets to selection of record formats, data set documentation, and the data verification process, a congenial interaction leads to a mutual cooperation, without which the RDM system cannot work. My background as an ecologist and my involvement with research on Konza Prairie prior to becoming data manager made it easier to develop an organized framework for managing the data and to assist the researchers in all phases of RDM.

CONCLUSIONS

Research data management in an ecological research environment requires an integrated approach that involves (1) a clear understanding of the purposes of the research program and the types of data expected, (2) an organizational scheme which facilitates retrieval and use of the data, (3) an awareness of technological tools which are available, and (4) an appreciation for the future value of the database. A data set inventory and catalog, and procedures for storage, quality assurance, archival, maintenance, and administration should be incorporated into RDM, regardless of the size of the research staff or databases. The RDM budget must balance costs for personnel, hardware/software, storage, and data exchange as appropriate to the research project, with careful consideration of existing facilities.

RDM should be institutionalized so that it will proceed smoothly even if the original data manager leaves. Perhaps the worst scenario is that federal support for ecological research may be so sharply reduced that ongoing projects would have to be suspended. If there were only enough money to support storage costs, could collection and management of data be resumed after 5 or 50 years? We owe it to future generations of ecologists, as well as those who seek solutions to society's future problems, to create a data legacy that can survive interruption.

ACKNOWLEDGMENTS

I thank John Briggs, Dick Marzolf, Susan Stafford, and several anonymous reviewers for their constructive comments on earlier drafts of this paper. The author is supported by the National Science Foundation under grant BSR-8410425 granted in 1984.

LITERATURE CITED

Gorentz, J., G. Koerper, M. Marozas, S. Weiss, P. Alaback, M. Farrell, M. Dyer, and G.R. Marzolf. 1983. Data Management at Biological Field Stations. Report of a workshop at W.K. Kellogg Biological Station, Michigan State University, May 17-20, 1982. Prepared for the National Science Foundation. 46 pp.

Kanter, J.   1981.   Taming Your Computer.   Prentice-Hall, Inc., Englewood Cliffs, NJ. 246 pp.

Lide, D.R., Jr.   1981.   Critical data for critical needs. Science 212: 1343-1349.

Sinclair, R.A.   1983.   Long-term Ecological Research Data Management Workshop. Urbana-Champaign, IL. November 22-23, 1982. Illinois State Water Survey Miscellaneous Publication 72. 24 pp.

Stockmayer, W.H.   1978.   Data evaluation: A critical activity. Science 201: 577.

CHARACTERISTICS OF SCIENTIFIC DATABASES AND DATABASE MANAGEMENT SYSTEMS

V. Komarkova and J.L. Bell

ABSTRACT

The scientist needs database management software to obtain effective use of data. The most complex scientific tasks could not be accomplished without advanced database management. Business-oriented database management systems (DBMS) have been successfully applied to some scientific databases, but there is a clear need for systems tailored to the special characteristics of scientific data.

Because of the diversity of scientific databases, it is not possible to design a generalized data management system which will satisfy all needs of scientists. To provide increased data handling efficiency, such systems will have to be specifically designed for individual classes of scientific data, databases, and tasks. A scientific database classification, based on the underlying structure of the described systems, either anthropogenic or natural, is proposed.

INTRODUCTION

Scientific data may be kept in unstructured flat files, but a structured database management system (DBMS) usually greatly increases the efficiency of data use since most scientific data are being modified and added to frequently (Gault, 1984) and are the subject of diverse and repeated analyses. A DBMS is a software system that handles all access to the database and accommodates a variety of different applications using the same data. It presents a consistent and well maintained logical view of the data to the user or application program, completely separated from the physical storage and access techniques (McCarthy, 1979; Bacon, 1982).

Databases and thus DBMS's vary widely, but DBMS features usually include (Ross, 1978; Huffenberger & Wigington, 1979):

- data definitions, describing the data relationships and content, that are entered by a user.
- data are automatically arranged in physical storage.
- data manipulation, including retrieval and update, is supported through a user language or host language interface.
- data integrity, security and privacy are controlled.
- restart and recovery facilities in case of system failure.
- utility programs to load, reorganize, and copy data.
- reporting mechanisms are provided.

While the need for database management in technical fields originated with space exploration and military command and control programs in the 1950's (Jones & Ries, 1979), most of the generalized DBMS's in use today were developed in business environments and influenced by the hardware, software, and other problems encountered there (e.g., Benkovitz, 1979). Most of these systems are oriented toward modification and retrieval of formatted data such as inventories or accounts. They have a wide range in size and are available for microcomputers and mainframes; capabilities and flexibility are usually sacrificed on microcomputers (Huffenberger & Schermer, 1984). The rapid growth of business DBMS's reflects the recognition of information as a critical asset. In 1982, an estimated 50 companies marketed 54 different DBMS packages, and there were more than 15,000 systems installed in the United States (Blasgen, 1982).

A number of lower-level data retrieval and automated search information systems have been applied to scientific databases. For example, bibliographic, abstracting, and indexing services were developed in the 1960's (Blasgen, 1982). Many business-oriented DBMS's have been successfully applied to scientific data, particularly for low-level tasks such as structured record keeping and report generation (Jones & Ries, 1979; Town et al., 1980). Our experience includes testing of several DMBS's on site-keyed data. Hierarchical and some network systems (System 2000,

DMS-1100, and ARX-Boeing) worked well in managing the data and keeping dictionaries; the only exception was the network system SEED which was inefficient and expensive to use for our type of data and tasks.

Systems developed primarily for statistical analysis of scientific data began incorporating data management features in the 1970's (Greenberg et al., 1981). In general, statistical packages offer only limited data management which is not adequate for analyzing large data sets (Bates et al., 1982).

Systems designed specifically for management of scientific data (i.e., SDBMS's) began appearing in the 1970's (Hampel & Ries, 1978). Most of these systems either do not have a full set of data management operations or are specialized. Portable, generalized SDBMS's are still scarce; most of these systems have been developed at the National Laboratories. The most complex and computationally intensive scientific tasks now do, and probably always will, require special processing.

The lag of scientific behind business data management (e.g., Felippa, 1979) may be due to
1.   lower level of support for data management in science, specifically the absence of a coordinated information dissemination system in federal information policy (Sherrod, 1977; Luedke et al., 1977).
2.   the specific characteristics and usually more demanding requirements of scientific databases.
3.   the lower level of knowledge about processes and systems subject to scientific investigations.

This lag may only be an inconvenience to the individual scientist, but on a national basis it may lead to large-scale loss of data which were often collected at great expense (Stevens et al., 1979; Freeman, undated). In addition, data often cannot be shared and verified independently (Hampel, 1973).

Management of scientific data is most advanced in well-funded areas such as defense, medicine, and industry-related research, and in those environmental sciences which are the focus of federal agency research. One of the worst situations exists in the chronically underfunded biological sciences. Olson (1984) reviewed environmental and natural resources databases in the US and concluded that ecological and biological data appear to be more dispersed and have less standardization than water, climate, and other abiotic data. Biological phenomena are generally less well known (e.g., due to the absence of a US Biological Survey), and the biological data entities and their relationships are more vague than, for example, physical data (see also Chayes, 1981). Lack of funding prevented the maintenance of project-wide, machine-readable databases even in large cooperative biological projects in the past (e.g., as in parts of the International Biological Programme).

The need for management of ecological data, particularly of the unpublished information, is well recognized (e.g., Oppenheimer et al., 1976; Armentano & Loucks, 1979; Altman & Fischer, 1981; Olson, 1984; Marzolf & Dyer, this volume; Risser & Treworgy, this volume). The establishment of regional or global information centers for environmental sciences has been proposed repeatedly (e.g., Oppenheimer et al., 1976; Armentano & Loucks, 1979). Suggestions have been made that a portion of all research project funds be dedicated to data management. Part of the system could be a depository of raw data, perhaps used as a prerequisite for publication after an article has been accepted by a refereed journal. Raw data are usually not accepted for publication by standard outlets (e.g., Rossmassler, 1980a) although they are inherently more valuable than one researcher's conclusions. Raw data should be easily available to other users, preferably in machine-readable form (Stevens et al., 1979). As SDBMS's become more widely used, such data collection and sharing will become easier, will become more cost-effective and accepted, and will facilitate networking in the future (Klopsch & Stafford, this volume).

This paper explores the characteristics of scientific databases and the implications for scientific database management systems (SDBMS's) that would provide data manipulation operations automatically. An understanding of the characteristics of scientific data is needed so that programs can be built which manage scientific data in an efficient way and which scientists can use easily.

DISCUSSION

Characteristics of Scientific Data

The characteristics of scientific data and databases are as diverse as the purposes and sources from which the data are collected. Scientific databases generally document or model natural systems and processes. The database is composed of data elements (e.g., chemical element weights) and descriptions of relationships among elements (e.g., molecular structures). Scientific data include measurements, counts, textual descriptions, and calculated or derived values. These values arise from a variety of data types, both numeric (predominantly) and non-numeric. Fixed point, floating point, double precision, complex numbers, and character string data types should all be provided for data display and storage in SDBMS's (Bestougeff, 1984). Numeric values can vary over several orders of magnitude in one database (e.g., Birss et al., 1978; Turner et al., 1979) and accuracy is very important (Fuja & Lindeman, 1978), so the SDBMS must provide for accurate representation of wide ranges. Among non-numeric data, variable length data are very common and must also be accommodated.

Descriptive and annotative information is an essential part of a scientific database (Hampel & Ries, 1978). For example, data that give measured values may be annotated with uncertainty or tolerance, units of measurement, normalization, method of measurement, measurement instrument, conditions or constraints, bibliographic reference, comments, proprietary status, classification, and cost (Hampel et al., 1977; Hampel & Ries, 1978). It is often prohibitive to store the data descriptions with the data values, since the descriptive information may itself require too much storage (Chan & Shoshani, 1981; Bates et al., 1982); however, the SDBMS must make some provision for it.

Scientific databases often contain a very high volume of data, both a large number of collection periods and a large number of attributes describing each entity; therefore data compression is very important. The SDBMS should allow storage of compact codes in a data field that can be expanded to a longer textual or numeric description when desired. This improves storage efficiency and maintains control of precise textual description of technical data (Szczesny & Gersbacher, 1978; Tubbs, 1979). Textual data in science and technology show little specificity as compared with data from other fields, so that standard checks, coding, and compression techniques can be used (Bestougeff, 1984).

Other special characteristics of scientific data include scientific notations and symbols (e.g., Hampel & Ries, 1978). The symbols may be application-specific, so the SDBMS should allow symbols beyond a predefined set. Automatic conversion from different units of measurement for comparison, retrieval or display should be provided by the SDBMS.

Some difficulties in database design reflect the very high complexity of the scientific environment. A database is an abstraction of the scientific system, which cannot be described in all its original detail. Problems with definition of abstract entities include delimitation of entities (from the continuous whole); distinguishing of like entities (e.g., how to assign unique identifiers to each pine tree or each patch of forest); and changes in the essential nature of an entity (Bell, 1982, 1983).

Some problems with modeling a scientific system result from insufficient knowledge of the system itself. Deciding what attributes to include can be a major database design problem because the types of questions to be analyzed are not yet fully known. Modeling of time is not easy in a database, especially when precise time intervals are not known.

In contrast, business-oriented databases usually store data about people or anthropogenic systems. There are well-known identifiers (e.g., social security number) and the questions to be answered by analyzing the data are well known. The data are stored as fixed format records consisting of fields containing integers or short character strings (e.g., "NAME" and "SALARY") (e.g., Brooks, 1978). Business databases do not include other data types (e.g., floating point), scientific notation, special symbols, or complex user-visible structures (Benkovitz, 1979).

Relationships in Scientific Data

The relationships among data elements can be described by vectors, matrices, arrays, graphs, and hierarchies (trees). Regardless of the underlying storage structures of the SDBMS, the user should be able to view all of these structures (Brooks, 1978; Fuja & Lindeman, 1978; Hampel & Ries, 1978; Richards, 1978; Jones & Ries, 1979; Tubbs, 1979; Ehrnsperger & Kelm, 1981; Gault, 1984).

The three most common underlying structures within computer storage are relational, hierarchical, and network (see Date, 1977 for a full description of each structure; structures are also discussed in Komarkova & Bell, 1985). The efficiency of a SDBMS depends in large part on the appropriateness of its storage structures for the "natural" structure of the data. Retrieval of related data can be accomplished much faster if the relationships are described explicitly in the storage structure. However, pointers to related data or other structural descriptions take up space, and deciphering them during retrieval takes computer time. Therefore, a general rule is to use the simplest structure that fits the natural relationships in the data: first relational, then hierarchical, and, lastly network (Brooks, 1984).

The structure of a SDBMS should be flexible so that the data can be used in ways and for applications which were not anticipated, the structure can be easily reorganized based on new knowledge, and new data and data handling require-ments can be integrated as the research continues (e.g., Birss et al., 1978; Jones, 1981). This is a major advantage of the relational approach; relationships need not be prede-fined since they are not explicitly used to organize the underlying storage (Haskin & Lorie, 1982). Other storage structures can be reorganized, but few existing DBMS's offer easy restructuring (Tubbs, 1979).

Data Manipulation Operations

Standard data manipulation operations are data entry and validation, retrieval, display, and update. Efficient and fast update and retrieval are emphasized in all DBMS's (Fuja & Lindeman, 1978; McCarthy, 1979). A major difference for scientific databases is the relative frequency of per-forming each operation, and hence decisions about which operations to optimize. Also, differences in data charac-teristics may influence the way that these operations are performed.

Data Entry and Validation. The data to be stored in a database is generally entered manually, usually in a batch mode. Helpful DBMS facilities include data validation (i.e., checking for errors), automatic conversion to dif-ferent units (Hampel & Ries, 1978; Szczesny & Gersbacher, 1978; McCarthy, 1979), and reformatting for transfer to/from other machines (Rumble & Seidenman, 1984). These types of facilities are available in most standard DBMS's. Scienti-fic users will need additional validation procedures for the extra data types in scientific databases. These procedures include checking individual values for maxima, minima, and precision; and statistical tests or preprocessing of groups of values (e.g., curve smoothing) (Bestougeff, 1984). Data standardization for local applications and for the exchange of information between sites (Berger & Tucker, 1980; Glad, 1981; Glad et al., 1981; Bestougeff, 1984) can be also done during data entry.

Retrieval. In scientific databases both sophisticated and simple queries are required. For some applications, data is retrieved using simple criteria and is always retrieved in the same pattern (e.g., by spatial region) (Bell, 1983). For other applications, on-line query of the database may combine data from many sources in unpredictable ways (Brooks, 1978; Jones & Ries, 1979; Ehrnsperger & Kelm, 1981; Haskin & Lorie, 1982).

For applications requiring an interactive interface, the query language must be powerful enough to express complex queries (i.e., it must provide Boolean, predicate calculus, and structured operands such as data ranges, matrix indices, and relational operands) (Brooks, 1978; Szczesny & Gersbacher, 1978). The SDBMS should also be flexible. Scientific data retrievals are typically of the form "Give me all instances of X that satisfy criteria Y." Many instances are retrieved at one time, based on several attributes. The scientific database, therefore, must have an underlying storage structure that can efficiently retrieve many instances based on any combination of attributes, regardless of pre-designated "key" attributes that were used to organize the physical storage.

Scientific users usually do not want to become programmers. There should be an interface for naive users that requires no "programming" (Hampel & Ries, 1978; Stevens et al., 1979; Chan & Shoshani, 1981; Kitagawa, 1981). This may take the form of non-procedural queries, a menu selection, natural language, or other alternatives. The ability to "browse" through a directory or index of information is very important because of the often large numbers of data elements (Chan & Shoshani, 1981).

Scientific users often perform the same type of query repeatedly. One should be able to store regularly used requests (Szczesny & Gersbacher, 1978), especially complicated ones, to avoid having to re-enter them. Similarly, one should be able to store the retrieved data, or summary data derived from the retrieved data, as a separate data set (or temporary file) for later analysis or further querying (Hampel & Ries, 1978; Chan & Shoshani, 1981; and others). This capability requires more disk storage but saves the expense of re-performing the same query at a later time.

Data Display. A database user must have a variety of options for the display of data once it is retrieved. An interactive display capability is required for most databases. The user should be able to selectively request display of all or any subset of the retrieved data (Stevens et al., 1979), either at a terminal or on paper. Report generation (i.e., complex output formats) and graphic display are highly desirable options (e.g., Szczesny & Gersbacher, 1978). Report capabilities for scientific databases should include formulas and scientific notations (Hampel & Ries, 1978; Rumble & Seidenman, 1984). Editing, composing, and typesetting that produce output acceptable for publication are desirable (Rumble & Seidenman, 1984).

Some display capabilities (e.g., report and graphics generation) may be provided by software outside of the SDBMS. Data must be transferred to a file for input to other software or for exchange of information between sites. The user should be able to output data from database storage to a file, and he should be able to specify the exact format (Glad et al., 1981; Bestougeff, 1984).

Update. Scientific databases are usually quite static because they typically contain observations, measurements, and facts. Most databases are infrequently updated, with only additions of new material and corrections to old data. If an update is to be performed, it is often a replacement of an entire data set, rather than changing individual instances or attributes (Fuja & Lindeman, 1978; Szczesny & Gersbacher, 1978; Bates et al., 1982). For example, atmospheric measurements at a given time may be updated by replacing them all with measurements from a later time. Therefore, efficient retrieval is more important than efficient update; concurrent updates and queries are not critical for scientific databases (Bernstein & Goodman, 1981; Kohler, 1981).

An interactive update capability can be a useful feature (e.g., for small corrections). Logging of whatever updates are performed and recovery of damaged files should be standard procedures (Tubbs et al., 1981).

Computation. Unlike commercial applications, scientific data are usually subject to considerable computation after retrieval (e.g., Fuja & Lindeman, 1978; Suich & Honeck, 1978). Many kinds of analyses are performed, from simple mathematical tasks through sophisticated statistical procedures. The SDBMS may provide a variety of options for computation, or the SDBMS may be callable from within a programming language. It is an open question as to how far a SDBMS should go in providing computational functions. However, to the extent that the SDBMS provides computational functions, the user is spared the extra step of transferring the data and reformatting it for processing by another program (Green & Atkisson, 1981; Ikeda & Kobayashi, 1981).

For example, statistical database users almost always require the ability to aggregate data and do statistical analyses. If these functions are available in the SDBMS, the user may stay in that environment for both retrieval and processing. Otherwise, the retrieved data must be outputed to a file and then inputed to a statistical package. The process of data analysis and data retrieval will continue interatively, necessitating much data transfer between the SDBMS and the statistical package. Therefore, many experts recommend that a SDBMS provide comprehensive computational facilities, including the ability to "program" application-specific analyses (Brooks, 1978; Hampel & Ries, 1978; Szczesny & Gersbacher, 1978; Jones, 1981; and others).

Classification of Scientific Databases

Because we were concentrating on common or typical characteristics of scientific data, the above discussion does not indicate the diversity of scientific databases. As we approach the questions of efficient implementation of SDBMS's, we see that it is very difficult to make a system that will efficiently satisfy the database manipulation needs of all scientific databases. For example, spatially-related data need an underlying structure which reflects the three-dimensional spatial relationships, but a simple hierarchical structure would suffice for biological classification data.

The best strategy for data management depends heavily on the nature of the data and its use (McCarthy, 1979); the influence of target applications is now being confirmed by formal studies. For example, Hawthorn (1981) compared business, bibliographic search, and statistical analysis data systems through the single metric of the effective instruction rate and found that the optimal organization of data manipulation processors was application dependent and a family of approaches was indicated. Similar studies could be extended to different classes of scientific applications.

Categories of scientific data have been listed, for instance, by Luedke et al. (1977), Rossmassler (1980b), Watson
(1980), and Rumble (1984). Specialized data management
systems have been designed for some of the categories (e.g.,
textual data).

Chambaud (1977) defined classes of scientific databases
by performing correlation, cluster and factor analysis on a
number of scientific databases. His analysis examined variables such as identity of the data subject; simplicity in
defining the circumstances of the data subject; number of
states taken by the data subject; desired precision; homo-
geneity; lifetime of the data; and frequency of updating.
He concluded that there were two principal classes:

1. Observational databases (e.g., weather observa-
    tions). These large databases have limited con-
    trol and store data that cannot be remeasured.
    The data can be aggregated or summarized.
2. Depositional databases (e.g., material proper-
    ties). This data can be obtained again, are
    smaller, have data validation problems, and are
    seldom aggregated.

The principal axes of the factor analysis represented ease
of database use, frequency of movement of data in the database, and the complexity of the data. Chambaud did not find
significant divisions according to scientific discipline or
technical field. Different disciplines can have very
similar databases, for example, the databases for the classification systems of biological organisms and of chemical
compounds. Similarly Chambaud did not find the division of
numerical vs. non-numeric data helpful as many databases
have both types of data.

Extending this classification, we propose a classification of scientific databases based on their conspicuous
systematic characteristics, reflected primarily in their
structure:

1. Anthropogenic systems. An example is the spatial
    arrangement of specimens in a museum, or a bibli-
    ography. These systems are close to business
    systems, in that they reflect human activity.
    They have a variety of underlying structures.
2. Anthropogenic systems reflecting, more or less,
    natural order. For example, taxonomy of organ-
    isms, evolutionary systems, and mineral systems.
    Some systems may be time- and/or space-dependent.
    These databases will usually have underlying

hierarchical structures. They will be relatively small in volume of data but may have many attributes.

3. Natural systems which are always time- and space-dependent. These databases contain observational data, in small or large amounts, or interpolated (via computer simulation) data, in unlimited amounts. These systems record data about many spatial locations and must also store the spatial relationships. The underlying structure is a multidimensional grid (array) or a network. This class can be further subdivided into

   a. Time element not included. These databases usually store observations at only one point in time and are very static.

   b. Time element included. Data include observations for many spatial locations and for many times per location, for example, satellite data or weather prediction data. These databases typically have very large volumes of data and are very dynamic, although often by adding data rather than by updating specific locations.

The proposed classification recognizes the diversity of scientific databases, and allows us to study each class separately. The classification is largely based on the underlying structure of the system being modeled, either anthropogenic or natural. To the extent that the storage structure can closely represent the structure of the natural system, the SDBMS can efficiently retrieve data from the stored structure. Thus SDBMS's for each class can be more efficient than a generalized SDBMS that tried to satisfy all classes of scientific databases. Similarly, the proposed classification may guide new users to SDBMS's which apply to their situations.

For example, scientific simulation models (e.g., weather prediction models) are part of the class of natural systems with time element included (3b. above). These models predict natural phenomena in some spatial domain by predicting local conditions at a very large number of contiguous spatial locations. For example, windspeed, humidity, etc. may be predicted at each grid point in a three-dimensional grid superimposed on the atmosphere over North America. A grid of size 100 by 100 by 15 is typical; with 40 attributes for each grid point, there are 6 million numbers being stored. Each number is re-computed and up-dated hundreds of times as the predictions evolve. Because of the combination of database size and computational inten-sity, efficient data retrieval is critical for making a realistic model. The data retrieval, computation, and up-date must be very closely coupled (Bell, 1982, 1983). Bell (1983) has proposed a specialized SDBMS, callable from FORTRAN, that is designed for efficiency with a large, multi-dimensional grid structure. It also economizes on retrieval time by allowing the user to sweep through all of the data in a pre-defined, often sequential, order.

Software for Scientific Data Management

The scientist, who is more concerned with science than with computer operation, would prefer to leave the imple-mentation of a scientific database to someone else. Unfor-tunately, this is usually not possible. The scientist must go through the following steps in a project which is data intensive: identify and define the data needs, plan the data project, design the database, obtain a suitable SDBMS if desired, link the database to users, and link the data-base to other databases (Rumble, 1984).

Often, the data are simply stored in a single file and processed sequentially. If the scientist does not desire more sophisticated processing, this is appropriate and very efficient (Brooks, 1984). However, whenever random access to the data or access to data in different orders is needed, repeated re-reading of a sequential file is very ineffi-cient. Also, the annotations, etc. about the data may not be recorded and are often lost after the project terminates.

Historically, scientists have been reluctant to use existing database management systems because they were too inefficient (Patterson, 1982; Bell, 1982, 1983). Most com-mercially available DBMS's were designed for business appli-cation, and hence optimize for quick random access retrieval primarily on the basis of identifiers. For anthropogenic scientific systems, these DBMS's may be efficient; but they are not efficient for many scientific applications.

Generally the user should pick a SDBMS that provides the desired functions within the constraints of efficiency for the particular scientific database. If a SDBMS is too general, it may decrease efficiency and be unnecessarily complicated to use (Bestougeff, 1984). On the other hand, a generalized SDBMS will be more flexible, be useful for many related projects, and have a longer life. Tubbs (1979) suggested criteria for projects that need a generalized SDBMS: the database is shared among several users, complex relationships among data items exist, queries are unpredictable, and the database is large.

Complete specification of the capabilities required in a generalized SDBMS is very difficult (Szczesny & Gersbacher, 1978), and no system meets all needs. Emphasis on specific features stems, in part, from differences between installations and work environments (Hampel & Ries, 1978). Major constraints on SDBMS design have included characteristics of phenomena and systems; scientific principles; numerical methods; types and intended uses of the data; computer environment, especially amount and type of storage; size of effort, such as funds and personnel available; and familiarity of designers and users with computer systems (McCarthy, 1979; Bates et al., 1982). Other considerations include the perceived desirability of portability to other environments and for growth (e.g., Jones & Ries, 1979; Stonebraker, 1981), security from unauthorized use (e.g., Bestougeff, 1984), tunability for optimal performance (Szczesny & Gersbacher, 1978), and ability to archive unused portions of the database (Szczesny & Gersbacher, 1978). Many of the above considerations are mutually inconsistent, so the SDBMS designer must carefully evaluate the tradeoffs for his eventual users.

The alternative to using existing software is, of course, to build your own, and most scientists have built their own data management systems from scratch (e.g., Bouillé, 1976, 1977; Bouillé et al., 1979; Patterson, 1982). The primary advantage is that these systems can be tailored to the exact database being constructed. However, this usually precludes extensive refinement of the SDBMS because of the extremely high initial cost of building a full system (Felippa, 1979; Tubbs, 1979). Many experts suggest that the SDBMS designer avoid the temptation to build his own system (Gindra, 1977; Farrell et al., 1982; Bestougeff, 1984). A newly developed product probably costs ten times more than a standard one (Huffenberger & Schermer, 1984). In addition, locally developed systems often lack needed functions and cannot be used for more than one project or for shared data.

CONCLUSIONS

There are significant special characteristics of scientific data, databases, and their uses. For example, a science application emphasizes use of information while a business application emphasizes its control (Gault, 1984). Scientific data and databases have been less well defined than business ones, and scientific tasks may be computationally and organizationally more intensive. Despite these differences, scientists can benefit from business experience with database management systems.

The scientist has the following data management options: use a DBMS not specialized for scientific databases; use a generalized SDBMS, if obtainable; develop his own system; or do without any software. Commercially available DBMS's are the most cost-effective solution if the scientific database will perform adequately on such software. Using a generalized SDBMS is advisable for a wide range of scientific databases. Developing your own system should be considered only for organizations with large resources or very simple databases. Doing without database management software altogether is reserved for extremely simple databases. Even the simplest data management greatly increases the efficiency and productivity of the scientific process. The authors of this paper encourage the development of specialized portable SDBMS's, to provide additional options for a wider range of scientific database users.

The development of scientific database management has been spontaneous and is continuing at this time. Experts differ about the "best" design for a SDBMS and these disagreements may often be the result of unrecognized diversity among scientific databases. Because of this diversity, a complete specification of the capabilities required in a generalized SDBMS may be impossible.

The preceding compilation of scientific database characteristics was based on the empirical insights of scientific database designers, administrators, and users. Only a few designs and applications have been subjected to formal analysis by information science and data modeling theory. As this work continues, the characteristics and classification of scientific databases will permit the construction of better general purpose data management software for scientific database users.

ACKNOWLEDGMENTS

The idea and much of the information for this paper originated with the Working Group on Scientific Data Management, organized by Dr. R.D. Hackathorn at the Division of

Information Science Research, College of Business and Administration, University of Colorado. We are grateful to Gaurangi Kamani and Martha Andrews who helped to obtain the literature. This work was supported by National Science Foundation grants DEB-8012095 (Long-Term Ecological Research in Colorado alpine) and DPP-7825748 (Vegetation of the sand region, Arctic Coastal Plain, Alaska).

LITERATURE CITED

Altman, P.L. and K.D. Fischer. 1981. Guidelines for development of biology data banks. Life Sciences Research Office, Federation of American Societies for Experimental Biology, 9650 Rockville Pike, Bethesda, MD. 74 pp.

Armentano, T.V. and O.L. Loucks. 1979. Ecological and environmental data as under-utilized national resources: Results of the TIE/ACCESS program. The Institute of Ecology, Indianapolis. 205 pp.

Bacon, G. 1982. Software. Science 215(4534): 775-779.

Bates, D., H. Boral, and D.J. DeWitt. 1982. A framework for research in database management for statistical analysis or a primer on statistical database management problems for computer scientists. pp. 69-78. In: Proceedings of the ACM SIGMOD International Conference on Management of Data. Orlando, FL, June 2-4, 1982.

Bell, J. 1982. Data modelling of scientific simulation programs. pp. 79-86. In: Proceedings of the ACM SIGMOD International Conference on Management of Data. Orlando, FL, June 2-4, 1982.

Bell, J.L. 1983. Data structures for scientific simulations programs. Ph.D. Thesis, University of Colorado, Boulder, CO. 250 pp.

Benkovitz, C.M. 1979. The use of SYSTEM 2000 in a scientific research environment. In: Meeting of the Association of SYSTEM 2000 users for Technical Exchange (ASTUTE). Austin, TX, April 1979.

Berger, P.W. and J.C. Tucker. 1980. Standards and guidelines for data. pp. 93-113. In: Data Handling for Science and Technology. S.A. Rossmassler and D.G. Watson (eds.). North-Holland, NY.

Bernstein, P.A. and N. Goodman. 1981. Concurrency control in distributed database systems. Computing Surveys 13: 185-221.

Bestougeff, H. 1984. Designing the database management system. pp. 119-166. In: Database Management in

Science and Technology. J.R. Rumble and V.E. Hampel (eds.). North-Holland, NY.

Birss, E.W., S.E. Jones, D.R. Ries, and J.W. Yeh. 1978. Scientific data base management at Lawrence Livermore Laboratory: Needs and a prototype system. pp. 132-144. In: Generalized Data Management Systems and Scientific Information. Organization for Economic Cooperation and Development (OECD), Nuclear Energy Agency, Paris.

Blasgen, M.W. 1982. Database systems. Science 215(4535): 869-872.

Bouillé, F. 1976. A model of scientific data bank and its applications to geological data. Computers and Geosciences 2: 279-291.

Bouillé, F. 1977. A model of protected, shareable and portable scientific data bank. pp. 523-531. In: Proceedings of the Fifth Biennial International CODATA Conference, Boulder, CO, June 28-July 1, 1976. B. Dreyfus (ed.). Pergamon Press, NY.

Bouillé, F., D. Pajaud, and M.-J. Roulet. 1979. Paleontological data processing with a HBDS data bank at the Universite Pierre et Marie Curie. pp. 381-391. In: Proceedings of the Sixth International CODATA Conference. B. Dreyfus (ed.). Santa Flavia, Italy, May 22-25, 1978.

Brooks, A.A. 1978. The structure of R & D information - Some observations. pp. 93-105. In: Generalized Data Management Systems and Scientific Information. Organization for Economic Cooperation and Development (OECD), Nuclear Energy Agency, Paris.

Brooks, A.A. 1984. Data types and structure in science and technology. pp. 15-37. In: Database Management in Science and Technology. J.R. Rumble, Jr. and V.E. Hampel (eds.). North-Holland, NY.

Chambaud, S. 1977. Classification of databanks. pp. 533-536. In: Proceedings of the Fifth Biennial International CODATA Conference. Boulder, CO, June 28-July 1, 1976. B. Dreyfus (ed.). Pergamon Press, NY.

Chan, P. and A. Shoshani. 1981. SUBJECT: A directory driven system for organizing and accessing large statistical databases. pp. 553-563. In: Proceedings of the Seventh International Conference on Very Large Data Bases (ACM). Cannes, France, September 9-11, 1981.

Chayes, F. 1981. Attitudes toward data in the hard and medium-hard sciences. pp. 116-121. In: Data for

Science and Technology. Proceedings of the Seventh International CODATA Conference, Kyoto, Japan, October 8-11, 1980. P.S. Glaser (ed).

Date, C.J. 1977. An Introduction to Database Systems. 2nd ed. Eddison-Wesley, Reading, MA. 536 pp.

Ehrnsperger, W. and R. Kelm. 1981. The new database system established in the German Geodetic Research Institute, Dept. 1 (DGFI). pp. 239-253. In: Proceedings of the International Symposium on Management of Geodetic Data. Geodetic Institute, Copenhagen, Denmark.

Farrell, M.P., R.H. Strand, J.C. Goyert, K.L. Daniels, D.H. Holzworth, J.R. O'Fallon, and A.D. Magoun. 1982. Review of research data management. pp. 457-467. In: Proceedings of the Fifteenth Hawaii International Conference on System Sciences. Vol. 1. Software, hardware, Decision Support Systems, Special Topics. W. Riddle, K. Thurber, P. Keen, and R.H. Sprague, Jr. (eds.). University of Hawaii Press, Honolulu.

Felippa, C.A. 1979. Database management in scientific computing. I. General description. Computers and Structures 10: 53-61.

Freeman, R.P. (undated). The national environmental data referral service: A publicly available online data catalog. 14 pp. (Unpublished manuscript; National Environmental Satellite, Data, and Information Service; National Oceanic and Atmospheric Administration, Washington, DC 20235).

Fuja, P.M. and A.J. Lindeman. 1978. Status of database management systems at Argonne National Laboratory. pp. 167-172. In: Generalized Data Management Systems and Scientific Information. Organization for Economic Cooperation and Development (OECD), Nuclear Energy Agency, Paris.

Gault, F.D. 1984. Database management systems for science and technology. pp. 39-73. In: Database Management in Science and Technology. J.R. Rumble, Jr. and V.E. Hampel (eds.). North-Holland, NY.

Gindra, A.W. 1977. Some experiences with remotely accessed databases. pp. 161-164. In: Proceedings of the Fifth Biennial International CODATA Conference, Boulder, CO, June 28-July 1, 1976. B. Dreyfus (ed.). Pergamon Press, NY.

Glad, A.S. 1981. System-wide data management for General Atomic's Doublet III distributed computer system. pp. 587-590. In: Ninth Symposium on Engineering Problems of Fusion Research. IEEE, NY

Glad, A., D. Drobnis, and B. McHarg. 1981. Data management in a fusion energy research experiment. General Atomic Project 3344. Contract DE-AT03-76ET51011 to San Francisco Operations Office, Department of Energy. 30 pp.

Green, R.S. and C.C. Atkisson. 1981. A model system for psychotherapy research data management. Behavior Research Methods and Instrumentation 13: 499-510.

Greenberg, E.A., W.M. Ivey, and B.R. Lewis. 1981. Comparison of some available packages for use in research data management. SigSoc Bulletin 13: 1-8.

Hampel, V.E. 1973. Problems and some solutions for the creation and utilization of large interdisciplinary computerized data banks. Lawrence Livermore Laboratory, Livermore, CA.

Hampel, V.E., E.A. Henry, R.W. Kuhn, and L. Lyles. 1977. Acquisition, storage, retrieval, display and utilization of computerized data in the LLL data bank of physical and chemical properties. pp. 553-566. In: Proceedings of the Fifth Biennial International CODATA Conference, Boulder, CO, June 28-July 1, 1976. B. Drefus (ed.). Pergamon Press, NY.

Hampel, V.E. and D.R. Ries. 1978. Requirements for the design of a scientific data base management system. pp. 111-131. In: Generalized Data Management Systems and Scientific Information. Organization for Economic Cooperation and Development (OECD), Nuclear Energy Agency, Paris.

Haskin, R.L. and R.A. Lorie. 1982. On extending the functions of a relational database system. pp. 207-212. In: Proceedings of the ACM SIGMOD International Conference on Management of Data. Orlando, FL, June 2-4, 1982.

Hawthorn, P. 1981. The effect of target applications on the design of database machines. pp. 188-197. In: ACM SIGMOD International Conference on Management of Data. Ann Arbor, MI, April 29-May 1, 1981. University of Michigan. Y.E. Lien (ed.). Association for Computing Machinery, NY.

Huffenberger, M.A. and C.A. Schermer. 1984. Selecting a suitable database management system. pp. 167-196. In: Database Management in Science and Technology. J.R. Rumble, Jr. and V.E. Hampel (eds.). North-Holland, NY.

Huffenberger, M.A. and R.L. Wigington. 1979. Database management systems. Annual Review of Science and Technology 14: 153-190.

Ikeda, H. and Y. Kobayashi. 1981. Additional facilities of a conventional DBMS to support interactive statistical analysis. pp. 25-36. In: Proceedings of the Workshop on Statistical Database Management. Menlo Park, CA, December 1981.

Jones, S.E. 1981. Using a relational DBMS in the scientific community. pp. 491-499. In: Proceedings of the Seventh International CODATA Conference. Kyoto, Japan, October 8-11, 1980. Pergamon Press, NY.

Jones, S.E. and D.R. Ries. 1979. A relational database management system for scientific data. pp. 353-357. In: Proceedings of the Sixth International CODATA Conference. B. Dreyfus (ed). Santa Flavia, Italy, May 22-25, 1978.

Kitagawa, T. 1981. Group study on database management system. pp. 36-42. In: Scientific Information Systems in Japan. H. Inose (ed.). North-Holland, Amsterdam.

Klopsch, M.W. and S.G. Stafford. 1985. The status and promise of intersite computer communication. pp. 115-123. In: Research Data Management in the Ecological Sciences. W. Michener (ed.). Belle W. Baruch Library in Marine Science, No. 16. University of South Carolina Press, Columbia.

Kohler, W.H. 1981. A survey of techniques for synchronization and recovery in decentralized computer systems. Computing Surveys 13: 149-183.

Komarkova, V. and J.L. Bell. 1985. Database management systems (DBMS's) in scientific applications: Use and bibliography. Long-Term Ecological Research in Colorado Alpine Data Report. Institute of Arctic and Alpine Research, University of Colorado, Boulder, CO.

Luedke, J.A., G.J. Kovacs, and J.D. Fried. 1977. Numerical databases and systems. Annual Review of Information Science and Technology 12: 119-181.

Marzolf, G.R. and M.I. Dyer. 1985. Summary and future directions of research data management in ecology. pp. 409-417. In: Research Data Management in the Ecological Sciences. W. Michener (ed.). Belle W. Baruch Library in Marine Science, No. 16. University of South Carolina Press, Columbia.

McCarthy, J.L. 1979. Data characteristics, application requirements, and database management tools: Matching statistical user's needs and systems capabilities. pp. 37-44. In: Proceedings of the Computer Science and

Statistics, 12th Annual Symposium on the Interface. University of Waterloo, Canada, 1979.

Olson, R.J.   1984.   A review of existing environmental and natural resource data bases. Report no. ORNL/TM-8928. Oak Ridge National Laboratory, Environmental Sciences Division, Oak Ridge, TN. Publication 2297. 71 pp.

Oppenheimer, C.H., D. Oppenheimer, and W.B. Brogden (eds.). 1976. Environmental Data Management. Plenum Press, NY 244 pp.

Patterson, G.S., Jr. 1982. Large scale scientific computing - Future directions. Computer Physics Communications 26: 217-225.

Richards, D.R. 1978. An overview of BDMS: The Berkeley Database Management System. pp. 145-150. In: Generalized Data Management Systems and Scientific Information. Organization for Economic Cooperation and Development (OECD), Nuclear Energy Agency, Paris.

Risser, P.G. and C.G. Treworgy. 1985. Overview of ecological research data management. pp. 9-22. In: Research Data Management in the Ecological Sciences. W. Michener (ed.). Belle W. Baruch Library in Marine Science, No. 16. University of South Carolina Press, Columbia.

Ross, R.G. 1978. Database systems: Design, implementation and management. American Management, NY. 229 pp.

Rossmassler, S.A. 1980a. Presentation of data in the primary literature. pp. 65-73. In: Data Handling for Science and Technology. S.A. Rossmassler and D.G. Watson (eds.). North-Holland, NY.

Rossmassler, S.A. 1980b. Introduction. pp. 1-7. In: Data Handling for Science and Technology. S.A. Rossmassler and D.G. Watson (eds.). North-Holland, NY.

Rumble, J.R., Jr. 1984. Data, computers and database management systems. pp. 1-13. In: Database Management in Science and Technology. J.R. Rumble, Jr. and V.E. Hampel (eds.). North-Holland, NY.

Rumble, J.R., Jr. and N.L. Seidenman. 1984. Analysis and display of data in science and technology. pp. 75-91. In: Database Management in Science and Technology. J.R. Rumble, Jr. and V.E. Hampel (eds.). North-Holland, NY.

Sherrod, J. 1977. Review of federal legislation establishing numerical databases. pp. 137-139. In: Proceedings of the Fifth Biennial International CODATA Conference. Boulder, CO, June 28-July 1, 1976. B. Dreyfus (ed.). Pergamon Press, NY.

Stevens, P.R., A. Rittenberg, F.D. Gault, and B.J. Read. 1979. The use of database management systems in particle physics. pp. 167-169. In: Proceedings of the Sixth International CODATA Conference. R. Dreyfus, (ed.). Santa Flavia, Italy, May 22-25, 1978.

Stonebraker, M. 1981. Operating system support for database management. Communications of the ACM 24: 412.

Suich, J.E. and H.C. Honeck. 1978. Extensions to the JOSHUA GDMS to support environmental science and analysis data handling requirements. pp. 151-159. In: Generalized Data Management Systems and Scientific Information. Organization for Economic Cooperation and Development (OECD), Nuclear Energy Agency, Paris.

Szczesny, K.F. and W.M. Gersbacher. 1978. The capabilities required in a generalized database management system for handling scientific and technical data. pp. 106-110. In: Generalized Data Management Systems and Scientific Information. Organization for Economic Cooperation and Development (OECD), Nuclear Energy Agency, Paris.

Town, W.G., J. Powell, and P.R. Huitson. 1980. Recent experience of the application of a commercial data base management system (ADABAS) to a scientific data bank (ECDIN). Information Processing & Management 16: 91-108.

Tubbs, N. 1979. Database management and scientific information. pp. 105-111. In: Proceedings of the Sixth International CODATA Conference. B. Dreyfus (ed.). Santa Flavia, Italy, May 22-25, 1978.

Tubbs, N., D. Johnson, L. Pellegrino, and W. Schuler. 1981. Application of a DBMS at NEA Data Bank: Problems and experience with a large database on a small computer. pp. 486-490. In: Data for Science and Technology. Proceedings of the Seventh International CODATA Conference. P.S. Glaser (ed.). Kyoto, Japan, October 8-11, 1980. Pergamon Press, NY.

Turner, M.J., R. Hammond, and P. Cotton. 1979. A DBMS for large statistical databases. pp. 319-327. In: Proceedings of the Fifth International Conference on Very Large Data Bases (ACM). Rio de Janeiro, Brazil, October 3-5, 1979.

Watson, D.G. 1980. Compilation and evaluation of data. pp. 83-91. In: Data Handling for Science and Technology. S.A. Watson and D.G. Watson (eds.). North-Holland, NY.

COMPUTER DATA ENTRY TECHNIQUES USED IN SCIENTIFIC APPLICATIONS

K.C. Zinnel and M.F. Marozas

ABSTRACT

Prolific data production by the Long-Term Ecological Research (LTER) program has demonstrated that the critical path for analysis and interpretation of biological phenomena involves the initial step of converting observers' measurements of the environment to a machine-readable format. A historical perspective describes the evolution of current data entry procedures used by LTER sites and emphasizes where new techniques allowed the assimilation of additional LTER data sets. Recent electronic technology has made it cost effective to minimize, and potentially eliminate, human involvement in data entry procedures. Critical factors to be considered when making decisions regarding technological advances in data entry are discussed within the framework of a scientific research organization.

INTRODUCTION

Research data management (RDM) encompasses a broad continuum of functions, which are outlined in Figure 1. Scientific research organizations involved in ecological research tend to become oriented toward long-term projects with multiple investigators from different disciplines (Culliton, 1984). This approach produces a need for a "database" system to ensure accessibility of project information. One reason why some large studies have difficulty producing breakthroughs may result from keeping data on paper or in filing cabinets rather than in a RDM system. The role of data entry within RDM is important because measurements made during the course of an experiment must be encoded into a

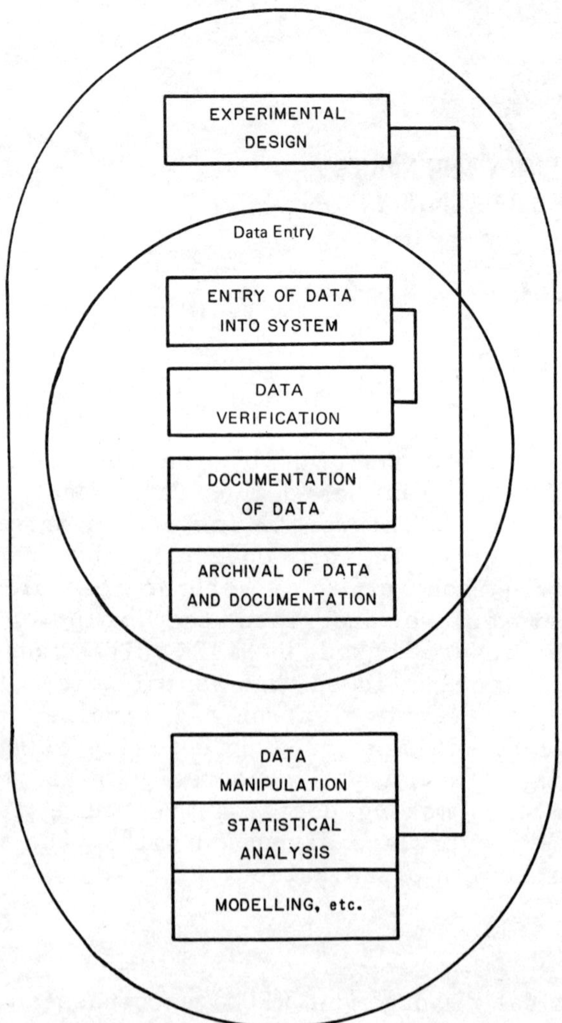

Fig. 1.   Capsule of research data management.

computer-compatible media before analysis can be performed. Analyses are necessary to answer the questions posed by the research design, thus producing publications on which applications for future funding are judged (Kennedy, 1985).

Scientists must be aware of the costs and benefits associated with data entry technology. Data entry encompasses several discrete steps, which are generally employed by most scientists (Fig. 2) and which are essentially the same steps used by corporate data-processing departments for

control over information input (Cerullo, 1983). Data entry
may consume more resources than any other RDM step. Documentation may take almost as much time and analysis can cost
more, but both areas are dependent upon successful data
entry as an initial step. Among the 11 LTER sites, the
actual mechanisms of data entry procedures have converged
over time.

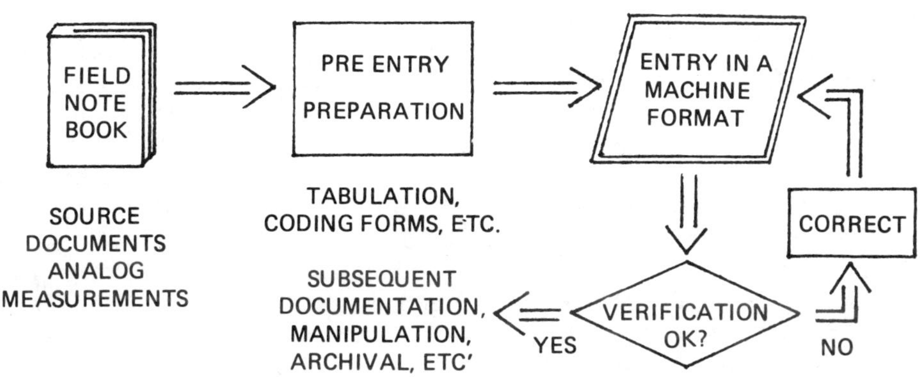

Fig. 2.   Steps within the data entry process.

Data entry cannot begin until the experimental design
phase has been completed. Time spent on designing appropriate field data recording forms will assure consistent and
compatible measurements of biological phenomena. According
to a pre-established and documented experimental design,
data are then recorded either by human observer or by instrumentation. After encoding the data into a machine-
readable format, the raw data are tabulated and statistically summarized. Data must then be verified to assure correct transcription from source form to machine-readable
file. Finally, each corrected data set must be associated
with documentation concerning origin and purpose before
entry into the project database is considered complete.
More than one set of documentation can exist, as projects
often change emphasis between inception and completion.
Tools for capturing documentation during data entry, which
can be automated, greatly enhance future usability of data
(Kellogg Workshop Report, 1983).

EVOLUTION OF DATA ENTRY TECHNIQUES

Precise definition of the steps for data entry first
requires the choice of the machine that will be used to

physically encode the data. Four phases of historical development can be recognized from the LTER group experience. Prior to 1978, most research projects keypunched 80-column cards from coding forms and submitted decks in a batch computer environment. Keypunch facilities were expensive to maintain and operate, and the turn around time was slow. For large data sets, researchers had to contend with a backlog of data and had no access to any part of the data while it was being entered. Another feature of large data sets was that errors introduced during the transcription process stood a greater chance of not being detected.

Because a high error rate was characteristic of the process, data tabulation was performed prior to the encoding process. Due to the high cost of double entry, verification was accomplished by visually checking printed listings of card decks against the raw data coding forms. Data were corrected by repunching cards, and summary programs were run using corrected data providing tabulations for comparison with pre-entry calculations to further reduce errors. Ultimate data bank disposition was generally a metal card catalog file of card decks with limited shelf life.

By the beginning of the LTER program in 1981, data entry at most sites had progressed to an interactive editing program and access to a mainframe computer through a time-sharing terminal for entering and correcting data (Table 1). Again, a printed listing of data was produced and visually verified against raw data coding forms. Computer centers subsidized this type of data entry by promoting interactive usage through low costs for connect time and permanent file storage. With many users sharing the cost, computer centers could afford to buy statistical packages and other software supported by vendors, thus sparing users the cost of developing their own software to tabulate data and flag outliers. A library of data files on magnetic tape, sometimes with documentation, was the usual data bank.

In many cases, methods of data entry employed prior to the start of LTER proved inadequate for handling the flood of information generated by LTER scientists. As university computer facilities increased their rates, a need for more efficient forms of data entry was apparent; resulting in the development of a distributed approach using off-line microcomputers for entry and verification of data (Table 2). Development of data entry procedures using microcomputer technology has been adopted by most LTER projects (Fig. 3). The use of microcomputers helped alleviate raw data backlogs, reduced computer usage charges, and increased error detection through full-screen data entry software featuring validity checking. Scientific research organizations, which employ multiple investigators, technicians, and students working toward common goals, also began to employ communications packages in much the same manner as corporate

Table 1.   Data entry techniques utilized by LTER sites when the program was initiated.

ANDREWS	Keypunch main encoding device, corrections made with interactive editor on mainframe, raw data stored on magnetic tape in mainframe library.
CEDAR CREEK	Microcomputer and commercially available software used for data entry and correction. Raw data archived on floppy disks and mainframe mass storage after upload.
COWETTA	Keypunch used to encode data, raw data maintained as card decks and archived on IBM 370 magnetic tapes.
JORNADA	Interactive entry/editing of data with mainframe system, raw data maintained on mainframe mass storage and archived to tape. Meteorological data logger read directly to mainframe disk file.
KONZA	Microcomputers and in-house generated, user friendly software used to enter/correct raw data. Meterological data logger entered as microcomputer disk file.
LARGE RIVERS	Interactive entry/editing or keypunch of raw data. RJE terminal transfer of raw data to mainframe for storage and archival.
NIWOT RIDGE	Keypunch, interactive terminal used for data entry/editing. Raw data archived as published reports and/or mass storage files on mainframe. Meteorological data logged onto cassette tape.
NORTHERN LAKES	Microcomputers and commercially available spread sheet software used to enter/edit data. Raw data uploaded and reformatted for archival as SIR mass storage files on mainframe.
NORTH INLET	Interactive software for full-screen management on mainframe system used for data entry/correction. Raw data archived as mass storage files on mainframe.
OKEFENOKEE	Keypunch used to encode data, raw data maintained as card decks, archived to IBM 370 magnetic tapes.
PAWNEE	Keypunches used for data encoding. Meteorological data logging on cassettes. Raw data archived on mainframe as magnetic tape files.

Table 2. Procedures evolved during the course of the LTER program.

ANDREWS	Microcomputers with commercially available software used for data entry, correction, and report generation. Documentation on microcomputer mass storage as well as archived to mainframe magnetic tape along with data.
CEDAR CREEK	Microcomputers used for automatic data acquisition in laboratory chemical analyses, weighing plant samples, and animal location by telemetry.
COWETTA	Individual PIs enter data with microcomputers and are responsible for maintenance. Database administrator uses microcomputer to create and maintain data set documentation.
JORNADA	Microcomputers used for data entry/editing. Raw data maintained on microcomputer mass storage, but archived with mainframe magnetic tape system. Minicomputer workstation also used for data entry/editing/access.
KONZA	Incorporated validity checking into in-house data entry software and established procedures for sending data files to mainframe.
LARGE RIVERS	Microcomputers with commercially available data entry and management software networked to database on in-house minicomputer.
NIWOT RIDGE	Microcomputers with commercially available software used for data entry, correction, and report generation. Multiple levels of data maintenance with data not necessarily in machine-readable form.
NORTHERN LAKES	Increased number of microcomputers and communication capabilities while maintaining procedural stability.
NORTH INLET	Meteorological data station and laboratory microcomputer used for automatic data acquisition. Microcomputers used for data entry and editing of data sets to supplement full-screen management software.
OKEFENOKEE	Multi-user microcomputer system used for data entry, correction, archiving, and documentation of raw data.
PAWNEE	In-house minicomputer, interactive entry/editing of raw data.

executives. Using microcomputers as replacements for remote job entry terminals or data entry stations, files of database additions were uploaded to a host mainframe and reports and/or summary statistics were downloaded to personal computers for display (Ferris, 1983).

A fourth phase in the historical development of data entry procedures has been the use of machines to take measurements and encode data directly into a machine-readable media. For experimentation requiring continuous monitoring, sampling for long periods of time, or high frequency sampling intervals, an automated data acquisition system is cost effective. Automation of measurement processes has lead to vast increases in the amount of data generated in every scientific discipline (Lide, 1981). Because instrumentation generates large volumes of data, only an automated data entry system can minimize the lag time betwen raw data collection and machine-readable data encoding. For these reasons, a majority of LTER sites now use data acquisition systems for meterological monitoring and/or chemical analyses (Fig. 3).

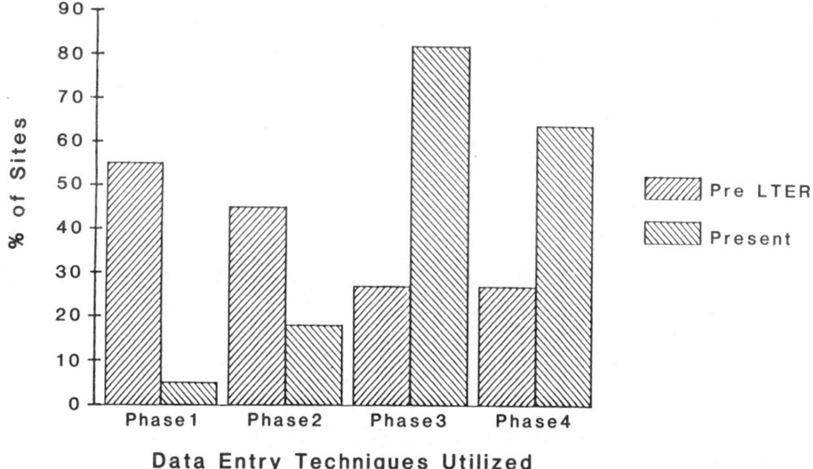

Fig. 3.   Evolution of data entry techniques employed within LTER.

ANALYSIS OF PRESENT DATA ENTRY TECHNIQUES

Selection of a microcomputer system for data entry involves weighing factors such as availability and cost of hardware and software, plus the time required to convert from an old procedure to a new system. The LTER sites selected different hardware, partially according to budgetary constraints. Most software packages selected, such as the STAR family by MicroPro, included characteristics to reduce transcription errors. Among these is full-screen data entry software, which allows designing a "form" that appears on the screen as an imitation raw data sheet. With cursor movement, the data entry technician is guided to fill in the blanks with the values from the data sheets. These packages also allow data formating controls and data validity checking which informs the operator that the data value does not meet specified criteria.

Other features include the ability to integrate repetitive information from other databases and generate reports and summaries. Raw data files can be listed for visual verification or another data entry technician can access each record and verify data entered. Previously, an expensive form of verification involved entering data twice, into two different files by two different people. Verifying the two files against each other with a "compare" program was an excellent, but expensive, way to locate transcription errors. Because of increases in speed of entry when utilizing full-screen software and off-line microcomputers, it currently is cost effective for sites to perform this type of double entry verification, which has the highest rate of error detection. A communications package capable of data file transmission between the microcomputer and a mainframe computer system is also needed for retrieval and archiving.

Data acquired by automatic instrumentation is cheaper on a per sample basis than data recorded by a human observer. If operating procedures are correctly followed, acquired data will be virtually error-free (Enke, 1982). Human involvement is required to ensure that sensors, signal conditioning, and recording equipment are performing accurately. Because automatic instrumentation is capable of generating far more data than can be manually checked by a human observer, elaborate software must be used to ensure data integrity. Currently, sophisticated validation software, developed in-house, is an essential requirement for any project utilizing data acquisition. Periodic checking of formats and data field ranges, plus testing for data accuracy, must be required as part of normal operating procedures. Final assessment of data integrity can be performed by an experienced technician through inspection of graphical data summaries.

Verification programs for data acquired by instrumentation should check for error conditions and extract usable data. Control totals such as record counts and average values per batch are used to evaluate machine performance during operation. Adequate operator training and supervision are essential to producing quality data. A specific technician should be assigned responsibility for determining if errors have occurred during operation, logging the errors, and reentering corrections. Recommendations have been made to record and preserve all auxilliary information (such as calibration standards or ambient temperature and pressure when the analyses were run) in order to use the data with confidence (Lide, 1981). Future possibilities for data entry techniques include the use of optical character recognition (OCR) for entering field notes directly and the use of voice recognition system for field data recording.

DISCUSSION

Tables 1-3 illustrate an evolving trend for LTER projects in the use of technological advancement, such as microcomputers, to meet the challenge of entering an increasing amount of source measurements into RDM systems. While the costs of computing hardware has continued to plunge, computing power and ease of interfacing different devices have increased (Enke, 1982). Although it is technically easy to put together hardware to perform complex operations, the software required for varying applications, or combinations of complex operations, is not as easily implemented. Testing is an especially critical requirement as each application is inherently different. There always will be modifications before a system becomes fully operational. As expectations for software capabilities have increased and hardware costs declined, software has rapidly become the major consideration and limiting factor in applying microcomputer technology to scientific research needs (Enke, 1982). Most LTER sites use commercially available software for data entry as well as word processing, but in-house programs for data reformatting and archival are common.

Within the last two years, direct application of microcomputer techniques for control of scientific instrumentation has promoted increased use of data acquisition involving one or more sensors without direct human involvement. Initially, research organizations desiring this capability had to invest in expensive laboratory minicomputer systems with expensive support software such as LIMS/2000 or RS/1 (Robinson, 1983). Within a year, half a dozen firms were offering controllers and data stations that would allow scientists to interface APPLEs and IBM PCs to numerous instruments (Robinson, 1984). Despite the great power of microcomputers to capture immense amounts of data, it is important to remember three caveats: (1) a machine is not a perfect observer, it is only as good as its electronic calibration, which is performed by humans; (2) larger amounts of data acquired by instruments take longer to analyze (software and human filters can reduce data to manageable levels); and (3) beware the salesman propaganda,

Table 3.   Decision basis for changes in data entry procedures and other pertinent factors.

ANDREWS	Using micros for data entry more cost effective than maintaining keypunch machines and personnel. Data to be entered at centralized facility growing steadily at 10% per year. Faster turnaround time with micros.
CEDAR CREEK	Cost is primary factor. Initial hardware investment less than maintaining technicians' salaries. Number of data items entered increased 170% from Year 1 to Year 2, and 250% from Year 1 to Year 3.
COWEETA	Data administrator maintains data documentation and data directory, but does not supervise data entry or maintain data library.
JORNADA	Use of micros and workstation facilitates data screening and error detection. Reduced use of mainframe computing increases flexibility and project control of computing resources.
KONZA	A reassessment of project needs necessitated redesign of data entry software and validity checking. File reformating needed because of procedural changes.
LARGE RIVERS	Specialized application package for geographical data required operating system stability not possible when using university mainframe system. Easier to predict and budget for computer costs of in-house system.
NIWOT RIDGE	Data dictionary maintained on microcomputer used to document location and status of data sets. All raw data published as CULTER data reports. No funds available for data entry and maintenance of machine-readable data library.
NORTHERN LAKES	Procedural stability most important factor. Data entry with micros adequate for task. Inefficient use of resources and time acquiring unnecessary technology.
PAWNEE	Backlog of preexisting data from International Biological Program. University mainframe costs out-of-control. Difficult to obtain funds for data management of pre-LTER data. In-house minicomputer stabilizes costs.

there always will be hidden costs. Troubleshooting problems, such as faulty grounds, are an inescapable part of normal maintenance. System performance criteria such as accuracy or linearity will be directly affected by the extremes of the operating environment (Chase, 1982). Unless care is taken when monitoring data acquisition system operation, there is the risk of producing large data sets (in nice neat formats) that are highly unreliable. This point concerning expert supervision of data acquisition systems is especially important in government-supported projects, which tend to be under-funded with regard to support (staff) personnel. Before any research project opts to use the latest technology, the question must be asked whether or not the benefits of using instrumentation to carry out experiments offset the costs of proper maintenance.

Intelligent tools require intelligent operators to supervise their operations effectively. Using microcomputers for data entry and data acquisition means that programming support is also required. Because in-house software support is affected by personnel changes, commercially available software packages are preferred by most sites. Gains in stability may be offset by losses in flexibility. Commercial software procedures tend to be less efficient than software custom designed for a specific application. Data managers involved with microcomputers also find that they have additional duties, including responsibility for maintenance and supplies, and consulting on the usage of various microcomputers and software packages.

CONCLUSION

Scientific and technical data are deemed necessary for the solution of societal problems such as energy supply, environmental quality, and industrial productivity (Lide, 1981). Since the late 1970's, ecologists have become increasingly aware that certain questions could be answered only by long-term data sets. Electronic tools have made mathematical analysis and management of such data feasible. One of the most labor-intensive RDM aspects resulting from this information explosion is the development of data entry procedures. A key consideration is the time lag between turning in data sheets and receiving access to a verified, corrected, and documented data file. Obviously, publications cannot be produced unless data are accessible for analysis or merging with another data set.

Many large-scale research projects, such as LTER, have found the critical path between measuring biological phenomena and interpreting experimental results to be the initial RDM step of data entry. In most cases, methods of data entry employed prior to LTER proved inadequate for handling the flood of information once the research was underway. As part of their strategy, many large research projects, including LTER, are automating as much of the data entry procedures as resources permit (Table 3). Reorganization of personnel and redefinition of procedures are usually necessary since data entry issues change as research data management in the ecological sciences evolves. Regardless of which combination of data entry techniques are chosen, any method will prove successful only if project leaders provide support for adequate software and hardware, as well as supervisory personnel.

LITERATURE CITED

Cerrullo, Michael.   1983.   Computer control systems in organizations. Data Base 14(4): 16-22.

Chase, D.H.   1982.   Many criteria affect the performance of a data acquisition system. Computer Technol. Rev. 3(1): 19-23.

Culliton, B.J.   1984.   NIH proposes extending life of grants. Science 226(4681): 1400-1402.

Enke, C.G. 1982. Computers in scientific instrumentation. Science 215(4534): 785-791.

Ferris, David.   1983.   The micro-mainframe connection. Datamation 29(11): 126-138.

Kellogg Workshop Report. 1983. Data Management at Biological Field Stations. May 17-20, 1982. W.K. Kellogg Biological Field Station, Michigan State University. 93 pp.

Kennedy, D.   1985.   Government policies and the cost of doing research. Science 227(4686): 480-481.

Lide, David R., Jr.   1981.   Critical data for critical needs. Science 212(4501): 1343-1349.

Robinson, Arthur L. 1983. LIMS is the next step in laboratory automation. Science 220(4593): 178-183.

Robinson, Arthur L.   1984.   Personal computers attract lab software. Science 224(4644): 40-44.

OPTIMIZING THE COMPUTATIONAL ENVIRONMENT FOR ECOLOGICAL RESEARCH*

S.G. Stafford, M.W. Klopsch,
K.L. Waddell, R.L. Slagle, and
P.B. Alaback

ABSTRACT

Computational tasks for ecological research--database management, statistical and graphical analysis, word processing, communications, and accounting--are diverse. Hardware and software options should be considered as a unit to create the most useful computational environment. Hardware includes mainframes, minicomputers (minis), microcomputers (micros), and supermicrocomputers (supermicros), and such attendant peripherals as printers, plotters, and tape drives. Ease of use, low cost, fast response time, and potential graphics capability make micros a popular choice for interactive computing (statistical analysis and word processing). Software, which must be reliable and easy to use, includes statistical analysis, graphics, data management, word processing, modeling, accounting, and communications. Graphics software can be a powerful medium for presenting research results. Restricting equipment use to specific tasks and appropriately scheduling machines, providing manuals and short courses, and carrying maintenance contracts are important to maximizing output. Resources should be shared through networking to make the best use of existing equipment. Hardware acquired should be adequate for the job intended, and choice of vendor and support level

---

*FRL 1951, Forest Research Laboratory, Oregon State University, Corvallis. The mention of trade names or products does not constitute endorsement or recommendation for use.

must be carefully evaluated. Developing in-house software should be avoided if commercial packages are already available; a table of statistical software for micros, minis, and mainframes is provided. Although computing systems have hidden costs, the potential gains remain high for all but the most modest research projects.

INTRODUCTION

Optimizing the computational environment means increasing the productivity and efficiency of researchers with existing hardware, acquiring compatible new hardware and software, and coordinating all hardware and software into a well-integrated, cost-effective system. Computational tasks in ecological research include database management, simulation modeling, statistical and graphical analysis, word processing, data entry, communications, automated data collection, and accounting. Each of these tasks places different demands on the available facilities, and some systems do the job more efficiently than others, saving time and money.

Database management generally requires considerable data storage capacity, computer memory, and speed. Large databases on mainframe computers and minicomputers commonly need substantial storage. Even small databases on microcomputers can quickly exceed the available floppy disk storage, requiring the addition of a hard disk. Data analyses, including modeling, matrix operations, graphics, and other statistical techniques, generally make heavy demands on computer memory, speed, and, to a lesser degree, storage. In production statistical analyses (in which a large number of predetermined analyses must be run), computational speed is crucial. Response time is critical for editing and word processing, but less so for analyses. For interactive statistics, simplicity of operation and integrated graphics are most important. Real-time tasks like data entry, laboratory instrumentation, digitizing, and communications are often more efficient on microcomputers than on large multi-user systems.

Staff at the Forest Science Data Bank, which serves the Department of Forest Science at Oregon State University, manage large quantities of data collected from ecological studies conducted throughout the Northwest, including data from the H.J. Andrews Experimental Forest, a Long-Term Ecological Research (LTER) site. The resulting data files and associated documentation are organized, stored, and manipulated within the Data Bank's database management system. In this paper, we describe hardware and software options for optimizing the computational environment on the basis of our experiences.

HARDWARE OPTIONS

Mainframe Computers

Data management and analysis have relied primarily on mainframe computers, which are fast, have large memory and random-access storage, and are very expensive. The peripheral devices (peripherals) for mainframes (tape drives, line printers, large plotters, high-resolution graphics terminals, and laser printers) add substantially to mainframe utility, but also to their cost. Maintenance is also expensive, and system support requires a large staff. The high cost has been justifiable to universities and other large research organizations because mainframes offer a uniform computational environment for a large number of users. To be cost efficient, however, these machines must carry heavy workloads, resulting in long waits for interactive applications such as word processing or data entry.

Minicomputers

Minicomputers recently have become important to ecological research. They have many of the characteristics of mainframes at a substantially lower price. Minis are sufficiently fast for most production statistics and modeling, but cannot serve as many users as a mainframe. Memory and storage generally are not limiting, and their cost is within reach of some large projects and departments. The end user has significantly more control over the software and peripherals than with a mainframe. Furthermore, most applications available on mainframes are also available on the major minicomputers. Minis might become a more cost-efficient alternative to mainframes as the need for computational capacity increases. For example, preliminary estimates suggest that the Forest Science Department could save at least half its annual computer budget if it bought and maintained a minicomputer rather than continuing to pay for the University mainframe.

Microcomputers

A revolution in computing has been brought about by

microcomputers, which have proliferated in homes and offices, putting many people in direct contact with computers for the first time. Micros are fast enough for most single-user applications, performing best on tasks requiring fast response time and only moderate computational capability (e.g., data entry, word processing, communications, and automated data acquisition). Their moderate-resolution graphics are useful for interactive statistics and many other applications.

Microcomputers today range from 8-bit microprocessors with 64K (K is about 1000 bytes, or characters) of memory to 16- and 32-bit microprocessors with up to 16,000K of memory. Floppy disks with large storage capacities (e.g., 1.2Mb; Mb is a megabyte, or 1000K bytes) have made more applications possible. A hard disk of 10Mb or greater is a powerful peripheral when users are working with many databases or multitasking software. Networks significantly increase the utility of microcomputers by linking them to shared databases, faster computers, and expensive, shared peripherals. Micros are relatively easy to interface with a wide variety of devices, making them ideal for laboratory instrumentation and other real-time data collection devices.

Supermicrocomputers

Recently, a new class of computers based on the latest 16- and 32-bit microprocessors has been developed. These supermicrocomputers bridge the gap between micros and minis, representing our closest glimpse of what the microcomputers of the next 5 to 10 years will be. They now cost between $10,000 and $50,000 and should become less expensive within the next few years.

Supermicros provide minicomputer architecture and capabilities to a small group of users. They are fast enough for several users or small-scale production statistical analyses. Most machines have between 0.5 and 4Mb of memory, but have operating systems that allow virtual addressing of up to several megabytes of disk memory. Disk storage ranges from 40 to 320Mb, which is less than that of many minicomputers, but adequate for most applications. Many supermicros use a version of Bell Laboratories UNIX multi-user operating system; however, so far, only a moderate amount of UNIX software has become available for the supermicros (Darwin, 1984). With the introduction of AT&T computers into the supermicro market, software may not be long in coming (Hunter, 1984).

Peripheral Devices

Over the past 5 to 10 years, the performance of computer peripherals has improved significantly, and their cost has decreased. The hard-disk drive has become physically smaller, yet can store orders of magnitude more bytes of information. Laser disks, providing in excess of 1000Mb (10^9 bytes) of random-access storage, are currently being tested (Hecht, 1984). Graphics displays have attained higher resolution with more colors and faster response time. Memory chips have become cheaper, permitting larger computer memories and making it practical to allocate memory to simulate ultrafast disk storage. Printers have become faster, producing higher quality text and graphics output. Quiet laser printers, which provide near letter-quality printing at speeds approaching those of the fastest dot matrix printers, are now available (Bernard, 1984). Color graphics printers have become inexpensive enough to permit the quick reproduction of medium-resolution color graphics. Communications are being revolutionized by local area networks (LANs), which allow data storage and peripherals to be shared among computers. Intersite communications are becoming faster and more reliable with microwave satellite links and better phone modems; modems for communication at 2400 to 9600 baud are becoming more common, and AT&T is experimenting with 56K baud modems.

SOFTWARE OPTIONS

General Statistical Software

Statistical computing has come a long way when we consider that computer technology is only a few decades old (Boardman, 1982). The reduced costs of devices for processing and memory (e.g., microchips) have added the personal computer to the computer environment and made it a practical research option (Chambers, 1980; see also Stafford et al., "Data Management Procedures in Ecological Research," this volume). The development of computer systems and programs is central to statistical analysis because it expands the quantity of data that can be analyzed quickly and accurately (Chambers, 1980).

Statistical software has been classifed into three basic categories (Chambers, 1980). Single programs are task-specific programs reading data in specific formats, performing calculations, and outputting results. The disadvantages are that single programs are not easily modified to perform larger, more general analyses, are often inadequately documented, and may unknowingly be duplicated at different institutions to solve similar problems.

Algorithm collections (statistical libraries) are sets of subprograms designed to supplement a main program, which the user must write. A good algorithm should be easy to run on many different computers, be reliable, and adapt well to a variety of problems. It should not be limited by the size of problems handled or make restrictive assumptions. Some large collections of statistical algorithms have been developed (e.g., the International and Mathematical Statistical Library [IMSL] [IMSL, Inc., 1980] and Biomedical Computer Programs [BMDP] [Dixon, 1983]); using algorithms simplifies programming and usually improves the quality of results. However, the attributes that make an algorithm machine efficient and statistically rigorous are seldom going to make it easy for the casual user to understand or apply. Often, an interface such as user-friendly documentation is needed between user and algorithm.

Statistical systems are collections of programs that combine the advantages of both single programs and algorithm collections. Normally, they are thoroughly tested major packages (e.g., SAS [SAS Institute, Inc., 1982], SPSS [Nie et al., 1975], and MINITAB [Elkins, 1971]), which include a strict organization of commands through which the user specifies the analyses to be performed; the underlying design incorporates a collection of well-chosen algorithms. Statistical systems must be flexible to be able to handle a great diversity of data.

However, the power of statistical systems often has more to do with their data management and graphical flexibility than with their statistical routines. A common dilemma is that the most powerful statistical systems, which allow for flexible, complex data mergers and support presentation-quality graphics, are the most machine-dependent ("nonportable"). For example, until very recently, SAS (SAS Institute, Inc., 1982), one of the best engineered statistical packages, ran only on IBM mainframe computers, though it is now available on several minicomputers and will soon be available for 16-bit micros. In contrast, SPSS (Nie et al., 1975) is available for most mainframes, minis, and micros, but its data management capabilities are much more limited. Clearly, well-documented statistical systems integrated with spreadsheets, graphics, and word processing will increase user efficiency and capability. For most researchers, attaining a powerful yet easy to use statistical system can be the most critical element in the computational environment.

Microcomputer Statistical Software

Many attempts have been made to describe criteria that might be useful in evaluating statistical software on microcomputers (Carpenter et al., 1984). Key features of certain major software packages, current as of November 1984, are listed in Table 1. Software packages need to provide an integrated system of routines, allowing the user complete flexibility in performing major operations. Input routines must allow direct data entry as well as entry from external databases. Manipulation routines should include transformations, sorting, merging, recoding, and editing. Data management routines need to provide several methods of storing and cataloging data sets. It should be possible to select portions of a data set on the basis of prespecified criteria and pass those portions to one or more of the data exploratory or statistical analysis routines.

Microcomputers have three important features for conducting statistical analysis: ease of use, low cost, and potential graphics capability. The industry has been moving toward increased user friendliness, higher computational speeds, improved color graphics, faster and more versatile peripherals, and more modern programming languages, all on highly capable, inexpensive machines (Chambers, 1980). The impact of increased computing options on micros has already affected software development for data analysis and graphics. Analyses that were too time consuming and costly are now manageable on micros. However, obtaining reliable statistical software for microcomputers can be a problem because the statistical accuracy and long-term reliability of commercial packages recently entering the market have yet to be proven (Lachenbruch, 1980; Carpenter, et al., 1984).

Graphics Software

The way in which analytical results are presented can be as important as the validity of the results themselves. In fact, few statistical tools are as powerful as a well-chosen graph (Chambers et al., 1983). Graphs have been used primarily to show the overall data structure or pattern, though exploratory graphical procedures are most often used

Table 1.   Statistical software for microcomputers, minicomputers, and mainframes (adapted from Carpenter et al., 1984), current as of November 1984.

Software package	Statistical capabilities[a]	Operating system	Minimum requirements		
			RAM[b]	DD[b]	Software/(Other)
Microcomputers					
ABSTAT 3.04	D/R/A/G-low	CP/M-80	56	2	
		CP/M-86	128	2	
		MS,PC-DOS	128	2	
ABSTAT 4.00	D/R/A/G-low	MS,PC-DOS	196	2	
AIDA	D/R/A/MV/G-high	Apple-DOS	48	1	
ASTAT	D/R/A/MV/G-low	CP/M-80	48	1	
		Apple-DOS	48	1	
		MS,PC-DOS	48	1	
ASYST	D/R/A/C/G-high	MS,PC-DOS	320	2	
BMDPC	D/R/A/MV/G-low	PC-DOS	640	1	(5Mb hard disk)
DYNACOMP	R/A/G-low	Apple-DOS	48	1	
HSD-STATS PLUS	D/R/A/G-high	Apple-DOS	48	1	
HSD-ANOVA II	A/G-high	Apple-DOS	48	1	
HSD-REGRESS II	R/G-high	Apple-DOS	48	1	
HSD-PC STATISTICIAN	D/R/A/G-high	MS,PC-DOS	128	2	
INTROSTAT 2.2	R/A/G-high	Apple-DOS	48	1	
		MS,PC-DOS	64	1	
		Atari-DOS	48	1	
MICROSTAT 4.1	D/R/A/MV/G-low	CP/M-80	64	1	APC-BASIC
		CP/M-86	64	1	
		MS,PC-DOS	128	1	APC-BASIC
MICRO-TSP	D/R/G-high	Apple-DOS	48	1	
		CP/M-80	64	1	
		MS,PC-DOS	128	1	
MICRO-TSP 4.0	D/R/A/G-high	MS,PC-DOS	256	1	
MINITAB	D/R/A/G-low	PC-DOS	320	1	(10Mb hard disk)
MSTAT	D/R/A/G-low	CP/M-80	64	1	
		CP/M-86	64	1	
		MS,PC-DOS	128	2	
NUMBER CRUNCHER	D/R/A/MV/G-low	CP/M-80	64	2	M-BASIC
		MS,PC-DOS	196	2	

Table 1. Continued.

Software package	Statistical capabilities[a]	Operating system	Minimum requirements		
			RAM[b]	DD[b]	Software/(Other)
Microcomputers (continued)					
NWA-Statpak	D/R/A/G-low	CP/M-80	64	2	M-BASIC
		CP/M-86	64	2	
		MS,PC-DOS	128	2	
RS/1	D/R/A/C/G-high	P-System[c]	128	2	(DEC PRO 300 series)
SAM	D/R/A/MV/G-low	Apple-DOS	48	2	
		CP/M-80	56	2	M-BASIC
		CP/M-86		2	
		MS,PC-DOS		2	
Speed STAT	D/R	Apple-DOS	48	2	
SPS	R/A/MV/G-high	Apple-DOS	48	1	
		CP/M-80	64	1	M-BASIC
		CP/M-86	64	1	
		MS,PC-DOS	64	1	
SPSS/PC	D/R/A/MV/G-high	PC-DOS	320	1	(10Mb hard disk)
STAN	D/R/G-high	P-System[c]	64	2	Apple-PASCAL
		P-System[c]	64	2	IBM-PASCAL
StatPac (Wallonick)	D/R/A	CP/M-80	64	2	
		MS,PC-DOS	128	2	
STATPRO	D/R/A/MV/G-high	P-System[c]	64	2	(Apple)
		P-System[c]	128	2	(IBM,DEC,Sage)
SYSTAT	D/R/A/MV/G-low	CP/M-80	56	1	
		MS,PC-DOS	256	2	
		UNIX	500	2	
TWG-ELF	D/R/A/MV/G-high	Apple-DOS	48	1	
		CP/M-80	64	2	
		CP/M-86	128	2	
		MS,PC-DOS	128	2	
Minicomputers					
BMDP	D/R/A/MV/G-low	UNIX, VMS			
MINITAB	D/R/A/G-low	UNIX, VMS			
PSTAT	D/R/A/MV/G-low	Most systems			
S	D/R/A/MV/G-high	UNIX			
SAS	D/R/A/MV/G-high	Most systems except UNIX			

Table 1. Continued.

Software package	Statistical capabilities[a]	Operating system	Minimum requirements
			RAM[b] DD[b] Software/(Other)
Minicomputers (continued)			
SPSS-X	D/R/A/MV/G-high	CMS,MVGS,VMS	
Mainframes			
BMDP	D/R/A/C/MV/G-low	Most systems	
S	D/R/A/MV/G-high	UNIX	
SAS	D/R/A/C/MV/G-high	IBM only	
SPSS	D/R/A/MV/G-high	Most systems	
SPSS-X	D/R/A/MV/G-high	Most systems	

[a]
D = Descriptive statistics
R = Regression
A = ANOVA
C = Nonlinear curve fitting

MV = Multivariate statistics
G-low = Graphics, low resolution
G-high = Graphics, high resolution

[b]
RAM = random access memory, in kilobytes (K); DD = minimum number of disk drives.

[c]
Provided with software by manufacturer.

to show departures from an expected structure. Graphs are
powerful diagnostic tools for confirming assumptions or,
when assumptions are not met, for suggesting corrective
action (Chambers, 1980). There has been significant development in graphics software for micros; most microcomputer
statistics software also makes extensive use of graphics.

How graphical results are presented is important.
Poorly arranged tables and figures can hinder the understanding of research results and lead to incorrect conclusions. Caution must be exercised, for instance, in using
color on statistical graphs. Experimental psychologists
have demonstrated that varying the intensity of colors can
cause optical illusions in the apparent size of objects,
thus affecting people's judgment of quantitative information
(e.g., Cleveland & McGill, 1983). "Chernoff faces"--
depicting clusters or natural groupings within a multivariate data set--also can bias interpretation of numerical
data. Variables are assigned different facial features; if
a relatively insignificant feature is assigned to a highly

significant variable (or vice versa), the researcher's conclusion can be directly influenced. In summary, the researcher should be aware that eye-catching presentation-
quality graphics, though attractive, must be used cautiously
to ensure that data are presented objectively.

USING HARDWARE AND SOFTWARE
TO THEIR BEST ADVANTAGE

Steps for Optimizing the Computational Environment

Despite the allure of new technological developments
and their potential applications to research, the top priority for data managers, who often have limited budgets and
tight deadlines, must be to optimize the efficiency of
existing equipment. The first task should be to inventory
all equipment and look for bottlenecks in data processing.
Research plans and usage trends should be examined so that
future demands on computational resources can be anticipated. We have found it effective to take the following
steps:

- Restrict the use of equipment to specific tasks to
  prevent a computer with capabilities in high demand
  from being tied up on a task which could be executed
  on a less critical machine. A busy microcomputer
  should not be used for time-consuming "number crunch-
  ing" if more powerful machines are available. Likewise, using a mainframe for data entry is not cost
  effective if micros are available.
- Schedule equipment to spread usage more uniformly
  over a larger portion of the day, reducing the "feast
  and famine" cycle typical of unscheduled equipment.
  Scheduling allows users to plan their time more efficiently and, in the case of mainframes, to take
  advantage of the lower traffic and reduced time-
  sharing rates available during certain shifts.
  Scheduling should be flexible enough to accommodate
  machine down time (time lost to repairs) and open
  time (time on unscheduled machines) for short-term
  "rush" jobs.
- Provide in-house handouts summarizing operating commands for each machine or analysis package; an easily
  accessible and well-stocked library of commercial
  user manuals; and, when possible, regularly scheduled
  short courses designed for specific users. Acquiring
  software that is easier to learn and more interactive

has also helped increase the productivity and efficiency of data bank personnel. Increased automation
of routine data retrievals, analyses, and archivals
using batch command languages on both mainframes and
micros has contributed to this process.
- Maintain the balance between batch and interactive
  service modes; that is, use whichever mode is best
  suited for the given task. Batch mode conserves
  system resources (e.g., allowing a user to submit a
  job which runs during less congested times), whereas
  interactive mode facilitates user involvement. This
  balance provides flexibility, which translates into
  increased productivity.
- Share resources to make the best use of all existing
  equipment. Computers not continuously used by one
  research team might be shared by others. A color
  graphics terminal and color printer/plotter are prime
  examples of expensive equipment that can be purchased
  cooperatively and shared among projects at a site.

The ultimate goal is a network system linking all
computers to shared software and peripherals (see
Klopsch & Stafford, this volume). For instance, at
Oregon State University, installation of a broadband
LAN with several thousand ports linking microcomputers, terminals, and larger computers is under way.
The LAN will use general-purpose communications software to allow any user to address any other willing
user. The campus-wide cable network can carry more
than 100 channels, of which the LAN will require only
one; the remaining channels will be available for
special dedicated uses (e.g., TV, sub-LANs). When
fully operational, the LAN will telecommunicate data,
documents, and electronic mail between cooperating
departments, including the University's Department of
Printing.
- Carry maintenance agreements and regularly schedule
  preventative maintenance to avoid one of the biggest
  potential costs to researchers: loss of productivity
  due to computer down time. Machines must be protected from electrical power surges, dust, smoke,
  temperature extremes, and humidity to minimize damage
  and time lost to repairs. Software for local diagnostic testing of computer problems and for recovering partially destroyed disks also can minimize these
  costs. In our experience, most microcomputer down
  time has been attributable to defective equipment;

defects have usually surfaced within 6 months of acquisition. To guard against these initial failures, equipment should be purchased from reputable vendors that test their products before shipping; it may be especially advantageous to get equipment that is "burned in" (i.e., tested for at least 100 hours).
- Provide adequate security. Misuse, theft, hardware vandalism, software piracy, and data file security (including problems associated with both public and internal network access) must be considered in system planning. If system security is inadequate, user passwords or numbers might be misused or stolen, and account funds subsequently depleted or data file contents damaged.

Optimizing Acquisitions

Understanding the technical and operational aspects of hardware and software helps the staff make the best selection; however, this knowledge alone is not enough. Current information on availability, compatibility, and plans to update programs and documentation of commercial software packages should be examined thoroughly before a product is purchased.

Hardware and software must be considered together because one without the other is useless. Avoid developing major in-house hardware and software if commercial products are available. Commercial software packages are cheaper and usually better documented. However, when no reasonable alternative is available, the decision to develop an in-house package must be made with full recognition of all the costs involved, initial and ongoing.

Some caveats for acquisitions apply:
- Do not delay purchasing hardware or software just because something better is coming; something better is always coming. The best way to avoid obsolescence is to purchase systems that can be expanded as new products become available and to choose systems which follow industry standards, use standard disk formats, and have widely used operating systems.
- Avoid outdated equipment unless it is the only available option. Little software is developed or maintained for older machines or operating systems.
- Be realistic about delivery times. Delivery times on newly released hardware and software are always optimistic. Vendors want users to know that new or

updated products are coming, but often cannot project when. Newly released products frequently have bugs which take considerable time for users to find and the company to correct. There may be a substantial time lag between new hardware releases and software availability, so wait for software before purchasing hardware. Several instances of firms advertising products that do not actually exist have been documented. Sometimes advance orders fund the basic research and development for the product!
- Establish a list of available products which meet current needs, and consult with other users whose requirements are similar. Considerable time and money can be saved by sharing software as well as solutions to common problems.
- Assess compatibility between hardware and software. Whenever possible, select similar software for all machines at a site to simplify the user environment and optimize equipment usage.

<u>Choosing Hardware</u>

Hardware, whether a mainframe or micro, should be designed to provide a full range of analytical tools for all sizes of tasks. Be sure the equipment is suitable for the job. Many microcomputers with 8-bit processors are restricted by 64K or less of memory, which may be inadequate for working with large arrays. Supermicros often come with virtual memory, allowing each user to access 16Mb of memory. Disk storage must be adequate; purchase all the storage you can afford. Hardware durability must also be considered; it is unfair to expect a product designed for occasional home use to stand up to production pressures.

Buy from a well-established, financially sound company that will continue to upgrade its products. But remember that brand names are not always a good indication of support. Many larger computer and software companies do not like working with end users and rely on dealers. If dealer support is poor, purchasing a brand name product can be worse than purchasing the product from a less well-known company whose dealer can provide the necessary support. Furthermore, distance from vendors is not always a good indication of support level. Local support often consists merely of sales personnel with poor technical training.

Some mail-order companies have excellent service departments, but, in general, local support is preferable. However, support must be evaluated in light of staff capabilities; the support adequate for one group may not be adequate for another. Ask vendors for the names of people who have purchased similar hardware and software and use those people as references.

Our experiences with vendors have varied. In one instance we found that servicing a standard IBM monitor was far more costly and time consuming than repairing any of our terminals bought through mail-order houses. In buying some widely available computers through local distributors, we have had trouble getting technical help; in the case of one popular computer company, we were expected to call a single phone number nationally (which was nearly always busy). Our biggest disappointment, however, was with a state-of-the-art multi-user microcomputer bought from a mail-order house. The unit as shipped was inoperable due to improperly designed memory boards which had been inadequately burned in. After many phone calls, patching the operating system, switching boards, and mailing vital components back and forth, we found the system was too slow to be useful for its original purpose. Our experience taught us that minimizing a purchase price through mail-order houses may be false economy when you consider ease of repair and availability of replacement parts. Although all repairs were done under warranty, by the time the system was finally fully operational (over a year later), prices and technologies had changed to such a degree that for less money we could have purchased a far more useful, capable, and probably reliable system. The moral to the story is this: although something better will always be coming, it is important to act decisively when a need arises.

Choosing Software

Software has been changing considerably. Operating systems, typified by UNIX and VMS (Digital Equipment Corp.'s minicomputer operating system), have become more flexible, permitting virtual memory, multitasking, and input/output redirection. Software for mainframes and minicomputers has become more interactive. Standards are being developed to make graphics displays more portable from machine to machine (McCune, 1984; Wong, 1984). Relational database management systems are superseding the older, more restrictive hierarchical systems and network systems. As microcomputers become more powerful and memory increases, analogs of many software packages previously restricted to minis and mainframes are appearing on micros.

A thorough evaluation of software before it is purchased will reduce disappointments from incompatibilities or unrealistic expectations of the product. LTER researchers, for instance, have the advantage of consulting research groups within and among LTER sites to determine which software best meets their needs. However, discussing statistical software is a problematical process. It is important to remember that software is constantly being improved; that is, it is easy to wind up evaluating an obsolete program (Hamer, 1981). We can safely infer only two things from papers evaluating statistical software: 1) specifics about the programs themselves, and 2) information about the attitudes and skills of those who write the programs. While the latter has longevity, the former is likely to be short-lived.

When first considering the acquisition of any software, define its objective and intended function. For example, if the software is to be used by a large number of people, a well-documented, user-friendly package is probably desirable. The most important, obvious advantages are 1) reduced personnel time training novices, and 2) broadly increased computer literacy within a research group because the software is easy to learn. Disadvantages may be less apparent, but are real and of concern. Software that is touted for its user friendliness may not be machine efficient, cost effective, adequate for specialized tasks, flexible enough to handle general tasks, or rigorous enough for sophisticated analyses.

When we first evaluated database management software for the Andrews LTER site, a bewildering array of options was available for our microcomputers. Logically, we thought that the most powerful package would be best. We eventually purchased two packages, one strictly for replacing our keypunch data entry system, the other for its highly regarded database management capabilities. After intensive use, we found that the package purchased for data entry was also the better one for database management because it was so easy to learn and use. When limitations arose, it was more efficient to develop some simple software to supplement the data entry package, using languages our staff already knew, than

to try to learn the system language of the more sophisticated database package.

It is useful to get a demonstration copy of the software (demo disk) before purchasing. Some manufacturers
provide these disks, which permit a trial of a "crippled"
system before final purchase. (A "crippled" version of a
program uses only a small data set or is in some other way
modified so that it cannot be applied to real analyses.)
Where demo disks are unavailable, try to test the software
on another system before purchasing. When copies of commercial programs are being considered for multiple machines,
additional licenses, usually available from the manufacturer
at a reduced cost, will be needed. Software vendors are
already incorporating protection devices into their programs
to prevent users from making unauthorized copies. Look for
complete and available documentation, independent training
manuals, ease of interfacing with other software, good
access to consultants, and a high degree of vendor support.

A generous amount of public-domain software already
exists for many operating systems. Unfortunately, documentation of public-domain software tends to be poor, and bugs
are not uncommon. Current periodicals and other users
should be able to point out both good and poor software
packages.

IMPLICATIONS

Chambers (1980) suggested that computing systems could
be arranged in three configurations to make less expensive
computing power available to more statistical users. The
first would be a _minicomputer center_ ($100,000 in 1984 dol-
lars) supporting about 10 users, with the same processing
capacity, program size, and data storage as formerly provided by the large mainframe computer systems. This facility would serve a more specialized user community and therefore require a less diverse selection of hardware and software. The second would be a _personal computer_ or microcomputer on every scientist's desk. And the third would be
a compromise between the first and second, a _distributed
system_ in which each user has a processor, but mass storage
is centralized. This is a forerunner of the previously
noted LAN system, which combines advantages of inexpensive
local processing and utilizes high-speed transfers from
system to system in the network.

Half a decade later, the future is not that different
from what Chambers envisioned; most sites currently use one

or more of the configurations described. Regardless of the arrangement, each computational environment must be configured so that users can gain the most advantage for the least cost.

Although people buy computers to save time and money, there are many hidden costs, both initial and ongoing, which need to be carefully evaluated. Software and hardware require time for installation and customization; the hardware itself may require additional furniture or wiring. Program bugs must be found and corrected, and users need learning time on new machines and with new programs. The annual costs of maintenance and supplies (e.g., maintenance contracts as well as personnel time for system management and preventative maintenance) are roughly 5 to 25% of the initial purchase price, depending on the level of support required. Nevertheless, the potential gains remain high for all but the most modest research projects.

The proportion of resources allocated to computing will doubtless increase with time as will the possibilities for more sophisticated research and increased productivity. Computer applications that were prohibitive with time-sharing on a mainframe are now highly affordable on micros or minis. Attempts in the past to conduct large-scale, integrated research suffered from the lack of computer flexibility and capability that we now routinely have at our disposal. We feel that the investment in computers has more than justified itself in terms of both education and analytical output. Clearly, the ability of the LTER program to realistically meet its goals of acquiring and managing data depends on the ability of data bank personnel to be innovative in applying the latest computer technologies to optimize the computational environment for ecological research.

LITERATURE CITED

Bernard, J. 1984. HP's new laser jet printer. Computers & Electronics 22(7): 36-39, 87.

Boardman, T.J. 1982. The future of statistical computing on desktop computers. The Amer. Statistician 36(1): 49-58.

Carpenter, J., D. Deloria, and D. Morganstein. 1984. Statistical software for microcomputers. Byte 9(4): 234-264.

Chambers, J.M. 1980. Statistical computing: History and trends. The Amer. Statistician 34(4): 238-243.

Chambers, J.M., W.S. Cleveland, B. Kleiner, and P.A. Tukey.

1983. Graphical Methods for Data Analysis. Wadsworth International Group, Duxbury Press, Boston, MA. 395 pp.

Cleveland, W.S. and R. McGill. 1983. A color-caused optical illusion on a statistical graph. The Amer. Statistician 37(2): 101-105.

Darwin, I. 1984. UNIX software directory. Microsystems 5(4): 45-58.

Dixon, W.J. (ed.). 1983. BMDP Statistical Software. University of California Press, Berkeley, CA. 733 pp.

Elkins, H. 1971. MINITAB Edit, MINITAB Frequencies and MINITAB Tables: A Set of Three Interrelated Statistical Programs for Small Computers. University of Chicago, IL. 100 pp.

Hamer, R.M. 1981. Papers that evaluate computer programs. The Amer. Statistician 35(1): 264.

Hecht, J. 1984. Store it with light. Computers & Electronics 22(7): 60-64, 94-96.

Hunter, B. 1984. AT&T enters the micro/mini market. Microsystems 5(6): 114-118.

IMSL, Inc. 1980. International Mathematical and Statistical Library Reference Manual. Eighth edition. Houston, TX.

Klopsch, M.W. and S.G. Stafford. 1985. The status and promise of intersite computer communication. pp. 115-123. In: Research Data Management in the Ecological Sciences. W. Michener (ed.). Belle W. Baruch Library in Marine Science, Number 16. University of South Carolina Press, Columbia.

Lachenbruch, P.A. 1983. Statistical programs for microcomputers. Byte 8(11): 560-570.

McCune, D. 1984. An introduction in NAPLPS. Microsystems 5(7): 54-65.

Nie, N.H., C.H. Hull, J.G. Jenkins, K. Steinbrenner, and D.H. Brent. 1975. SPSS: Statistical Package for the Social Sciences. Second edition. McGraw-Hill, NY. 675 pp.

SAS Institute, Inc. 1982. SAS Users Guide: Basics. 1982 edition. SAS Institute, Inc., Cary, NC. 923 pp.

Stafford, S.G., P.B. Alaback, K.L. Waddell, and R.L. Slagle. 1985. Data management procedures in ecological research. pp. 93-113. In: Research Data Management in the Ecological Sciences. W. Michener (ed.). Belle W. Baruch Library in Marine Science, Number 16. University of South Carolina Press, Columbia.

Wong, W.G. 1984. Graphics portability with Digital Research's GSX. Microsystems 5(7): 74-82.

DATA MANAGEMENT PROCEDURES IN ECOLOGICAL RESEARCH*

S.G. Stafford, P.B. Alaback,
K.L. Waddell, and R.L. Slagle

ABSTRACT

Ecological research requires a flexible, organized system for acquiring, documenting, and managing data. Careful documentation of data, best done through close collaboration of the researcher and data manager, is important if all users are to benefit from a centralized database management (DBM) system. A systematic approach to data management and analysis comprises four key steps: comprehension, planning, execution, and evaluation and interpretation. Statistical consulting at the beginning of a project helps scientists plan well-designed, efficient research strategies. Data collection forms should be designed to encourage recorders to enter all essential identifying information, thereby minimizing errors. Consistent, high-quality data verification, through either the visual or double-entry method, allows most data collection errors to be caught and corrected before analysis. Standardized forms help maintain uniformity in data documentation. Documentation and corresponding data files should be linked in a carefully organized relational DBM system in which all information may be easily stored and retrieved. Using commercial software for data management and analysis is usually more cost effective than developing and maintaining customized software. Scarce DBM resources should be invested in data validation, equipment

---

*FRL 1944, Forest Research Laboratory, Oregon State University, Corvallis. The mention of trade names or commercial products does not constitute endorsement or recommendation for use.

acquisition, system design, and data documentation, not programming. Database maintenance is costly, but critical, to keeping credibility with users. The key to future advances in ecological data management lies in the ability of data managers and scientists to move into a more cooperative, integrative mode through which comprehensive databases are established and more fully used, increasing overall research efficiency.

INTRODUCTION

Most ecological studies produce a wide variety of data over a period of years. These data are expensive to collect. The better they are documented, the more likely they are to be useful to other scientists presently and in the future. Research data management is a serious attempt to anticipate future needs.

Different types of data are normally stored in separate data files, all of which constitute the data set for a given study. Each data file is documented in a systematic process that identifies and records its structure and origin; this information is entered into the data catalog. The result is the organized storage of data files and corresponding documentation in a logical, easily retrievable format. All data and documentation can then be permanently archived into a central data bank so that a scientist, even one unfamiliar with the research, can readily obtain data for comparison with current, related studies, support of a new hypothesis, or synthesis with other studies. The methods used to organize, store, and retrieve data in this fashion make up the local database management (DBM) system.

The Forest Science Data Bank (FSDB), which serves the Department of Forest Science at Oregon State University, has evolved over more than a decade. Faculty and students have collected sizable amounts of data from ecological studies conducted throughout the Northwest. In addition, research at the H.J. Andrews Experimental Forest, a National Science Foundation (NSF) Long-Term Ecological Research (LTER) site, has produced vast quantities of data now stored with documentation in the FSDB (Stafford et al., 1984). The prime reason for NSF's emphasis on sound, reliable data management in its LTER program has been to help overcome the difficulty experienced in the past of finding appropriate data sets and coordinating research from multiple institutions or geographic regions. Success or failure of an LTER DBM system may well rest on its ability to provide a uniform standard of documentation and facilitate accurate transfer of experimental data among all 11 LTER sites. Thorough study documentation enhances the understanding and accessibility of valuable research results and allows for greater research efficiency and opportunities.

In this paper, we give an overview of the systematic approach we have developed for retrieving past data and managing current data. Specific DBM procedures, as they relate to the Andrews LTER site, are described. Procedures at other LTER sites are referred to when they are known to differ from those at the Andrews LTER site.

OUR SYSTEMATIC APPROACH: AN OVERVIEW

The FSDB personnel include a consulting statistician, data managers, data analysts, and a microcomputer specialist, all with a combination of statistical, ecological, and biological training. Through time, we have seen how some research strategies succeed and others fail. From this experience we have evolved our own philosophy, a systematic approach to data analysis and management, comprising four key steps: comprehension, planning, execution, and evaluation and interpretation (Fig. 1). The approach itself is a modification of a more general schema currently used in some teaching programs (Chervany et al., 1980).

The first step involves recognizing a specific problem as one instance of a general category of problems and organizing the approach to find a solution. The experimental design needed to test the problem must be understood so that statistically valid data will be available for final analysis. Problems that might have been avoided had a statistician been consulted before the experiment was initiated often develop during analysis of a poorly designed experiment.

During the second step, the strategy (plan) for analysis should be formulated (Fig. 2) and then critically reviewed. The data bank's involvement at this point encourages careful thinking and precise planning relative to experiments that are long-term and expensive. A complete set of documentation should be maintained as a permanent record for future reference and to ensure consistency over time. As research progresses, changes in the design should be reviewed and documented; some changes are inevitable because

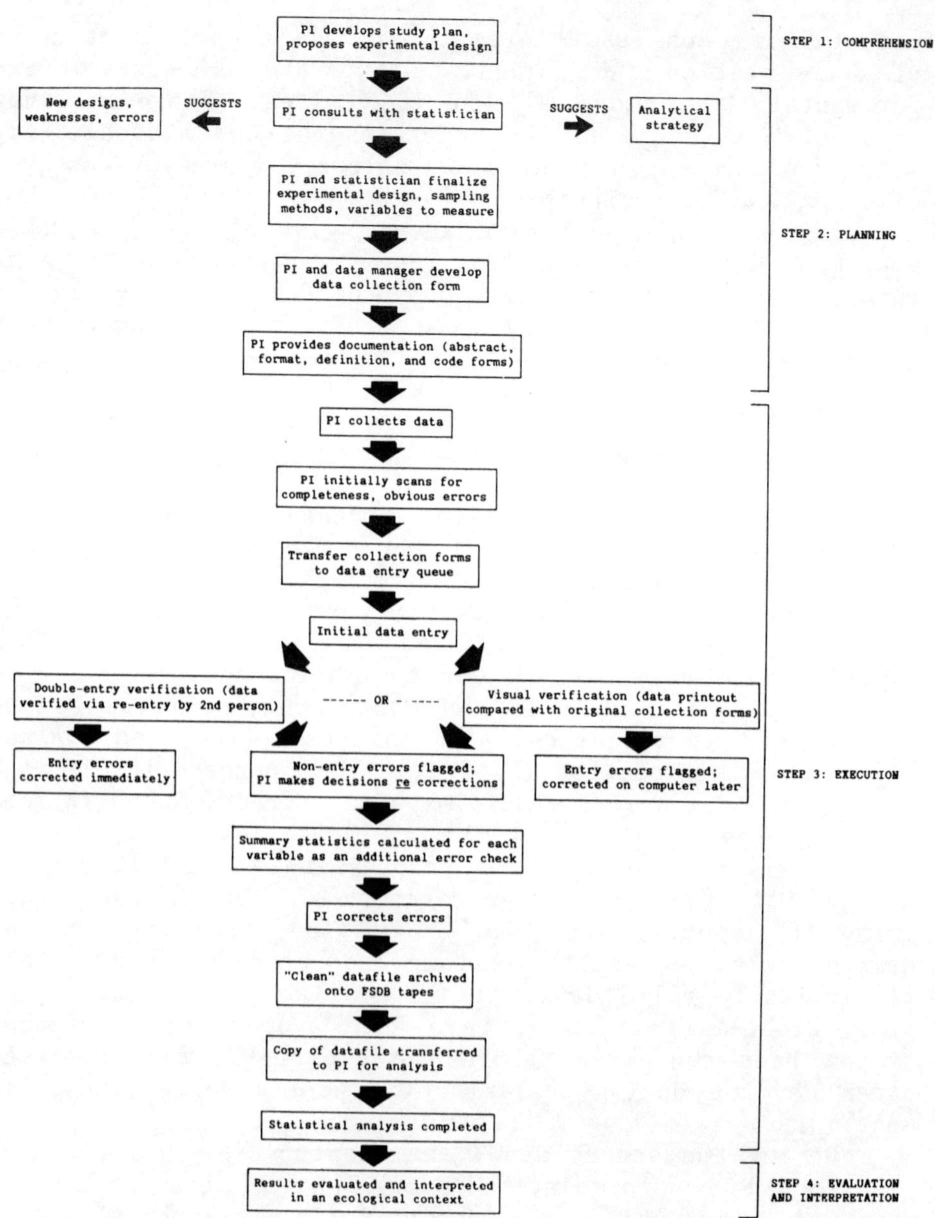

Fig. 1. Flow chart for efficient data collection, documentation, and analysis at the Andrews LTER site. The procedures are blocked according to the four steps specified by the FSDB's systematic approach to data analysis and management.

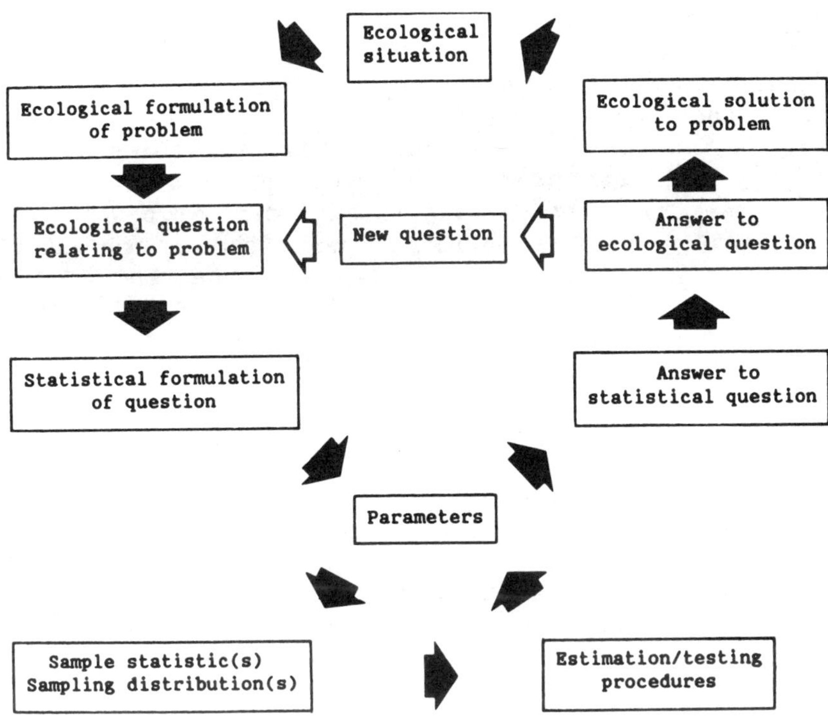

Fig. 2.   A strategy for sound, computer-aided analysis to resolve ecological problems (adapted from Chervany et al., 1980).

many facets of ecological experimentation are impossible to foresee or are beyond direct human control.

Once data have been collected and entered, data files should be validated and summarized to cross-check data file structure with the documented experimental design. Many problems can be caught at this stage, before extensive, complex, and costly analyses have begun. The clean data file is then analyzed with a statistical procedure (or group of procedures) to test the experimental hypotheses of interest; this process is the heart of the third step.

Often, the data file structure and sequence of analysis procedures are influenced by the statistical analysis package used; for example, the MANOVA procedure in SPSS (Nie et al., 1975) requires the data to be more structured than does SAS (SAS Institute Inc., 1982). The experimental analysis may be so complex and lengthy that it should be broken apart into a series of smaller analyses, completed individually.

We recommend starting simply, working with one component at a time and reviewing the initial results, before undertaking the entire analysis.

In the fourth and final step, each set of results should be scrutinized to ensure that variables were read correctly and that procedures were appropriate for the intended statistical tests, and to determine if new procedures are suggested by the results. Once the final output is obtained, a statistician can help the researcher interpret analytical results in an ecological context. This interpretation should be the basis for discussion in the final presentation of the data.

Long-term ecological research poses different types of statistical problems than traditional agronomic or agricultural experiments. Most ecologists recognize the difficulties of specifying appropriate statistical criteria for replication and determining appropriate error rates in a world full of unique sites and situations (Hinds, 1984; Hurlbert, 1984). Statistical consultation through all four steps of the systematic approach will help ensure that statistically sound, credible inferences be drawn.

THE STATISTICIAN'S ROLE

Data produced from any research project must be collected according to a sound experimental design. Whether the design is simple or complex, thorough analysis can be completed only with statistically valid data.

We recommend consultation with a statistician to confirm that the field or laboratory procedure is consistent with the statistical design and study objectives (Fig. 1). Suggestions for analysis, study weaknesses, or potential sources of error can be pointed out and dealt with early on. Consultation usually benefits both parties: the researcher learns how all phases of experimental research and attendant assumptions of statistical inference are interrelated and how they will most likely affect the project's outcome, and the statistician becomes more familiar with the research project as well as unavoidable constraints on design. Data bank personnel and scientists should work closely together to chose variables, design data collection and documentation forms, and anticipate future analyses so that data can be properly documented as permanent, logically arranged files in the data bank.

DESIGNING DATA COLLECTION FORMS

Careful design of data collection forms fosters accurate, efficient data collection and a smooth transition from field or laboratory to the computer. Experimental design, sampling intensity, periodicity, and variables to be measured should be determined before forms are designed, and forms should be developed before data are collected to encourage the recorder to enter all essential information. Many items are neglected when no form exists, which can compromise quality control and later prevent complete statistical analyses.

Media used to record data in the field or laboratory include analog, digital, or audio electronic devices and paper forms. The most common are project-specific paper forms. Although these forms cannot be completely standardized, data bank staff and scientists at each site should develop a prototype form whose general structure can be extended to every study and which should take into account the following:

* The most efficient order for collecting data for each variable.
* All pertinent experimental information (e.g., treatment level, location, site, plot or subplot, block, sample date).
* Simplicity and legibility; areas for recording observations should be clearly labeled and should contain adequate space to accommodate the maximum value expected.
* Additional space for recording pertinent comments.

The real test of how well a data collection form can capture the needed information comes during the active data-collecting phase, which tests all the predictions, assumptions, and formats imposed on the data by that form. The form's success will partly depend on how well the researcher and data manager communicate; an "open line" between both is crucial. Terminology often is misunderstood. For example, a forest mensurationist traditionally classifies a tree 7.6 to 8.5 inches in diameter as an 8-inch tree; the study documentation must distinguish this value as within an 8-inch diameter class and not a true decimal measurement. The question of precision must be posed. For instance, a variable allocated three spaces as an integer value, but which really demands two or three additional decimal places for greater precision, may be squeezed into too small an area, compromising legibility, or, even worse, rounded off to fit the space provided. Poorly designed forms increase the likelihood of errors in the data and unnecessarily complicate the data entry procedure. Moreover, transcribing values onto a more appropriate form introduces additional potential for error. If possible, briefly test a new form before deciding on its final design.

A data collection form is only a tool--a means, not an end. Occasions will arise when experimental conditions change or when assumptions about the data are proven incorrect. For example, mortality of a tree may entirely eliminate a species, or the number of places allocated to a particular variable may need to be increased due to unexpected extreme values. Both of these situations require only a simple modification of the form, but severe changes during data collection may require that an entirely new form be created.

Direct electronic input from analog or digital devices is becoming increasingly common for meteorological and laboratory data. These instruments can automatically record large quantities of data at set time intervals on digital media such as cassettes or EPROM (erasable programmable read-only memory) microchips. The hardware-dependent, limited storage capacity of most of these devices produces data files in a condensed format that often cannot be read by the data collector. Documentation for these types of data should include a textual header, entered at the beginning of each data file, that specifies, for example, the type of experiment, date, time, people involved, measurement interval, sensors or instruments used, and output data format. If calibration data are required to convert raw data to useful measurement units, the source of this calibration data and the calibration function should also be included.

Special problems occur when re-measurement data (e.g., annual tree height and diameter growth measurements) are being collected during a long-term research project. If a new or updated form is designed for the re-measurement sample, the existing data and documentation should be reviewed to assure that the subsequent (new) data file will be compatible with the older one. It is often useful to have the original data on the collection sheet in the field to serve as a baseline for verifying the new data. Commercial or custom software can efficiently produce a collection sheet containing the previous measurement, with adjacent spaces to record re-measurements.

ENTERING AND VERIFYING DATA

Before data are entered into the computer, the researcher reviews every data collection sheet for completeness and other obvious errors (Fig. 1). If the collection forms are unacceptable for data entry, the data collector or researcher may have to transcribe the data onto new forms. Occasionally, if the collection forms are not too disorganized, transcription can be avoided by entering the data as it is and reformatting later. Though time consuming and costly, this route reduces the number of transcription errors. Easy-to-use DBM software tends to minimize the need for transcribing poorly formatted data because data from new or older data sets can be efficiently restructured, reorganized, and reformatted after it is entered. A variety of commercially available software packages are designed specifically for data entry. Many simplify the data entry process by allowing the user to reproduce the actual paper collection form on the computer's video display.

The researcher is responsible for the accuracy of recorded and computerized data. For many LTER sites, a range or code check is built into the data entry software for each data set. For others (including much of the Andrews LTER), collected data are rapidly entered into the computer and checked for errors later, in the second (verification) stage of data entry. In any case, the person entering the data should <u>not</u> be responsible for correcting faulty information, even when it may be obviously wrong (<u>e</u>.<u>g</u>., an alpha value in a numeric field).

Two systems are commonly used to verify the data being entered: double-entry verification and visual verification. Double entry requires that the data set be entered completely by one person, then reentered (<u>i</u>.<u>e</u>., verified) by a second person. Two different people are used because one person may be more likely to make the same mistake twice. This is the system used by the FSDB for virtually all new data and is frequently used by commercial data-entry operations. If a value typed during the second entry does not match that of the first entry, the computer signals a mismatch and the error can be corrected. However, the second data-entry person will have to consult with the researcher to know how to make the correction.

Visual verification is widely used on LTER sites, especially when initial costs and time prevent double entry. A data file printout is compared with the original collection sheets; any errors or changes are flagged and edited later. In our experience, and that of many large businesses, double-entry verification is the most straightforward, accurate system and is preferred for critical data sets.

DEVELOPING DATA DOCUMENTATION FORMS

Standardized documentation containing all pertinent information describing data collection is an essential component of a DBM system. A standard set of documentation forms, which may be created by DBM software or hand drawn on paper forms, should be developed for each research organization. Most LTER sites currently use the paper forms.

Documentation initially includes variable definitions, formats, measurement units, codes, and research methods and should be permanently recorded long before data collection begins. All information contained on the finalized data-collection form should be documented as soon as possible after actual data collection to avoid the problems incurred when the researcher cannot recall details as time passes. Summary handouts describing how to fill out each form make it easier for newcomers to the system to provide complete documentation and understand its rationale. Again, the most desirable forms usually are produced when the researcher and data manager collaborate.

Every study should be assigned a code or brief title (often called a data code) to uniquely identify its documentation and data files. If multiple files are produced because data types are diverse, each file should be documented with a unique name easily related to the data code. The Andrews LTER site uses a format-type identifier after the data code number to name each file in the study. To develop a successful relational database (a coding system linking all data files and supporting documentation with data location referenced within a catalog or directory), all documentation pertaining to each file should be labeled with the data code and format-type identifiers. Ideally, before the data manager agrees to assume responsibility for permanent data storage, documentation should be complete.

Every format type within a data code should be documented with a complete set of the following: (1) forms recording variable names and formats, with units of measurement; (2) forms explaining variable definitions and their acronyms; (3) forms specifying the meaning of each code for a given variable; (4) forms identifying the names of all stored data files under one data code and their description; and (5) a form abstracting the study purpose, goals, locations and availability of supporting documentation, and methodology for additional background.

For example, a long-term study investigating the effects of fertilization, crown pruning, and stand density on the population dynamics of a young forest stand at the Andrews LTER site was begun in 1981 (Perry, 1982). Various measurements are already stored within several different data files, each with a unique format type. Complete documentation, filed with the FSDB, is described here to demonstrate the database documentation system. We anticipate that the resulting database will be continually updated. The data will be analyzed periodically and will serve as a baseline for other studies and comparative analysis. For example, a researcher deciding to study the population dynamics of a forest 10 years from now could retrieve this data set and its documentation to aid in the planning phase of that new study. The study plan underwent critical statistical review, and the data collection forms, with supporting documentation, were developed before data collection was begun.

The data code assigned to this study is TP88, a unique identifier recorded on every documentation form associated with this project. Relevant details are summarized on the abstract form (Fig. 3). The permanent file name and a brief description of the data stored in each file are listed on the data file description form (Fig. 4). Each file contains a different set of measured variables; thus, variable formats (Fig. 5), definitions (Fig. 6), and code specifications (Fig. 7) are documented on separate sets of forms. To save space here, documentation for only one format type is provided. Although data collectors and analysts will come and go during the course of this long-term study, the fully documented files should pose no major problems to the successors.

STORING DATA AND DOCUMENTATION

Data files and study documentation can be permanently stored on a variety of media: paper files, microfiche, floppy diskettes, punched cards, hard disks, laser disks, EPROM chips, mainframe disk packs, and magnetic tapes or cards. For example, the FSDB currently stores all data on magnetic tapes on the University mainframe computer and has

RESEARCH STUDY ABSTRACT

DATACODE: TP88
PRINCIPAL INVESTIGATOR: PERRY, D.A.
STUDYID: —
PROJECT FUNDING: LTER GRANT
STUDY TITLE: POPULATION DYNAMICS OF YOUNG FOREST STANDS AS AFFECTED BY DENSITY AND NUTRIENT REGIME
OTHER RESEARCHERS INVOLVED: SCHROEDER, P.; CHOQUETTE, C.
CONTACT PERSON: PERRY, D.A.
DATA COLLECTION PERIOD: Begin 81/08/01 End __/__/ ONGOING
KEYWORDS: POPULATION DYNAMICS, FERTILIZATION, PRUNING, FEEDBACK MECHANISMS, LOGISTIC GROWTH, STAND DENSITY, LTER
PARAMETERS/MEASURED VARIABLES: SPECIES, DBH, HEIGHT, SAPWOOD AREA, WOOD INCREMENT
VEGETATION ZONE: TSHE
PLANT COMMUNITIES IN STUDY AREA: TSHE: RHMA:GASH, TSHE:BENE: GASH, TSHE:ACCI:BENE
TAXA (List Scientific Abbreviations of Plant Species studied): PSME, TSHE, THPL
SOIL TYPE: NA
DETAILS OF SITE CHARACTER: 2000-3050 FT ELEVATION RANGE, MESIC SITES, 20-50% SLOPE RANGE, BROADCAST BURNED IN 1959 FOR SITE PREP., 2 SITES ON N-FACING AND 2 SITES ON S-FACING SLOPES.
RESEARCH AREA/REGION: OR-WESTERN CASCADES, H.J.ANDREWS EXPERIMENTAL FOREST, BLUE RIVER RANGER DISTRICT, WILLAMETTE NATIONAL FOREST.
PERMANENT PLOT NAMES: HJA STAND UNIT NUMBERS L107, L405, L701, L111

DATACODE: TP88
STUDY PURPOSE AND GOALS: TO TEST HYPOTHESES CONCERNING FEEDBACK MECHANISMS (COMPETITION AND MORTALITY) WHICH AFFECT FOREST STAND DEVELOPMENT. SITE QUALITY WILL BE MANIPULATED VIA FERTILIZATION, AND LEAF AREA WILL BE ALTERED VIA PRUNING. GROWTH WILL BE MODELED AS STANDS APPROACH SELF-THINNING.
EXPERIMENTAL/SAMPLING DESIGN: A SPLIT-SPLIT PLOT DESIGN—WITH 4 REPS., 3 DENSITY LEVELS (WHOLE PLOT), 2 FERTILIZATION (SPLIT PLOT) AND 2 PRUNING TREATMENTS (SPLIT-SPLIT PLOT). ONE PERMANENT PLOT WILL BE ESTABLISHED WHERE HEIGHT, DIAM., AND SAPWOOD AREA WILL BE SAMPLED.
EXPERIMENTAL METHODS: DENSITY TREATMENTS THINNED TO: (1) 15% OF MAX. STAND DENSITY; (2) 25% MAX. STAND DENSITY; (3) NO THINNING. FERTILIZATION TREATMENTS: (1) 50 kg/ha/yr OF UREA ADDED UNTIL CURRENT FOLIAGE REACHES 2% N; (2) NO UREA. PRUNING: TREE BRANCHES REMOVED TO ATTAIN CERTAIN CROWN WIDTHS.

REFERENCE CITATIONS--(Only those that cite data, relate to methods, etc.):

Num.	Year	Authors	Title	Journal and Vol.
1	1981	PERRY, D.A.	LTER STUDY PLAN—POPULATION DYNAMICS OF YOUNG FOREST STANDS	AVAILABLE FROM AUTHOR

Fig. 3.   The research study abstract form documenting ecological studies may request items such as study location, habitat type, site and soil characteristics, experimental design, analytical methodology, and keywords. The contents of this form should provide a reasonable overview of the experimental procedures and study objectives.

DATAFILE DESCRIPTION FORM
(FSDB)   841010

DATACODE _TP88_   DATE _83_/_10_/_11_   ONE-TIME COLLECTION ___
                              yr mo da   LONG TERM STUDY _✓_
STUDYID _—_   RECORDER _KW,CC_   DATA HEADERS EXIST ___

STUDY TITLE _POPULATION DYNAMICS OF YOUNG FOREST STANDS_   CRAFTS _✓_
                                                          LTER _✓_
            AS AFFECTED BY DENSITY AND NUTRIENT REGIME   FIR ___

PRINCIPAL INVESTIGATOR _PERRY, D.A._

OTHER PERSONS INVOLVED _SCHROEDER, P.; CHOQUETTE, C._

DOCUMENTATION STATUS (✓)

FORMAT TYPES →	1	2	3	4	5	6	7	8	9	10
FORMATS	✓	✓	✓	✓	✓	✓	✓	—	—	—
DEFINITIONS	✓	✓	✓	✓	✓	✓	✓	—	—	—
CODES	✓	✓	✓	✓	✓	✓	✓	—	—	—

FORMAT TYPE	DATAFILE NAME	BRIEF DESCRIPTION OF DATAFILE CONTENTS
1	TP8801	PRE-AND POST-THINNING DENSITIES ON PLOTS
2	TP8802	INITIAL TREE DIMENSIONS
3	TP8803	DELAYED HEIGHT MEASUREMENTS AND CHANGES
4	TP8804	TREE GROWTH MEASUREMENTS
5	TP8805	EXTRA TREE DATA OUTSIDE EXPERIMENTAL AREA
6	TP8806	COMPETITION DATA FOR SAMPLE TREES
7	TP8807	MINERALIZABLE NITROGEN SAMPLES

Comments: _YEARLY MEASUREMENTS ARE EXPECTED_

Fig. 4.   The data file description form contains an organized list of all data files pertaining to one data code (study). Individual data file names and a brief description of the type of data stored in each file are recorded.

recently converted storage of paper file documentation to floppy and hard disks on a microcomputer. The storage location and media chosen depend primarily on computer accessibility, available software, ability to store text as well as numbers, and cost of storing and retrieving information. In any case, it is important to have adequate, secure backups in the event that originals are lost, stolen, burned, or otherwise destroyed. We maintain four copies of our FSDB tapes, one of which is stored at a location outside the University computer center. The durability of media also is an important consideration. Magnetic tapes and disks deteriorate after extensive use. Cards, paper tapes, and magnetic tapes become brittle with time and require periodic use, cleaning, or replacement.

Initially, we store most study documentation in paper files, where completed forms, sample field collection sheets, the study plan, and other related information are maintained. Some DBM systems store much of their documentation as a header at the beginning of each data file. The storage device for this type of file must be able to handle text efficiently and search for and retrieve specific information. Many systems have specialized software for relational databases, but in our experience, mainframe software is too expensive to run, and disk storage costs are too high to leave documentation continuously online. If files must

841010   **VARIABLE FORMAT FORM**   Page _1_ of _1_
(FSDB)

DATACODE TP88   DATE 10/11/83   RECORDER KW, CC
                     mo da yr

FORMAT TYPE 4

STUDYID —   DATA TITLE: TREE GROWTH MEASUREMENTS

VAR. NUM.	VARIABLE NAME	COLUMNS OCCUPIED	FORTRAN FORMAT	CODED (√)	UNITS	MISSING VALUE CODE
1	DATACODE	1-4	A4			ALL "BLANK"
2	FORMTYPE	5-6	I2			
3	SITE	7-11	1X,A4			
4	DENSITY	12-13	1X,A1	✓		
5	TMT	14-16	1X,A2	✓		
6	YRMODA	17-23	1X,I6			
7	TREENUM	24-27	1X,I3			
8	DBH	28-31	1X,F3.1		cm	
9	HT	32-35	1X,F3.1		m	
10	SAPW1	36-39	1X,I3		mm	
11	GRINC1	40-42	1X,I2		mm	
12	PRERING1	43-46	1X,I3		mm	
13	SAPW2	47-50	1X,I3		mm	
14	GRINC2	51-53	1X,I2		mm	
15	PRERING2	54-57	1X,I3		mm	
16	TREECOND	58-59	1X,I1	✓		
17	BARK1	60-62	1X,I2		mm	
18	BARK2	63-65	1X,I2		mm	

CONTINUED, Reverse side ___

Fig. 5.   The variable format form delineates the exact structure of a data file by identifying the columns occupied by each variable, the precise format, and related information. Each variable is assigned an acronym (variable name no greater than eight characters) that remains constant on every documentation form.

first be retrieved from magnetic tape, time delays are
inevitable and additional costs accumulate before the software is used. We have relied on publicly accessible microcomputers for organizing, storing, and retrieving data documentation and the University mainframe for storing the data
files themselves. Interested users can scan our documentation files on the microcomputers, note the data code or data
file name, and retrieve the file(s) from our magnetic tape
library on the mainframe.

CATALOGING AND RETRIEVING DATA AND DOCUMENTATION

Paper files may be suitable for storing completed documentation forms, but a computerized relational database is
invaluable for quick search and retrieval tasks. For example, with a computerized database, a file containing study
abstracts of all projects currently being managed may be

841010   **VARIABLE DEFINITION FORM**   Page 1 of 1

(Complete one form for each format type)

DATACODE TP88   DATE 10 / 11 / 83   RECORDER KW, CC
mo   da   yr

FORMAT TYPE 4

STUDYID —

VAR. NUM.(1)	BRIEF DESCRIPTION OF EACH VARIABLE	PRECISION
1	FSDB DATASET CODE	
2	FSDB FORMAT TYPE	
3	HJ ANDREWS SITE IDENTIFICATION CODE	
4	STOCKING - DENSITY LEVEL FOR WHOLE PLOT TRTMT.	
5	TREATMENT - FERTILIZED OR PRUNED	
6	YEAR - MONTH - DAY OF SAMPLING	
7	TREE TAG NUMBER	
8	DIAMETER AT BREAST HEIGHT	$\pm$ 0.1 cm
9	TOTAL TREE HEIGHT	$\pm$ 0.1 m
10	SAPWOOD WIDTH - DIRECTION #1	$\pm$ 1 mm
11	GROWTH INCREMENT - LAST TWO YEARS, #1	$\pm$ 1 mm
12	PRE-THINNING RING WIDTHS - RINGS 4-8 COMBINED, #1	$\pm$ 1 mm
13	SAPWOOD WIDTH - DIRECTION #2	$\pm$ 1 mm
14	GROWTH INCREMENT - LAST 2 YEARS, #2	$\pm$ 1 mm
15	PRE-THINNING RING WIDTHS - RINGS 4-8 COMBINED #2	$\pm$ 1 mm
16	TREE CONDITION - DAMAGES	
17	BARK THICKNESS - DIRECTION #1	$\pm$ 1 mm
18	BARK THICKNESS - DIRECTION #2	$\pm$ 1 mm

(1) Use exact numbers from the corresponding Variable Format Form.

Fig. 6.   The variable definition form briefly describes
each variable and specifies the desired measurement precision.

841010		VARIABLE CODE FORM	Page 1 of 1

(Complete one form for ALL format types)

DATACODE TP88   DATE 10/11/83 (mo/da/yr)   RECORDER KW,CC

STUDYID —

VARIABLE NAME	CODE VALUE	BRIEF DEFINITION OF EACH CODE VALUE
DENSITY	C	CONTROL
	L	REMOVAL OF 15% MAX. STAND DENSITY
	M	REMOVAL OF 25% MAX. STAND DENSITY
TMT	C	CONTROL - NO UREA OR PRUNING
	F	FERTILIZED WITH UREA
	P	PRUNED TO SPECIFIC CROWN WIDTHS
	FP	FERTILIZED AND PRUNED
TREE COND	0	LIVE
	1	DEAD
	2	BROKEN TOP AND DEAD
	3	STEM DEFORMITY DUE TO SNOW
	4	ROOT ROT (PHELLINUS) INFECTION
	5	BROKEN TOP - WILL LIVE
	6	LEANING TREE
	7	ANIMAL DAMAGE - BASAL SCARS
	8	ANIMAL DAMAGE - IN CROWN

CONTINUED, Reverse side ___

Fig. 7.   The variable code form describes the codes allocated to constant nonmeasurable variables or cate-
gorical variables (used where each observation is
classified into a predefined category). Specific
codes should be identified before active data collection begins.

searched for a particular topic. Once the appropriate
abstract is found, the data code or name is extracted and
used as a key variable in the database. The most efficient
method of referencing paper files, if used, would again be
by data code or name. If documentation can be centralized
on a computer, commercial DBM software packages can be used.
Because these packages are more user friendly, DBM personnel
need be only minimally involved in routine searches and
retrievals. Once the desired information has been located,
the computer can produce a printed report fully describing
the data sets.

Ease of retrieval depends on the existence of a relational system for cataloging information for a given data

code and on the access mode of the storage medium: random access (disks) or sequential access (magnetic tapes). Software that simplifies data retrieval is available for data files permanently stored on magnetic tape. Currently, researchers at the Andrews LTER site use a program that requests the name of the data file to be retrieved, automatically creates a batch file that mounts the proper tape, retrieves the file, and stores a copy on a requested computer account, all without DBM staff assistance. If data are stored on disk or a similar random-access device, the optimal strategy is to tie data and documentation together so that retrieving both is one basic process. Researchers using the SIR system (Robinson et al., 1979), for example, can retrieve data-structure documentation easily with a query language. Data files can also be stored with a documentation header, which allows documentation and data to be retrieved as one unit, but this approach may require more custom software development or be incompatible with some software.

SOURCES OF ERROR

A popular misconception is that errors first occur when the data are collected. Actually, problems potentially begin much earlier, with the design and documentation of data collection forms. Detectable errors may still exist even after data files are in final form. A preliminary data screening analysis, designed to further validate the data, will identify these errors. Validation analysis serves two purposes: first, to validate the correspondence between data and associated documentation; second, to identify data collection and entry errors where possible. We have found that data screening always pays. Locating and correcting as many errors as possible before a full statistical analysis is undertaken prevents costly reruns and wasted time.

During validation analysis, the documented data file format is compared with the actual data file format. For discrete variables, a list of frequencies for each code is generated. For continuous variables, the following statistics are produced: minimum, maximum, mean, number of valid observations for each variable, and total number of observations in the data file. The analysis reveals where incorrect codes were used or where data were out of the expected range. These errors, which may have occurred during field collection, data entry, or recording of the documentation, are then identified and located on the collection forms so that the researcher can decide how to correct them. Validation checks have been incorporated into DBM systems at several LTER sites to catch errors when the data are entered. At other sites, graphical displays identify possible errors. The most elaborate systems produce printouts of raw data with annotated comments, which point to possible errors. Once again, responsibility for changing faulty values lies with the researcher.

After data entry, data files frequently need to be reformatted and then transferred to another medium or storage device. These procedures should be done by data bank personnel to avoid the possibility of introducing "unprotected errors" (errors created by a program or by another logical error that could produce a major, but virtually undetectable, error in the data file). For example, merging two data sets for which there is no one-to-one matching variable (which exactly identifies the matched cases) can create unprotected errors. A successful merge requires that the sorted order of each data file be identical. If cases are out of order or missing, the completed merge may appear to be normal, but actually may have produced undetectable errors in the data file. Where reformatting and manipulating data are unavoidable, exercise extreme caution.

Though documentation and data entry errors are usually easily corrected in the data file, field or laboratory errors must be handled quite differently. One approach, though not widely recommended, is to leave the data exactly as recorded on the collection forms. In many cases, however, the data file cannot be processed with these errors; for example, an alpha value in a numeric field could cause a program failure. Another approach is to try to update the data where the correction is obvious (and we do mean obvious), for instance, a misspelling. Such corrections, once approved by the researcher, can be made efficiently without running the risk of introducing extraneous or incorrect information. We have found that the best approach concerning faulty information is to set the observation equal to a "missing" code to avoid guessing the correct value; these observations are then excluded from all analyses.

Data collection errors usually are not rectified easily. Continuous variables that are out of the expected range often require field or laboratory validation. If no other validation method is available, the value should be set to "missing." Losing information is always unfortunate, but a value that is out of range can be far more damaging to analysis results than missing information. Similarly, data that are illegible should be entered as "missing" and that observation noted for later evaluation if no other validation method is available.

In a few circumstances, data may be modified purely on a "best approximation" basis. This is not often recommended, but can sometimes have merit. For example, suppose tree diameter on a particular plot had been measured annually for many years, but the value for an intermediate year was found to be completely out of range. That diameter could be approximated according to the values recorded for the adjacent years, and the probability of error would be very small. The alternative is entering "missing." The consequences of each approach should be judged for each instance. Problems like this arise when long-term ecological data are not analyzed yearly (in which case the error would have been identified and corrected at the time) or when data were obtained from other sources. In any case, situations do occur when recorded data must be modified, and those modifications must be documented. Systems for documenting changes range from merely jotting them down on the collection forms to elaborate notation in the actual data files.

Another type of error common to ecological research is the failure to record unusual sampling situations, which may lead to atypical data values. Unfortunately, there is no way to plan for these when designing a data collection form except to provide space for comments. Although the comments are not generally archived with the data file, they help the researcher interpret the outliers in the data. We encourage the archiving of raw data collection forms (usually in microfiche) to preserve this information. In some long-term studies in which re-measurements are frequent, certain comments are encoded in the data files themselves; however, this must be done so that the comments can be separated easily and excluded during analysis.

Sites in the LTER network have functionally similar data validation (quality assurance) systems. The most sophisticated systems produce graphical analysis and annotated output, usually via the SAS system (SAS Institute Inc., 1982). Data bank personnel verify the entry of data at least once at all sites and provide researchers with information on illegal codes and possible outliers. Decisions on what to do about outliers always rest with the researcher. Problems sometimes arise because scientists do not adequately proof and validate data; at some sites, they must sign or initial approval forms after reviewing entered data. Providing a uniform standard of quality assurance for all archived data remains the most critical test of the usefulness of an ecological data bank.

The issue of data integrity must be emphasized to assure that the data analyzed are a valid, accurate representation of the facts. In this process we often take advantage of the "Law of Large Numbers," which basically states that the larger the sample size, the closer the sample mean will be to the true population mean. To require every bit of data to be an exact representation of reality is unrealistic. A more reasonable aim is to record measurements as accurately as possible, given the experimental constraints and overall research perspective.

One additional source of error remains after documentation, data entry, and validation are completed: errors in the execution of the statistical analysis. These can range from relatively minor mistakes to completely misleading results that may be invalid if the data are not analyzed with the statistical design initially chosen.

CONCLUSIONS

Evolution of hardware and software for data management in recent years has made possible great advances in the flexibility and utility of ecological data management. In our experience, developments in the computer industry have far outstripped ecologists' training to use them. The most substantial barriers to improving data management are our own failures of imagination. New ways of thinking and co-operating in multidisciplinary, long-term research lay the groundwork for future, more fundamental advances in the field. Increased availability of computers, more powerful software that is easier to use, and support from DBM personnel should augment scientists' abilities to improve database quality and utilization. Once credible, reliable, comprehensive databases are established, more integrative data-intensive research projects will be a practical reality for answering the key questions of systems ecology.

LITERATURE CITED

Chervany, N.L., P.G. Benson, and R.K. Iyer. 1980. The planning stage in statistical reasoning. Amer. Statistician 34: 222-226.

Hinds, W.T. 1984. Towards monitoring of long-term trends in terrestrial ecosystems. Environ. Conserv. 11(1): 11-18.

Hurlbert, S.H. 1984. Pseudoreplication and the design of ecological field experiments. Ecol. Monogr. 54(2): 187-211.

Nie, N.H., C.H. Hull, J.G. Jenkins, K. Steinbrenner, and D.H. Brent. 1975. SPSS: Statistical Package for the Social Sciences. 2nd edition. McGraw-Hill, NY. 675 pp.

Perry, D.A. 1982. Population and dynamics of young forest stands: Establishment report for Long-term Ecological Research at H.J. Andrews Experimental Forest. Appendix D. In: Proposal to National Science Foundation Long-Term Ecological Research on the Andrews Experimental Forest. Unpublished. 10 pp.

Robinson, B.N., G.D. Anderson, E. Cohen, and W.F. Gazdzik. 1979. SIR: Scientific Information Retrieval. 2nd edition. Northwestern University, Evanston, IL. 321 pp.

SAS Institute Inc. 1982. SAS User's Guide: Basics. 1982 edition. SAS Institute Inc., Cary, NC. 923 pp.

Stafford, S.G., P.B. Alaback, G.J. Koerper, and M.W. Klopsch. 1984. Creation of a forest science data bank. J. Forestry 82: 432-433.

THE STATUS AND PROMISE OF INTER-SITE COMPUTER COMMUNICATION[1]

M.W. Klopsch and S.G. Stafford

ABSTRACT

One major goal of the National Science Foundation's Long-Term Ecological Research (LTER) program is to facilitate collaborative research among sites. To achieve this goal, intersite communication of both data and documents will be required. Most of this information can be handled electronically. Data exchange via magnetic tapes, floppy disks, or telecommunications is available at all LTER sites; as communication speeds increase, nearly all data will be telecommunicated using error-checking protocols to improve reliability. Information currently communicated on paper could be sent as electronic mail; parts of collaborative manuscripts could be prepared on word processors and merged without extensive retyping or delays; electronically exchanged data summaries could easily be transformed for analysis or merged with other information. Shared databases could be established on either a central computer or an information service. Although the information revolution brings with it certain hazards, linking LTER sites into a telecommunication network is feasible and has great potential to enhance future ecological research.

INTRODUCTION

One major goal of the Long-Term Ecological Research (LTER) program is to facilitate collaborative research among

---

[1] FRL 1957, Forest Research Laboratory, Oregon State University, Corvallis. The mention of trade names or products does not constitute endorsement or recommendation for use.

sites. To achieve this goal, intersite communication of both data and documents is required. We surveyed LTER sites regarding their communications capabilities; the survey results suggest a considerable amount of unutilized communications potential.

Traditionally, most intersite communication has been by telephone conversations and paper documents--without the aid of electronic media. With the advent of the microprocessor, this situation has changed dramatically: most communications will soon be handled electronically at some point in their preparation. This fundamental change in operations provides the opportunity to speed and enhance intersite communication. In this paper we review current communication capabilities and explore some short-term and long-term possibilities.

THE CURRENT STATUS OF INTERSITE COMMUNICATIONS

Data exchange via magnetic tapes, floppy disks, or telecommunications is available at all LTER sites. Once a standard for tape interchange, a list of acceptable disk formats, and a knowledge of communication protocols have been established, data exchange between sites will be simply a matter of matching the method to the need.

Magnetic Tapes

Magnetic tape, a standard medium for mainframes and minicomputers, is useful for transferring large data sets and programs. Unless another format is arranged, the standard for magnetic tape interchange should be an "unlabeled," 9-track tape recorded in ASCII at 1600 bpi (bits per inch) with an ANSI format and a block size less than 5000 characters. Although the suggested format is somewhat bulky due to the large number of interblock spaces, it could be easily created and read by all LTER sites. The tape must be "unlabeled" because labeling installs a system-dependent password that makes reading the tape at another site nearly impossible. The block size must be small because some systems, such as our CDC CYBER computer, are unable to efficiently buffer larger blocks. Thorough documentation of tape formatting is required even with this standard. Our biggest problems with tape exchange have been caused by inadequate documentation of tape formats and the inadvertent exchange of "labeled" tapes.

Floppy Disks

Floppy disks are more commonly used with microcomputers, word processing systems, and to a lesser extent supermicrocomputers and minicomputers for medium-size data sets. Although disks are cheaper and simpler to mail than magnetic tapes, their use as a communication medium will primarily depend on the ease of moving the information to and from the disks at the origin and destination sites. The type of disk and format to use in disk exchanges should be determined by convenience. Although every LTER site can use 5¼-inch double-sided, double-density, MS-DOS formatted disks, many sites prefer other types of disks and formats (Table 1).

Table 1. Disk formats usable at LTER sites.

Site	Computer	Operating System	Disk Format	Format Acceptability
Andrews	Compupro	CP/M-80	8" SSSD	Preferred
Andrews	IBM PC	PC-DOS 2.1	5 ¼" DSDD	Preferred
Cedar Creek	IBM PC	PC-DOS 2.1	5 ¼" DSDD	Preferred
Cedar Creek	Xerox	CP/M-80	8" SSSD	Acceptable
Cedar Creek	Apple IIe	AppleDOS	5 ¼"	Possible
Cedar Creek	Apple IIe	ProDOS	5 ¼"	Possible
Coweeta	IBM PC	PC-DOS 2.1	5 ¼" DSDD	Preferred
Coweeta	MacIntosh	MacIntosh OS	3 ½"	Preferred
Coweeta	Apple IIe	AppleDOS	5 ¼"	Possible
Coweeta Hydro Lab	Apple IIe	AppleDOS	5 ¼"	Preferred
Illinois Rivers	IBM PC	PC-DOS 2.1	5 ¼" DSDD	Preferred
Illinois Rivers	Apple IIe	AppleDOS	5 ¼"	Possible
Jornada	IBM PC	PC-DOS 2.1	5 ¼" DSDD	Preferred
Jornada	MacIntosh	MacIntosh OS	3 ½"	Acceptable
Hobcaw	IBM PC	PC-DOS 2.1	5 ¼" DSDD	Acceptable
Hobcaw	Dec Pro 350	POS	5 ¼" RX50	Preferred
Konza Praire	IBM AT	PC-DOS 3.0	5 ¼" DSDD	Preferred
Konza Praire	IBM AT	PC-DOS 3.0	5 ¼" DSQD	Preferred
Konza Praire	CBM 8032	CP/M-80	5 ¼" SSSD	Acceptable
Niwot Ridge	IBM PC	PC-DOS 2.1	5 ¼" DSDD	Preferred
Niwot Ridge	RS Model 4	TRSDOS	5 ¼" DSDD	Preferred
Niwot Ridge	RS Model 16	XENIX	8" DSDD	Acceptable
Niwot Ridge	RS Model 3	TRSDOS	8" DSDD	Acceptable
Niwot Ridge	Apple IIe	AppleDOS	5 ¼"	Acceptable
Northern Lakes	IBM PC	PC-DOS 2.1	5 ¼" DSDD	Acceptable
Northern Lakes	Apple IIe	ProDOS	5 ¼"	Preferred
Northern Lakes	Apple IIe	P-System 1.2	5 ¼"	Acceptable
Northern Lakes	Apple IIe	AppleDOS	5 ¼"	Acceptable
Northern Lakes	MacIntosh	MacIntosh OS	3 ½"	Acceptable
Okefenokee	IBM PC	PC-DOS 2.1	5 ¼" DSDD	Preferred
Okefenokee	Viasyn Sys 10	CCPM	5 ¼" DSDD	Acceptable
Pawnee	IBM PC	PC-DOS 2.1	5 ¼" DSDD	Preferred
Kellogg	IBM PC	PC-DOS 2.1	5 ¼" DSDD	Preferred
Kellogg	VAX	VMS	8" files-II	Preferred

<u>Telecommunications</u>

Most communications among sites require rapidly transferring small amounts of data, for which telecommunications is best suited. In the future, as communication speeds increase, nearly all data will be handled by telecommunication.

Telecommunications can be either synchronous or asynchronous. Synchronous communication is the transmission of compressed blocks of data "in sync" with a clock signal. Standardized protocols are used to detect and correct transmission errors and to unblock the data. Synchronous modems are faster, but more expensive, than asynchronous modems and usually require specially conditioned phone lines. Only two sites reported the capability to telecommunicate synchronously with other sites. In contrast, asynchronous communication is based on the transmission of individual characters; each byte or character is transmitted by sending a starting bit, the data bits, sometimes a parity bit, and stop bits at a predetermined rate. Because asynchronous modems, software, and phone lines are cheaper than synchronous, most microcomputer communications are done asynchronously. Computers communicating asynchronously must use the same data configuration (i.e., communication speed, data width, parity, stop bits, and half or full duplex). Most communication packages store these configurations along with the phone numbers to automatically configure the computer before dialing. The asynchronous configurations for LTER sites are reported in Table 2. Most asynchronous communication is conducted at a speed of 1200 baud (bits per second, about 120 characters), although some communication is still at 300 baud, and 2400-baud modems are becoming available. The most common 1200-baud configuration is eight data bits, no parity bit, and one stop bit, requiring ten bits to transmit a character.

The primary problems with telecommunications are comparatively high costs, relatively slow speed, and fairly high error rate. Since the reliability of telecommunications depends on the quality of the phone connection, it varies with location and time. The reliability problem can be overcome by using some form of error-checking protocol for important information. A standard protocol for intersite communication is desirable. The diversity of equipment among LTER sites requires that a protocol supported on a

Table 2. Intersite asynchronous communications capabilities.

Site	Telecommunications Contact Person	Phone	CS NET	Baud	Parity Data	Stop	Duplex	Error-Checking Protocols
H.J. Andrews	Mark Klopsch	(503) 757-4427	N	300/1200	7 E	1	Full	DATLINK, XMODEM, KERMIT
Cedar Creek	Robert Buck	(612) 376-9455	N	300/1200	7 E	1	Half	XMODEM
Coweeta	Polly Casale	(404) 542-2968	N	300/1200	7 E	1	Full	KERMIT (soon)
Coweeta Hydro Lab	Bryant Cunningham	(704) 524-2128	N	300/1200	8 N	1	Half	
Illinois Rivers	Frank Brookfield	(217) 333-6006	N	300/1200	8 N	1	Half	PLOT-10
Jornada	Walt Conley	(505) 646-2541	Y	300/1200	8 N	1	Half	Send twice and compare
Hobcaw	Robert McLaughlin	(803) 546-6219	N	300/1200	7 N	2	Full	
Konza Praire	John Briggs	(913) 532-6629	N	300/1200	8 N	2	Full	Hardcopy check
Niwot Ridge	Jim Halfpenny	(303) 492-6241	N	300/1200	8 N	1	Full	
Northern Lakes	Carl Bowser	(608) 262-8955	Y	300/1200	8 N	1	Full	Xon/Xoff
Okefenokee	Joe Schbauer	(404) 542-2968	N	300/1200	7 E	1	Full	KERMIT (soon)
Pawnee	Tom Kirchner	(303) 491-1986	Y	300/1200	7 E	1	Full	KERMIT
Kellogg	John Gorentz	(616) 671-5117	N	300/1200	7 O	1	Full	BTRANS, XMODEM

CS NET: Y = Currently Accessible, N = Not Currently Accessible
Parity: N = none, E = even, O = odd

wide variety of computers be selected. Only two asynchronous communication protocols currently deserve serious consideration: XMODEM and KERMIT. The most common, XMODEM, developed by Ward Christianson for his public domain MODEM7 package, is available on most microcomputers running CP/M-80, CP/M-86, MS-DOS, and ProDOS operating systems. Software providing this protocol include CROSSTALK, PC-TALKIII, MITE, and ASCOM (Helliwell, 1984). The KERMIT protocol, designed specifically for asynchronous communication between microcomputers and mainframes (DaCrus & Catchings, 1984a & b), is attractive because it is sufficiently simple to be easily implemented on most machines. Source code is available at cost from Columbia University for the IBM PC (PC-DOS), IBM 370 (VM/CMS), VAX-11 (VMS, UNIX), SUN (UNIX), PDP-11 (UNIX, RT-11, RSX, RSTS), 8080 (CP/M-80), 8086 (CP/M-86, MS-DOS), and Apple II (Apple DOS).

Information Services

Information services such as CompuServe and Source provide shared databases, electronic mail, bulletin boards, and forums on topics of general interest. The subscription, time-sharing, and telephone charges make the services relatively expensive ($10 to $50 per hour). Most of the features useful to the LTER sites could be approximated at a lower cost with a centrally located multi-user system.

THE POTENTIAL OF INTERSITE COMMUNICATIONS

Information currently communicated on paper could be sent more efficiently through an LTER electronic mail system, which could guarantee arrival in less than 24 hours for about the same cost as regular mail. For example, letters and memos, important news items, warnings, or requests for assistance could be routed electronically on the basis of names or keywords--perhaps to all principal investigators or to groups of people sharing a common interest. Parts of collaborative manuscripts could be prepared by their authors on word processors and merged without extensive retyping or delays. Electronically exchanged data summaries are easily transformed into spreadsheets or merged with other information into data files for future comparison and evaluation.

AT&T has recently been testing a new modem which takes advantage of the phone system's time compression multiplexing to permit full-duplex communication at 56K baud on ordinary phone lines (McDonnel, 1984). When the circuit-switched digital capacity (CSDC) system is universally available in late 1985, it should revolutionize telecommunications, replacing all other forms of communication, even for large data sets. The 20 cents it costs to telecommunicate a 2-page letter today should pay for the transfer of a 90-page manuscript a year from now. Facsimile transmission (the transmission of pictures and graphics), which also requires the transmission of large amounts of information, will be vastly enhanced. The CSDC system should make it possible to install reasonably efficient telecommunication bridges between local area networks (LANs) to achieve a functional regional, national, and international intercomputer communication network.

As the quality and speed of communications improve, certain graphics devices, specialty programs, and other network resources could be shared among sites. Because running a program on a distant time-sharing system is no more difficult than running it locally, many applications not requiring special display devices could be run remotely, reducing the need to replicate some expensive specialized equipment and software at each site. Shared databases, including lists of hardware and software with evaluations, data catalogs or dictionaries, and studies in progress, could be established on either a central computer or an information service like CompuServe.

Most LTER sites have access to a computer with the UNIX operating system. About 1500 large UNIX systems and many smaller computers are connected to UNIX communications program (UUCP) to form USENET (Emerson, 1983; Darwin, 1984). This program and associated utilities allow automatic file transfer between machines, electronic mail, and a subscription service for several hundred categories of news items. Each participant creates a subscription list identifying the news information categories desired. Because the UNIX system has provisions for scheduling unattended transfers, all USENET activities are accomplished with little or no supervision--a desirable feature.

However, as a network for LTER sites, USENET has several disadvantages. First, a message may pass through many computers before reaching its destination. For example, to get a message from Oregon State University to Emory University in Georgia, the message would have to pass through about ten machines. Though useful for passing on news and information, this is too slow for urgent messages and "rush" jobs. Second, USENET is so broadly based that,

when it is used as a bulletin board, only general categories
of news (such as physics, politics, and religion) are provided. Consequently, ecological researchers would have to
wade through a morass of unnecessary information to find
their topics of interest.

An alternative approach would be to set up a system
among LTER sites in which the information categories could
be specific to LTER needs. It might be possible to "piggy-
back" a smaller network of LTER sites onto the existing UNIX
systems using the same software. An alternative would be to
create an analogous network linking existing computer systems at each site. The disadvantage of developing an analogous network is that some software would have to be created
to handle file forwarding and selection. A rudimentary
system could be initiated using an asynchronous protocol
with file-forwarding software located on only one central
computer. In this star arrangement, users at each site
would either manually or automatically contact the central
computer node on a regular basis to exchange information.
Later, the system could be enhanced with several interconnected central computers to reduce the telephone distances
and to take advantage of faster or less expensive communications systems.

Some potential hazards accompany the proliferation of
electronic publications and the use of electronic mail, data
banks, and information services. First, users must remember
that telecommunications is a tool, not a goal in itself, and
must not be pursued to the exclusion of other projects
("network addiction"). Second, although freer exchange of
information is a desirable goal, effective mechanisms must
be developed to protect proprietary data and properly
acknowledge individuals for their contributions. What
constitutes a "publication," for example, may have to be
redefined (Zientara, 1984). Finally, the possibility that
we will all be overwhelmed by the sheer amount of information is a real danger. It is especially important for ecologists to learn the new tools for information access to
avoid the narrowing of focus that would otherwise be almost
inevitable.

The potential of efficient intersite communications
is staggering. In the future, "idea exchanges" might be
created in which hypotheses and supporting evidence could
be displayed, comments or extra information added by other
contributors, and an "electronic publication," with acknowledgments and authorship as appropriate for contributors,

ultimately produced. The exchange could reduce research overlap and double as a registry for ideas to prevent piracy by associating an author and date with each entry. A new "gray" literature could be created if authors post papers for comment before publication. Increasingly, manuscripts for submission to journals or magazines will be requested in electronic form to speed galley preparation--a prelude to the electronic journals of the future, which promise a more timely exchange of data and ideas. Subscriptions to such journals could provide access to a database already cross-referenced by keywords and authors. The net result of these new procedures would be the rapid, organized dissemination of protected information, enhanced research opportunities, and improved journal quality.

Most LTER sites could be linked today through a simple communication network with existing equipment. Although current communication among sites might not seem to justify establishing a network, this is not a good indicator of future usage. In addition, implementing and standardizing communications within an LTER telecommunications network would enhance local communications at some sites. The LTER program is designed to provide leadership in ecological research. An LTER network would be a concrete example of this leadership and serve as a model or provide impetus for a global network of ecological researchers.

LITERATURE CITED

Darwin, I.F. 1984. The UNIX file. Microsystems 5(5): 32-38.

DaCrus, F. and B. Catchings. 1984a. Kermit: A file-transfer protocol for universities. Part 1: Design considerations and specifications. Byte 9(6): 225-278.

DaCrus, F. and B. Catchings. 1984b. Kermit: A file-transfer protocol for universities. Part 2: States and transitions, heuristic rules, and examples. Byte 9(7): 143-145, 400-403.

Emerson, S.L. 1983. USENET: A bulletin board for UNIX users. Byte 8(10): 219-236.

Helliwell, J. ·1984. Guide to communications software. Computers & Electronics 22(8): 56-59, 86-89.

McDonnel, P. 1984. AT&T breaks the speed barrier. Computers & Electronics 22(9): 66-69, 100.

Zientara, M. 1984. Watch your words. Infoworld 6(32): 33-34.

LOCATING MACHINE-READABLE ECOLOGICAL DATA

Richard J. Olson and Paul Kanciruk

ABSTRACT

Increasing numbers of machine-readable ecological and environmental data files are becoming available from federal agencies specifically for public dissemination. Many agencies or projects with specific thematic orientations compile descriptions of databases, but usually do not maintain the actual data. These directories provide enough information to determine which data files have a high probability of fulfilling a user's need and include an individual's name to contact for obtaining more information (or the actual data file). This paper describes directories and clearinghouses that assist users in locating and acquiring data files.

Twenty-four directory/clearinghouse activities maintained by federal agencies are described and grouped into five classes: I) clearinghouses not dealing with actual data files, II) data repositories containing data files, III) data centers with data and some data evaluation, IV) data analysis centers with extensive data evaluation, and V) integrated database systems with evaluation, standardization, and integration. Published inventories of machine-readable data are reviewed. The Acid Deposition Data Network, developed at Oak Ridge National Laboratory by the National Acid Precipitation Assessment Program, is described as a project using data directories and as an example of an integrated database system that provides evaluated, standardized, and integrated information on acid-deposition-related data.

INTRODUCTION

Concerns about cumulative and regional-scale impacts of man's activities on human health and the environment have created a demand for environmental data. Acid rain is a good example of a current issue requiring extensive ecological data. Ecological data do exist; however, it is often a challenge to locate, acquire, and apply these data for specific purposes. This paper describes clearinghouses and inventories that assist users in locating and acquiring data files. The development of the Acid Deposition Data Network (ADDNET) database, which is utilizing these resources to compile an integrated database for acid deposition assessments for the National Acid Precipitation Assessment Program (NAPAP), will be used as an example. NAPAP is a multiagency, multiyear, federal program created by the Acid Precipitation Act of 1980 to evaluate potential impacts of acidic deposition and alternative control strategies.

CLEARINGHOUSES

Clearinghouses appear to be poorly known and underutilized resources maintained by many federal agencies for locating data. Although we use the term clearinghouse to describe these data inventory activities, this is only one of several terms commonly used (e.g., data inventories or data directories). "Clearinghouse" may also be used to describe activities included as one of the tasks of information centers, data referral centers, data centers, integrated database systems, or other similar operations.

Locating Data

We use the term clearinghouse to refer to those activities specifically oriented to compiling directories of machine-readable data files (MRDFs). These directories contain descriptions of the data, not the actual data files. A directory may often be an on-line, searchable database with specific data elements or components that describe each MRDF. Most directories contain the data file name or title, an abstract, the spatial and temporal attributes, institutional responsibilities, and the name of a contact person. They thus provide enough information to indicate which data files have a high probability of fulfilling a user's need, and furnish an individual's name if more information or the actual data file is needed.

Clearinghouses usually limit their inventories to their own agency's data. The National Oceanic and Atmospheric Administration's (NOAA) National Environmental Data Referral Service (NEDRES) (see Freeman, this volume) has a much broader scope: it is actively compiling data-set descriptions from many sources, with major emphasis on climatic, atmospheric, and physical environment themes (less emphasis is placed on water resources, vegetation, forests, wildlife, or other biologic components). Currently, NEDRES contains descriptions of over 15,000 data sets that can be searched interactively for specific data types.

Information centers usually focus on specific subjects and provide specialized services such as searching bibliographic databases for citations. Clearinghouses extend these bibliographic database techniques to repackaging information or numeric data for wider dissemination (Frank & Gandy, 1983). A recent report on health data (PHS, 1982) described 19 public health service clearinghouses. However, the majority of these information resource organizations did not include maintaining directories of MRDFs within their operation.

Data centers may have a clearinghouse or directory component, but they differ from information centers in that they are also a repository for data. Data centers vary as to the amount of data processing done: some centers may provide only archival storage, others perform data evaluation and extensive repackaging prior to distribution. Centers providing evaluated data files often rely on an established user community that determines criteria for data standards. Integrated database systems are another level of data compilation that entails maintaining a directory of the diverse data files within the system.

1983 SURVEY

For the 1983 Integrated Data Users Workshop (Olson & Millemann, 1984), federal agencies were contacted to identify clearinghouses or similar directory activities that provide a focal point for locating and/or distributing data within each agency. This limited survey emphasized major agencies compiling socioeconomic, demographic, environmental, and natural resource data because these were the types of spatial data emphasized at previous Integrated Data Users workshops.

Implied in this discussion on MRDFs is that these data files are readily available to all potential users. In practice, MRDFs may be restricted to internal agency use, to other federal agency use, or to public use only as special tabulations or selected subfiles. Similarly, the directories of MRDFs may be developed for internal use and are accessible to outside users only through special requests to the agency clearinghouse for specific data. The intent of this discussion is to provide information on clearinghouses and directories, leaving issues of data accessibility and exchange to be resolved by data users on a case-by-case basis.

Table 1.   Classes of activities related to machine-readable data file directories.

Class	Activity functions			
	Directory	Data	Evaluation	Analysis
I. Clearinghouse	Yes			
II. Repository	Yes	Yes		
III. Data center	Yes	Yes	Maybe	
IV. Evaluated data center	Yes	Yes	Yes	
V. Integrated database system	Yes	Yes	Yes	Yes

The 1983 survey classified the various directory activities according to five broad categories (Table 1). Class I includes those activities oriented toward maintaining directories that do not offer direct access to databases. Class II consists of archival repositories that provide public access to data acquired from federal agencies. Class III is a more diverse group of activities, but generally represents agency data centers that provide MRDFs for public distribution and, in the process, maintain inventories of their MRDF holdings. Some of these data centers limit their activities to within their specific agency. Class IV represents data centers that evaluate and package databases prior to distribution. Class V consists of integrated database systems such as those described by Merrill (1983); these were not covered in this survey.

Information in Tables 2 through 5 is based on mail questionnaires and follow-up telephone calls. Additional questions were asked concerning the level of use of the directories, typical search costs, updating procedures, etc.; however, this information was generally unavailable or incomplete. A draft of this report was sent for review to all clearinghouses listed in Table 5, although the authors assume responsibility for errors and omissions.

SURVEY RESULTS

Twenty-four clearinghouses that maintain directories of machine-readable data files (MRDFs) were identified. The following information is based on contacts with these clearinghouses; blanks in the tables indicate either that no response was received from the clearinghouse or that the information was not available at the time of the survey. Table 5 lists the assigned acronyms or initialisms used in Tables 2 to 4, the full clearinghouse name, the agency, and other contact information. The primary scope of each clearinghouse is given in Table 2, with their characteristics summarized in Table 3. The "published directory" column usually refers to a formal publication; however, some directories are simply computer listings. Three clearinghouses that allow searching of the directory by the public are identified. Those organizations that distribute MRDFs are also indicated. Finally, the number of databases described within each directory is given in the last column. Many of the clearinghouse contacts indicated that their directories are being expanded and the number of entries will increase.

Although some clearinghouses may have used computerized text processing systems to compile their directories, only a third of the clearinghouses have computerized systems that allow on-line searching. Table 4 indicates whether clearinghouse directories contain selected key elements to describe MRDFs. This does not suggest style or syntax consistency between clearinghouses. If a clearinghouse does not have a computerized database, the information refers to their published directory.

The selected elements for Table 4 include abstract, keywords for searching and indexing, description of variables, spatial characteristics, temporal characteristics, and quality assurance or sampling variability associated with a file. A "yes" for a given descriptor element indicates that a particular clearinghouse directory contains a field describing that specific attribute for each MRDF.

While some of the directories do not explicitly describe temporal and spatial characteristics of a file, this information is often embedded in the title, keywords, or abstract. Some directories include many additional descriptor elements as indicated by the "number of elements" listed in the last column of Table 4.

Table 2. Primary scope of clearinghouses.

Acronym	Scope
I. Directories of machine-readable data files	
EPA/CLEAR	Environmental data and models
DOE/NEIC	Energy resources and consumption
NOAA/NEDRES	Climate, oceanography, geologic, geographic, hydrologic data
USGS/ESIS	Earth sciences data and analysis systems
USGS/NAWDEX	Organizations and sites concerned with water data
USGS/NCIC	Geodetic, cartographic, remote imagery data
NASA/DND	Aeronautics and space, remote imagery data
II. Repositories of machine-readable data files	
NTIS	Agency files provided to NTIS for public distribution
NARS	Statutory responsibility to identify, preserve, and distribute permanently valuable federal records
III. Agency data distribution activities	
CENSUS	Economics, population, housing, etc.
BEA	Economic activity accounts, economic projections
NOAA/NCDC	Climate
NOAA/NODC	Oceanographic
NOAA/NGDC	Geophysical, ionospheric, solar-terrestrial space
DHHS/SSA	Retirement, socioeconomic, disability
DHHS/NCHS	Health and vital statistics
DOL/BLS	Labor economics, social statistics (LABSTAT)
DOT/CTI	Highway, rail, water, and air transportation statistics
ED/NCES	Education
USDA/SCS	Soils, land use, water, conservation
NSF/NCAR	Climate
IV. Data evaluation activities	
ORNL/CDIC	Climate and carbon data for use in global carbon dioxide studies
ORNL/RSIC	Radiation properties of materials

INVENTORIES

Inventories of MRDFs have been compiled nationally for the United States (NTIS, 1977; OFSPS/NTIS, 1981; Yardas et al., 1982; Olson, 1984), and internationally (Emptoz, 1982). The international survey was conducted for the Committee on

Data for Science and Technology (CODATA) of the International Council of Scientific Unions. The preliminary survey identified over 650 data centers and data referral centers in 94 countries, covering the general areas of energy resources, fertilizers, water resources, nutrition, pesticides, and soils. The 1981 US directory contains descriptions of approximately 500 files from 44 federal agencies and organizations. Preparation of the directory was coordinated by the Office of Federal Statistical Policy and Standards (OFSPS) and funded by the National Technical Information Service (NTIS). Much of the support and input for the compilation was provided by the Interagency Committee on Data Access and Use. Current fiscal constraints have prevented the updating of the directory.

In 1981, the OFSPS was abolished, and statistical policy functions were transferred to the Office of Management and Budget (OMB). The Interagency Committee on Data Access and Use continues to function, and its technical subcommittee recently proposed a "voluntary standard" for use in preparing abstracts to describe MRDFs (OIRA, 1983). The standard is based on the experiences of several agencies, the 1981 directory compilation, a style manual for MRDFs (Roistacher, 1980), and compatibility with widely used bibliographic systems. The report includes extensive descriptions and examples of components of the abstract. The style manual (Roistacher, 1980) gives syntax and stylistic standards for documenting MRDFs and also includes extensive examples.

Inventory of Sources of Computerized Ecological Information, which contained descriptions of 73 information centers or data systems providing access to numeric or bibliographic information, was complied by Tucker and Huber (1980). The majority of the entries (64%) were bibliographic abstracting projects that provided keyword retrievals of citations to published ecological literature.

A variety of national environmental data sets are readily available from various federal agencies as well as other sources. Over 130 of these were identified in a recent review (Olson, 1984). Criteria for selecting data sets were 1) national or regional coverage, and 2) availability in computer-readable form. The report provides data-set titles, general thematic orientation, information on spatial and temporal characteristics, and the name, agency affiliation, and telephone number of a contact person. The list was compiled from many sources, including data inventories (Watts, 1983), survey responses, and personal contacts.

Based on this inventory, it became apparent that many federal, state, and other agencies collect water resource data and that there have been several well-established efforts to index and integrate these data. The Office of

Table 3. Clearinghouse characteristics.[a]

Acronym	Published directory	On-line directory	Searchable	Data access/ distribution	No. of entries
I. Directories of machine-readable data files					
EPA/CLEAR	1982	1980	Agency		370
DOE/NEIC	1983			NTIS	25
NOAA/NEDRES	1984	1983	Public		13,500
USGS/ESIS	1984	1979			370
USGS/NAWDEX		1976	Public		350
USGS/NCIC		1978	Public		48,000
NASA/DND	1983	1983	Agency		144
II. Repositories of machine-readable data files					
NTIS	1982		Agency	Public	>1,000
NARS	1983		Agency	Public	>300
III. Agency data distribution activities					
CENSUS	1980			Public	
BEA	1983			Public	
NOAA/NCDC				Public	
NOAA/NODC				Public	
NOAA/NGDC				Public	
DHHS/SSA	1981			Public	27
DHHS/NCHS	1980			Public	23
DOL/BLS	1981			Public	24
DOT/CTI	1983	1983	Agency	Public	1,400
ED/NCES	1981			Public	50
USDA/SCS					
NSF/NCAR	1982				>100
IV. Data evaluation activities					
ORNL/CDIC	1983	1983	Agency	Limited	83
ORNL/RSIC					

[a]Description of headings: Acronym or initialism - See Table 5 for complete names. Published directory - Most recent year of a published directory of listing of the MRDFs. On-line directory - Year that a computer directory became available. Searchable - Signifies accessibility of the directory for searching. Data access/distribution - Indicates if the clearinghouse provides access to data described in its directory. No. of entries - Number of data files described in the directory.

Water Data Coordination (OWDC) was created in 1964 to coordinate and standardize water data collection and exchange (USGS, 1979). The USGS/NAWDEX (National Water Data Exchange Program) was established in 1976 to link data collectors and users for more efficient use of the nation's water data. Another reason for the well-coordinated system of water data is the established classification system and maps for the nation's streams.

In contrast to water data, ecological resource data appear to be less abundant, more dispersed, and poorly indexed. Reasons include the lack of federal regulations requiring the monitoring of biota, the lack of a widely accepted classification framework, and the various forms of ownership of land and biotic resources. A recent report identified information needed for regional and national assessments of fish and wildlife resources (Whelan, 1982). In documenting the available information to meet these needs, the study concluded that 1) there is no national habitat inventory, 2) there are many different resource inventories being conducted by federal agencies, and 3) there appears to be a proliferation of fish and wildlife planning efforts. Whelan (1982) also pointed out the lack of an effective mechanism to prevent duplication and overlap of these efforts.

Cushwa and Tunstall (1983) surveyed sources of national-level wildlife (flora and fauna) data in response to the 26-member international Organization for Economic Co-operation and Development (OECD). Their findings, based on contacts with 56 individuals representing 12 private and 14 government organizations, indicate that responsibilities for wild fauna are spread among several major US government departments (Agriculture, Commerce, Defense, and Interior) and agencies (Council on Environmental Quality [CEQ], Environmental Protection Agency [EPA], National Science Foundation [NSF], Tennessee Valley Authority [TVA], and Water Resources Council [WRC]). These groups are directed to collect information about wildlife under several legislative mandates; however, no agency or organization within the federal government is responsible for overall coordination. Cushwa and Tunstall (1983) recommended that 1) the federal government designate a national office to prepare a periodic report for Congress from data being compiled, and 2) standard common classifications and definitions be promoted in all programs.

Table 4. Clearinghouse descriptor elements.[a]

Acronym	On-line directory	Abstract	Keywords	Variables	Spatial	Temporal	Quality	No. of elements
I. Directories of machine-readable data files								
EPA/CLEAR	1980	Yes	Yes	Yes	Yes	Yes	Yes	84
DOE/NEIC								
NOAA/NEDRES	1983	Yes	Yes	Yes	Yes	Yes	Yes	15
USGS/ESIS	1979	Yes	Yes	Yes	Yes	Yes	No	41
USGS/NAWDEX	1976	Yes	Yes	Yes	Yes	Yes	No	110
USGS/NCIC	1978							
NASA/DND	1983	Yes	Yes	Yes	Yes	Yes	No	16
II. Repositories of machine-readable data files								
NTIS		Yes	Yes	No	No	No	No	
NARS	1983	Yes	Yes	No	Yes	Yes	No	
III. Agency data distribution activities								
CENSUS BEA	1983							
NOAA/NCDC								
NOAA/NODC								
NOAA/NGDC								
DHHS/SSA		Yes	No	No	No	Yes	No	1
DHHS/NCHS		Yes	No	Yes	Yes	Yes	No	5
DOL/BLS		Yes	No	Yes	No	Yes	No	6
DOT/CTI	1983	Yes	No	Yes	Yes	No	Yes	12
ED/NCES		Yes	No	Yes	No	Yes	No	
USDA/SCS								
NSF/NCAR								
IV. Data evaluation activities								
ORNL/CDIC	1983	Yes	Yes	Yes	Yes	Yes	Yes	17
ORNL/RSCI								

[a]Description of headings: Acronym or initialism - See Table 5 for full name. On-line directory - Year the computer directory became available; if blank, the table refers to elements in the published directory. Abstract - Textual description which often defines spatial and temporal characteristics. Keywords - List of keywords available for searching and indexing. Variables - Description of variables within the file. Spatial - Definition of geographic coverage and spatial resolution (state, county, etc.). Temporal - Definition of period of record and temporal resolution (annual, monthly, etc.). Quality - Description of sampling variability, potential errors, quality assurance procedures. No. of elements - Total number of descriptor elements within the directory.

ADDNET -- A SPECIFIC DATA CENTER

ADDNET compiles data to conduct assessments and policy analyses as required by NAPAP. The integrated assessments require emissions, deposition, air quality, aquatic, agricultural, forestry, materials, and other data with common spatial and temporal attributes that can be easily and rapidly accessed for statistical analysis, modeling, and display needs. The project uses many of the resources described earlier in this paper to locate necessary data from dispersed sources. It has incorporated data and methods developed by the GEOECOLOGY project (Olson et al., 1980), which compiled a national, county-level database of environmental data. ADDNET assembles selected files, verifies contents, conducts quality assurance, maintains documentation, and distributes data to the NAPAP community. Although specific projects and task groups within NAPAP are responsible for collecting and analyzing individual files, ADDNET is concerned with locating data not being produced within the NAPAP program and the integration of the various files as required by the NAPAP assessment.

The dispersed files must be located and acquired in usable forms and within the overall assessment schedule. Often schedules require data before researchers or agencies have had sufficient time to analyze and publish their results. While researchers may endorse the need to use the most current data in the assessment, experience indicates that many factors hamper their ability to provide documented data to a central database on time. Therefore, the development of a database such as ADDNET requires continual contacts with data producers to monitor and encourage the flow of data in a timely manner.

Once data files are submitted to ADDNET, the data are checked for consistency, the documentation is reviewed for completeness, and the data are entered into the ADDNET database. Typically, a new file is maintained as a unique SAS data set within the database. However, data sets may be partitioned or combined for the most efficient and logical representation of the data. Variables are assigned unique names and labels. Necessary editing, such as converting to standard state and county codes, date formats, missing value codes, metric units of measure, and latitude/longitude, is performed. Thematic maps may be generated to check for inconsistent patterns and errors. Similar data sets may be

used to check for inconsistencies; for example, new files
with county areas will be compared with existing files having this same variable. The goal is to develop accurate,
well-documented, consistent data sets that can be readily
integrated for assessment needs.

These data sets become the NAPAP "certified" data.
They provide a common set of data for deposition, air
quality, atmospheric transport modeling outputs, land use,
and other subjects for use in all parts of the assessment.
The certification process involves investigators and task
groups verifying substantive contents, with ADDNET staff
reviewing the contents and verifying formats. ADDNET provides copies of the data files on magnetic tape to researchers and offers on-line access or data analysis support. Data are formatted as SAS files or selectively as
ASCII or EBCDIC files. Documentation accompanies the files,
including references to associated published reports and an
appropriate reference citation for secondary users of the
data to acknowledge the primary source.

CONCLUSIONS AND RECOMMENDATIONS

More machine-readable data files are being produced now
than in the past by federal agencies, specifically for public dissemination or as by-products of research and statistical report compilations. Concurrently there is an increasing use of and demand for data in machine-readable
formats. With increasing costs of collecting data and
decreasing budgets for data collection, it is imperative
that we make maximum use of existing data resources. Directories and clearinghouses of MRDFs, especially those maintained as updated, searchable databases, are available in
many agencies to assist users in locating and acquiring data
files. However, this service varies greatly among agencies.
Currently, automated directory systems that utilize biblio-
graphic-based systems exist. In addition, a proposed voluntary standard is available for use in describing MRDFs. To
maximize the use of MRDFs, all agencies need to develop and
maintain a directory of MRDFs. As data producers, we must
willingly provide descriptions of our data files. As
secondary data users, we should use clearinghouses and support their operations. Finally, if a clearinghouse helps
you locate data, be sure to give it credit.

Table 5.   Clearinghouses and numeric data referral centers for machine-readable data files, October 1983.

Acronym or initialism	Name, agency, and address	Contact[a]
I. Directories of machine-readable data files		
1. EPA/CLEAR	EPA Information Clearinghouse US Environmental Protection Agency 400 M Street, SW, PM-211A, FM 2903 Washington, DC 20460	Irvin Weiss COM: 202/382-5918 FTS: 382-5918
2. DOE/NEIC	National Energy Information Center Energy Information Administration US Department of Energy EI-22, RM 1F048 Washington, DC 20585	COM: 202/252-8800 FTS: 252-8800

3. NOAA/NEDRES	National Environmental Data
	  Referral System
	National Oceanic and Atmospheric
	  Administration
	3300 Whitehaven Street, NW
	Washington, DC 20235
		Robert R. Freeman
		COM: 202/634-7722
		FTS: 634-7722

4. USGS/ESIS	Earth Sciences Information Service US Geological Survey National Center, MS 806 Reston, VA 22092	Ted Albert COM: 703/860-6086 FTS: 928-6086
5. USGS/NAWDEX	National Water Data Exchange Water Resources Division US Geological Survey National Center, MS 421 Reston, VA 22092	Owen Williams COM: 703/860-6031 FTS: 928-6031
6. USGS/NCIC	National Cartographic Information Center US Geological Survey National Center, MS 507 Reston, VA 22092	Clair Ketch COM: 703/860-6508 FTS: 928-6508

7. NASA/DND	Directory of Numeric Databases
	Science and Technology Information
	  Branch
	National Aeronautics and Space
	  Administration
	NIT-42
	Washington, DC 20546
		John Wilson, Jr.
		COM: 202/755-3465
		FTS: 755-3465

II. Repositories of machine-readable data files		
8. NTIS	Office of Data Base Services National Technical Information Service 5285 Port Royal Road Springfield, VA 22161	Stuart M. Weisman COM: 703/487-4808 FTS: 737-4808

Table 5. Continued

Acronym or initialism	Name, agency, and address	Contact[a]
9. NARS	Machine-Readable Archives Division US National Archives and Records Service Washington, DC 20408	Ross Cameron COM: 202/724-1080 FTS: 742-1080
III. Agency data distribution centers of activity		
10. CENSUS	Customer Services Branch Data User Services Division US Bureau of the Census Washington, DC 20233	COM: 202/763-4100 FTS: 763-4100
11. BEA	Public Information Office, BE-53 US Bureau of Economic Analysis Washington, DC 20230	COM: 202/523-0963 FTS: 523-0963
12. NOAA/NCDC	User Services National Climatic Data Center National Oceanic and Atmospheric Administration Asheville, NC 28801	COM: 704/258-2850 FTS: 672-0683
13. NOAA/NDGC	National Geophysical Data Center National Oceanic and Atmospheric Administration 325 Broadway, E/GC2 Boulder, CO 80303	Joe H. Allen COM: 303/497-6323 FTS: 320-6323
14. NOAA/NODC	National Oceanic Data Center National Oceanic and Atmospheric Administration 2001 Wisconsin Avenue, NW Washington, DC 20235	Albert M. Bargeski COM: 202/634-7500 FTS: 634-7500
15. DHHS/SSA	Office of Research and Administration Social Security Administration 1875 Connecticut Avenue, NW Washington, DC 20009	Henry Patt COM: 301/594-0324 FTS: 594-0324
16. DHHS/NCHS	National Center for Health Statistics Statistice 3700 East-West Highway HHS Center Building Hyattsville, MD 20782	Sandra Smith COM: 202/436-8500 FTS: 436-8500
17. DOL/BLS	Office of Systems and Standards Bureau of Labor Statistics US Department of Labor Washington, DC 20212	Tony DiFillipo COM: 202/523-1975 FTS: 523-1975
18. DOT/CTI	Center for Transportation Information Transportation Systems Center Kendall Square Cambridge, MA 02142	Santo J. LaTores COM: 617/494-2429 FTS: 837-2429

Table 5. Continued.

Acronym or initialism		Name, agency, and address	Contact[a]
19.	ED/NCES	Statistical Information Office National Center for Education Statistics 400 Maryland Avenue, SW Washington, DC 20202	COM: 202/254-6057 FTS: 254-6057
20.	USDA/SCS	Soil Conservation Service US Department of Agriculture 1000 Aerospace Road Matland Building 2 Lanham, MD 20706	George Bluhm COM: 202/475-4548 FTS: 475-4548
21.	NSF/NCAR	Data Support Section National Center for Atmospheric Research P.O. Box 3000 Boulder, CO 80307	Roy Jenne COM: 303/494-5151 FTS: 322-5526
22.	USDA/ESS	Data Administration Section Economics and Statistics Section US Department of Agriculture Washington, DC 20250	COM: 202/447-7017 FTS: 447-7017
IV.		Data evaluation centers	
23.	ORNL/CDIC	Carbon Dioxide Information Center Oak Ridge National Laboratory P.O. Box X, Bldg. 6025 Oak Ridge, TN 37831	Coordinator COM 615/574-0390 FTS: 624-0380
24.	ORNL/RSIC	Radiation Shielding Information Center Oak Ridge National Laboratory P.O. Box X, Bldg. 6025 Oak Ridge, TN 37831	Betty Maskewitz COM: 615/574-6176 FTS: 624-6176

V.   Integrated database systems

See Merrill, D. 1983. Overviews of Integrated Data systems: Context, Capabilities, and Status, pp. 3-24. In. R.J. Olson and N.T. Millemann (eds.). Proceedings of the 1982 Integrated Data Users Workshop. CONF-820120. National Technical Information Service, Springfield, VA. 186 pp.

[a]COM = commercial telephone number, FTS - Federal Telecommunications System telephone number.

ACKNOWLEDGMENTS

This research is sponsored by the US Environmental Protection Agency under Interagency Agreement No. 40-1441-84 with the US Department of Energy under Contract No. DE-AC05-840R21400 with Martin Marietta Energy Systems, Inc. Publication No. 2513, ESD/ORNL. This research has been funded as part of the National Acid Precipitation Assessment Program

by the US Environmental Protection Agency. The research
described in this report has not been subjected to EPA's or
NAPAP's required peer and policy review and therefore does
not necessarily reflect the views of these organizations and
no official endorsement should be inferred. Contributions
and comments of individuals providing information to compile
this report are acknowledged and greatly appreciated.

LITERATURE CITED

Cushwa, C.T., and D.B. Tunstall. 1983. Wildlife in the
     United States: The U.S. response to the Organization
     for Economic Co-operation and Development (OECD) 1982
     Wildlife Questionnaire. pp. 43-48. In: Renewable
     Resource Inventories for Monitoring Changes and Trends,
     Proceedings of an International Conference, Corvallis,
     Oregon, August 1983. J.F. Bell and T. Atterbury
     (eds.). College of Forestry, Oregon State University,
     Corvallis, OR.
Emptoz, G. (coordinator). 1982. Inventory of Data Sources
     in Science and Technology, a Preliminary Survey.
     CODATA, Committee on Data for Science and Technology of
     the International Council of Scientific Unions, Paris,
     France.
Freeman, R.R. 1985. The National Environmental Data Refer-
     ral Service: A publicly available on-line data cata-
     log. pp. 143-154. In: Research Data Management in
     the Ecological Sciences. W. Michener (ed.). Belle W.
     Baruch Library in Marine Science, No. 16. University
     of South Carolina Press, Columbia.
Frank, N.D. and J.V. Gandy. 1983. Data Sources for Busi-
     ness and Market Analysis, Third Edition. The Scarecrow
     Press, Inc., Metuchen, NJ. 470 pp.
Merrill, D. 1983. Overview of integrated data systems:
     Context, capabilities, and status. pp. 3-24. In:
     Proceedings of the 1982 Integrated Data Users Workshop.
     CONF-8210120. R.J. Olson and N.T. Millemann (eds.).
     National Technical Information Service, Springfield,
     VA.
National Technical Information Service (NTIS). 1977.
     Federal Environmental Data, A Directory of Selected
     Sources. PB-275 902. NTIS, Springfield, VA. 136 pp.
Office of Federal Statistical Policy and Standards, National
     Technical Information Service (OFSPS/NTIS). 1981. A
     Directory of Federal Statistical Data Files. PB81-
     133175. NTIS, Springfield, VA.

Office of Information and Regulatory Affairs (OIRA). 1983.
    Procedures for Preparation of Abstracts for Public Use
    of Statistical Machine-Readable Data Files. Proposed
    Voluntary Standard. Office of Management and Budget,
    Washington, DC. 49 pp.

Olson, R.J. 1984. Review of Existing Environmental and
    Natural Resource Data Bases. ORNL/TM-8928. Oak Ridge
    National Laboratory, Oak Ridge, TN.

Olson, R.J., C.J. Emerson, and M.K. Nungesser. 1980.
    GEOECOLOGY: A County-Level Environmental Data Base for
    the Conterminous United States. ORNL/TM-7351. Oak
    Ridge National Laboratory, Oak Ridge, TN.

Olson, R.J. and N.T. Millemann (eds.). 1984. Proceedings
    of the 1983 Integrated Data User's Workshop. CONF-
    831117. National Technical Information Service,
    Springfield, VA.

Public Health Service (PHS). 1982. Report of the Task
    Group on the Role of the Private Sector in the Collec-
    tion, Analysis, and Distribution of Health Data. US
    Department of Health and Human Services, Washington,
    DC.

Roistacher, R.C. 1980. A Style Manual for Machine-Readable
    Data Files and Their Documentation. Report Number
    SD-T-3, NCJ-62766. Bureau of Justice Statistics, US
    Department of Justice, Washington, DC. 75 pp.

Tucker, C.S. and E.E. Huber. 1980. Inventory of Sources of
    Computerized Ecological Information. ORNL-5441/R1.
    Oak Ridge National Laboratory, Oak Ridge, TN. 61 pp.

US Geological Survey (USGS). 1979. Water-Data Coordina-
    tion. US Geological Survey, Office of Water Data
    Coordination, Reston, VA. 12 pp.

Watts, J.A. 1983. Recent inventories of spatial data
    bases. pp. 116-130. In: Proceedings of the 1982
    Integrated Data Users Workshop, CONF-8210120. R.J.
    Olson and N.T. Millemann (eds.). National Technical
    Information Service, Springfield, VA.

Whelan, J.B. 1982. Information Needed to Determine the
    Regional and National Status of Fish and Wildlife
    Resources. FWS/OBS-82/W58. US Department of the
    Interior, Fish and Wildlife Service, Fort Collins, CO.
    25 pp.

Yardas, D., A.J. Krupnick, H.M. Peskin, and W. Harrington.
    1982. Directory of Environmental Asset Data Bases and
    Validation Studies. Resources for the Future, Inc.,
    Washington, DC.

THE NATIONAL ENVIRONMENTAL DATA REFERRAL SERVICE: A PUBLICLY AVAILABLE ON-LINE DATA CATALOG

Robert R. Freeman

ABSTRACT

The National Environmental Data Referral Service (NEDRES) is a database that provides a rapid and easy way to find out about the existence, location, and conditions for using environmental data held by people and organizations in the United States and Canada. NEDRES catalogs, abstracts, and indexes descriptions of data files in a standard format and makes these descriptions searchable by direct access to an on-line information retrieval system from the user's own terminal. NEDRES is also a network of organizations, coordinated by the National Oceanic and Atmospheric Administration (NOAA), that cooperate to make environmental data more readily accessible. This paper describes the structure and status of NEDRES and provides information on how to use NEDRES to catalog environmental data files.

INTRODUCTION

One of the motivating forces for this symposium was the recognition that there is a lack of accepted methods for managing research data in the ecological sciences. The organizers indicate that scientists are becoming increasingly aware of the need for collective stewardship of data and unpublished information related to ecological research sites. In this paper I describe an existing vehicle that permits an important step to be taken in the management of scientific data: preparing an organized record of the data possessed by an individual or organization and sharing that record with others who need similar data.

THE ROLE OF CATALOGING AND
INDEXING IN DATA MANAGEMENT

In an earlier meeting of data managers of the Long-Term Ecological Research sites and others interested in ecological data management, there was a lengthy discussion of the rationale, principles, and processes of documentation, cataloging, and indexing (Lauff, undated). The report of that 1982 meeting at the Kellogg Biological Station reflects a growing consensus on the importance of applying the same basic principles to scientific data that have served so well in making accessible the published record of science as it has grown prodigiously throughout the present century.

These principles include the development of commonly accepted standards for describing concisely the nature, organization, and circumstances of the collection and processing of data files held by a particular center or individual. They also include the development of methods for organizing these descriptions, permitting others at later times and different locations to discover that certain data exist and determine who has these data.

Bibliographical services and abstracting and indexing services for published scientific findings have existed for a long time. The data upon which these findings were based often could be included in the paper or stored in notebooks or paper files that the scientist could find if needed. Over the past 25 years, most of the services for locating published information have become computer-searchable databases in order to keep up with the growth of the literature.

Advances in computer technology, instruments, and instrument platforms have led to great increases in the amount, accuracy, and timeliness of data. Almost inexplicably, the application to data management of the same principles that have worked so effectively for published information has lagged. The results of this lag may be seen by individual scientists as an inconvenience at most: it may be a little more difficult to locate a particular data file.

On a national scale, however, the effect of these small inconveniences may be magnified many times over. Large-scale programs began to collect data in the early 1970's to determine environmental baselines and make environmental assessments of the actual or potential impact of actions. Millions of dollars were spent. Results were published and decisions made. But 10 years later, when there are still more decisions to be made, it has become evident that the data that cost so much to collect are gradually becoming dispersed and lost as people move on and organizations change.

Now some efforts have begun to locate, archive, and index the data. For example, the Minerals Management Service of the Interior Department has mounted a substantial program to archive and index the marine data collected in its Gulf of Mexico outer continental shelf environmental assessment programs.

The importance of cataloging and indexing environmental data in order to make it available to later users was recognized at the national level in the early 1970's when Congressman John Dingell sponsored a series of bills to establish better control over environmental data. The first of these bills envisaged the centralization of all environmental data in one giant computer system. After extensive testimony, Dingell's concept evolved into a national network of data centers linked by a central catalog to identify who had which data. One of the bills ultimately was passed by Congress, but pocket vetoed by President Nixon.

More recently, legislation has been passed that promotes better access to data which supports programs of national interest. For example, the National Climate Program Act of 1978 called for improved dissemination of climatic information and data. The National Ocean Pollution Research, Development, and Monitoring Planning Act of 1978 led to the development of an Ocean Pollution Data and Information Network.

Renewed interest in environmental data during the build-up of energy programs in the late 1970's resulted in another proposal for a network of environmental data centers. The network would "provide a comprehensive, frequently updated catalog of databases from current or recently completed research and environmental monitoring programs. For each database there would be an abstract which describes the nature and goals of the data collection, the research methods, the availability of the data, and other information needed for a potential user to determine whether/how he should obtain the data" (Armentano & Loucks, 1979).

Others in fields closely related to ecological and environmental data have called for inventories, catalogs, and indexes of available data. Documentation of data was discussed extensively at a workshop on fish and wildlife data and computer applications in 1983 (National Workshop,

1984). Oak Ridge National Laboratory proposed steps to be taken that would ensure better coordination of and access to environmental and national resources data (Olson, 1984). A survey of state natural resources information systems revealed a host of activities at the state-government level oriented toward management of data, much of it environmental (Caron, 1984). Most recently, a study of alternatives for improving access to water data and information called for linking existing data catalogs (such as NEDRES and the National Water Data Exchange [NAWDEX]) as a step toward making sure that individuals in broad, interdisciplinary fields (such as water research) can locate data quickly and easily (Lynch & Sullivan, 1984).

Many of these same goals and requirements have been apparent in NOAA, which was given the mission to maintain archives and make available to the public a broad range of environmental data, covering the spectrum from the solid earth to ocean resources, oceans to the solar environment and its relationship to the earth. The sheer magnitude of this interdisciplinary mission led us as early as 1970 to begin planning for ways to catalog and index our own data holdings as well as those of other centers. The result was the first service, the Environmental Data Index (ENDEX), to provide flexible, computerized cataloging of data. Operating from 1974 to 1980, it was the forerunner of the service described below and it yielded many useful concepts for organizing and producing a service of this type.

REQUIREMENTS FOR A FIRST STEP TOWARD
BETTER ACCESS TO ENVIRONMENTAL DATA

These diverse requirements for improved access to environmental data suggest a number of criteria for the design of a data referral service:

- The scope of the service should be <u>interdisciplinary</u>. The environmental sciences are comprised of a number of related disciplines with no clear borders separating one from another. They share the common feature of deriving some or all of their data from field observations and monitoring. Consequently, an environmental data referral service must be able to describe data, methods, and instruments used in a wide variety of scientific fields.

-   Another common feature of environmental data collections is the <u>significance of time and location</u>. Unlike the laboratory sciences, environmental scientists cannot repeat a set of observations because they are sampling from an environment that is changing. The identification of time and place thus become critical factors in the description of existing data sets.

-   In addition to the <u>content</u> of a referral service, the design of <u>access methods</u> is important. A service for an interdisciplinary, geographically dispersed community of users needs to be <u>convenient</u>, <u>easy to use</u>, <u>and flexible</u>. Our experience indicates that information services should be quickly available and able to accommodate the normal vocabulary of the user in order to be used. Fortunately, on-line, interactive information retrieval systems that use simple commands and natural language terms have been developed to deliver this kind of service to any location in North America, as well as to Europe and a few locations elsewhere, at reasonable costs.

A NATIONAL ENVIRONMENTAL DATA REFERRAL SERVICE AS A SOLUTION

We have attempted to use these design criteria in developing a National Environmental Data Referral Service (NEDRES). Since October 1983, NOAA has offered public access to this computer-searchable catalog and index of the environmental data holdings of organizations and individuals throughout North America.

NEDRES is designed to identify the existence, location, characteristics, and availability conditions of environmental data sets. A search of NEDRES gives a complete description of available data files that satisfy the request criteria stated by the user.

Users can specify desired environmental parameters, methods and instruments, geographic locations, time spans, or the names of people, organizations, or projects that were involved in the data collection activities. Geographic locations can be stated in terms of named places, rectangles of latitude and longitude, river basins, and ecological regions. NEDRES gives the requestor a listing of potentially relevant data files, their physical storage media, the address of the holder of the data, and the conditions of availability. This information generally is adequate for the user to decide whether to contact the data holder for specific details or to arrange to acquire a copy of the data.

The NEDRES database contains descriptions of approximately 15,000 environmental data files as of the fall of 1984. This number is constantly growing as new data files are located and described. Despite the large number of data file descriptions already included in NEDRES, we are convinced that only a small portion of the relevant data is in NEDRES as yet. Several years of additional effort are needed to make the service reasonably comprehensive. Nevertheless, users already are finding the service valuable.

NEDRES contains descriptions of data files from across the spectrum of environmental sciences (Table 1). Recognizing the lack of clear boundaries between sciences, we have made this range flexible in an effort to meet the needs of users of environmental data.

Table 1. Types of environmental data in NEDRES.

Climatological and meteorological
- standard surface and upper atmosphere
- atmospheric radiation, physics, and chemistry
- air quality

Oceanographic
- physical, chemical, biological
- ocean mineral and energy resources
- ocean pollution

Geophysical and geological
- geomagnetic and seismological
- marine geological and geophysical
- solar-terrestrial
- glaciological

Hydrological and limnological
- precipitation
- surface and ground water
- aquatic ecological
- water quality

Geographic
- geodetic
- cartographic
- land use/ground cover

Table 2. Types of information contained in NEDRES.

- data centers, services, and programs

- unpublished data files

- serial data publications

- published data files

- atlases or other published data in graphic form

- publications containing extensive data compilations

- manuals, user guides, or documentation of data files

- data catalogs or inventories

In addition to the subject or discipline scope of NEDRES, it is also useful to review the criteria for the types of data files and other source material included in NEDRES. As may be seen from Table 2, NEDRES includes more than just descriptions of machine-readable files. In many cases, environmental data are found only in published form or in analog representations on paper, film, or other recorded media. Data are increasingly available in digital form, but a major portion of the data are still found in these other media.

NEDRES also includes other forms of information useful in locating data, including published forms of data documentation, data inventories and catalogs, and descriptions of data centers and their services. Another useful representation of environmental data is the special data products or data syntheses, which were derived from basic data sets to serve some special purpose. For example, environmental assessment or impact studies frequently pull together all of the known data about a particular location of interest. References to these data products are included in NEDRES.

The NEDRES database resides in the on-line information retrieval system operated by BRS, Incorporated. This system is host to many databases, ranging from bibliographical citations to the full text of certain journals, reference works, and newspapers. The strength of the system is in providing an easily-learned means for pinpointing information in large volumes of text.

Although the purpose of NEDRES is to direct users to data, NEDRES itself is made up of text describing data files. For this reason, a system such as BRS is optimal because of its speed of operation and ease of use. A further advantage is that BRS provides access to its system via three public data communication networks, Telenet, Tymnet, and Uninet. This means that no user in North America is more than an intra-state or intra-provincial telephone call away from the system. In hundreds of cases, a local telephone call is all that is required to connect one's computer terminal to NEDRES.

Access to the NEDRES database is gained by entering a password when logging on to the BRS retrieval system. Users obtain their passwords from the NEDRES Office of NOAA. All that is required to obtain a password is the signing of a user agreement, which acknowledges responsibility for paying the modest user charges. These charges are based on recovering the cost of providing the information retrieval service. The cost of compling the NEDRES database is paid by NOAA.

USE OF NEDRES

The concept of a National Environmental Data Referral Service encompasses the provision of information about these data sources to all who need to locate data. NEDRES can and does serve three distinct types of user: (1) scientists who need data for research projects; (2) planners, managers, consultants, and engineers who need data for applications and decisions; and (3) data managers who need tools for controlling large volumes of data that are under their responsibility.

There are several ways in which NEDRES can be used to meet some of these needs:

To Find Data

Most users of NEDRES want to locate data that may be useful to them. A data search is analogous and complementary to a literature search, with the objective of avoiding duplication and advancing more rapidly in the research or application problem at hand. Users interested only in locating data may search NEDRES from their own computer terminals or by obtaining the assistance of organizations that specialize in information service. NEDRES is designed to be as easy to use as possible (within the capabilities of current computer systems) in order to encourage users to take advantage of it themselves.

<u>To Index Data</u>

Organizations that manage significant amounts of environmental data sooner or later find that their data holdings need to be cataloged and indexed in order to be of use to anyone. Unless the knowledge of the existence and location of an organization's data is readily accessible to any qualified person who needs it, an organization cannot be said to have control over this valuable resource.

In some cases, the answer is an in-house system based on locally developed software or on commercial software, often sold as part of a database management package. One example of this approach is the Oregon State University Forest Science Data Bank, which uses a microcomputer system with the Datastar and Reportstar software to maintain a data catalog (Stafford <u>et al</u>., this volume).

Another alternative is to use NEDRES as a tool for indexing data. Organizations that choose this alternative can take advantage of a ready-made system that is specially designed for describing environmental data. Access to listing of the data files is readily available via computer terminal, high speed laser printer listings, or by requesting delivery of a magnetic tape which then can be used locally.

<u>To Make Data Known To Others</u>

An important advantage to using NEDRES for cataloging and indexing an organization's data holdings is that the existence of these data files automatically becomes known to any other user of NEDRES. Thus an organization not only gains control over its own data holdings, but simultaneously gains a way of sharing its data with others. Each organization retains complete control over the use of its data. NEDRES permits statements of terms and conditions of availability to be given as part of each record so costs or limited access conditions may be made clearly known.

NEDRES AS A COOPERATIVE NETWORK OF ENVIRONMENTAL DATA SOURCES

It is a necessity that the effort required to develop a successful NEDRES is shared among the many organizations that are major holders and disseminators of environmental data. Even if far greater resources were available to develop a NEDRES, the nature of the problem (_i_._e_., the diversity of locations and disciplines involved in environmental data) would dictate that a cooperative effort would be more efficient.

If a large enough group of the major holders of environmental data agree to cooperate in developing NEDRES, the result will be more valuable than the result of independent individual efforts. This principle of cooperative effort is gradually beginning to work as more organizations see the results. It is already apparent that this cooperation can cross national borders, with both Canadian and US centers participating.

The idea of a cooperative network of organizations interested in the dissemination of environmental data has been a part of the planning of NEDRES from the start. Developing this type of network is expected to take many years to familiarize interested organizations with the concept and benefits of a cooperative approach.

The basic tools for the formation of an environmental data network are membership agreements. The essential elements of a NEDRES membership agreement are

- Member contribution. A member organization agrees to play a role in building the data referral service, normally by preparing descriptions of a significant collection of data files, which may be the member's own, but also may reflect a commitment to cover a geographic area or a larger organization of which the member is a part.

- Member benefits. In return for this contribution, members receive benefits negotiated individually, which may include a number of passwords and a certain amount of free use of the NEDRES database, tapes or special printouts of the member's records in the database, or copies of NEDRES catalog publications. The very existence of a membership agreement can benefit an organization that needs to show that it is making reasonable efforts to make its data known and available to a broad user community.

- Member participation in planning. Since NEDRES is intended to be a service that meets the needs of numerous organizations, members can participate in deciding the course of development of NEDRES. Although the availability of resources places an obvious limit on the practice of joint decision making, cooperating organizations already have influenced decisions as to the content of the database.

- <u>Coordination</u>. As part of the membership agreement, NOAA accepts the role of coordinator and the responsibiltiy of maintaining the NEDRES database in a publicly accessible on-line system. The coordination role involves promotion of the service, development of training materials and sessions, a newsletter, a substantial portion of the work of input to the database, and the administrative work involved in maintaining and servicing user accounts.

FUTURE DEVELOPMENTS

The concept of a network of environmental data centers need not stop with a referral service. In the future, NEDRES could become the "front end" for automated access to and delivery of data files. Already there are individual data centers that are developing systems to allow users to (1) search an on-line catalog of their data holdings, (2) switch automatically to a detailed inventory of one or more data files to explore detailed questions of data availability, and (3) access the referenced data files themselves. Graphic display and data manipulation capabilities may be a part of such systems.

Typically such developing systems are limited to the data holdings of one data center. But the data communication networks that are already available today are capable of permitting users to switch readily from one computer system to another. With careful planning, a network could be developed to permit users to enter at the level appropriate to their interest and knowledge. NEDRES could serve as the master index to these holdings, allowing users to switch automatically to the inventories and data files of suitably equipped centers and referring users to others not equipped for on-line access. Whether in a partially or completely automated network of the future or in today's disjointed network, NEDRES can play an important role in bringing users and holders of environmental data together.

For more information, inquiries may be directed to: National Environmental Data Referral Service Office, NOAA/NESDIS/AISC, 3300 Whitehaven Street, N.W., Washington, DC 20235.

LITERATURE CITED

Armentano, T.V. and O.L. Loucks. 1979. Ecological and Environmental Data as Under-utilized National Resources: Results of the TIE/ACCESS Program. The Institute of Ecology. Indianapolis. 207 pp.

Caron, L. 1984. An Inventory of State Natural Resources Information Systems. Final Report, Grant No. NAG2-201. Submitted by the Kansas Applied Remote Sensing Program, University of Kansas, Lawrence, Kansas, to the Office of University Affairs, Ames Research Center, National Aeronautics and Space Administration, Moffett Field, CA.

Lauff, G.L. Undated. Data Management at Biological Field Stations: Report of a Workshop, May 17-20, 1982, W.K. Kellogg Biological Station, Michigan State University. 45 pp.

Lynch, M.P. and J.K. Sullivan. 1984. Alternatives for a National Water Resources Research Center and Information Clearinghouse. Report submitted to the Council on Environmental Quality by the Chesapeake Research Consortium. Chesapeake Research Consortium Publication 120. Gloucester Point, VA. 111 pp.

National Workshop on Computer Uses in Fish and Wildlife Programs: A State-of-the-Art Review, December 5-7, 1983. 1984. Virginia Polytechnic Institute and State University, Blacksburg, VA. 304 pp.

Olson, R.J. 1984. A Review of Existing Environmental and Natural Resource Data Bases. Report No. ORNL/TM-8928. Oak Ridge National Laboratory, Environmental Sciences Division, Oak Ridge, TN. 71 pp.

Stafford, Susan G., Paul B. Alabach, Karen L. Waddell, and Rodney G. Slagle. 1985. Data Acquisition, Documentation, and Cataloging Procedures in Research Data Management. pp. 93-113. In: Research Data Management in the Ecological Sciences. W. Michener (ed.). Belle W. Baruch Library in Marine Science, No. 16. University of South Carolina Press, Columbia.

HISTORIC DATA SETS: LESSONS FROM THE PAST, LESSONS FOR THE FUTURE

Carl J. Bowser

ABSTRACT

Interest in research on long-term phenomena in ecology has prompted concern for collection of well documented, uniform quality data sets that will be useful to researchers long after initiation of such measurements. Lessons learned from previous investigations are valuable for proper planning of future studies. Types of historic data sets yielding valuable information include data generated by well planned, long-term research studies; data collected over many years by researchers committed to continued study of an area; long-term records that are a serendipitous product of data originally collected for short-term purposes; government and private agency data collected without prior knowledge of its value to long-term research; and naturally preserved historic records.

An anecdotal approach is used to illustrate the variety of data sets available and the different ways various types of historic data sets have been used in current research, and to draw lessons from these experiences that will aid in the design of long-term research experiments. Specific examples are drawn from work on lakes in northern Wisconsin.

INTRODUCTION

Data sets collected in the past are proving valuable to a large range of current studies in ecology and environmental science. Our interest in historic data sets has been heightened by the initiation of projects such as the National Science Foundation's (NSF) Long-Term Ecological

Research (LTER) program and the need to understand factors leading to long-term changes in ecosystems. Questions about changes in CO_2 levels and the effect of burning fossil fuels, concern for the effects of acid deposition on terrestrial and aquatic habitats, and the need for greater knowledge of natural levels of metals and organic constituents in the environment before pollution occurs in various regions of the world have increased interest in whatever sources of long-term data are available.

In nearly all papers presented in this symposium considerable attention was given to data management, software and hardware for data collection and management, and uses of such data sets. The common assumption is that such databases are devised for current use, as well they should be. Yet, little note was taken of the ways such data sets might be useful 20-50 years from now. Clearly, good databases are needed for current problems faced by the field of ecology; however, properly designed databases that pay particular attention to documentation, quality control, and long-term archiving will be even more valuable in the future. The NSF-sponsored LTER program has sharpened our concern for databases with present and future value.

For the past five years I have been closely associated with research at the LTER-Northern Lakes site in Vilas County, WI. The lakes in the surrounding region had been studied intensively by E.A. Birge, C. Juday, and several collaborators from 1926 to 1941. The studies range from biology to chemistry, geology, and physical limnology, and constitute one of the most historically valuable limnological data sets in existence. Although many collaborators shared in the research in the Trout Lake area, the data have become known collectively as the "Birge and Juday" data set.[1]

Experiences in dealing with these data sets have been valuable to researchers at the site and have seriously affected our thinking about our data collection and data management strategies. The following comments are designed to share our experience in dealing with historic data sets

---

[1] I use the expression "data set" to denote any collection of data representing a record of studies over several years in the same area, regardless of the manner in which the data are catalogued, tabulated, and stored. Use of the word database implies a similar record of data collected from a region over some period of time, but organized in a systematic manner and in a form that allows easy access to any part, or all, of the data (such as a computerized database or an organized, archived, paper file with well described documentation).

at the Northern Lakes LTER site. A chronicle of this experience with the Birge and Juday data set will be useful to others concerned with the integrity of long-term research data and the design of modern data collection and archiving systems that will be valuable to future researchers.

The focus on this paper is threefold: 1) to categorize the various types of data sets available from historic studies and contrast their potential value in ecosystem studies, 2) to draw on examples of the use of historic data sets to better understand changes in ecosystems that have taken place over the past years; and 3) to draw lessons from the use of existing historic data sets so that we might formulate ways to ensure that our current long-term data sets will have value to future generations.

HISTORIC DATA SETS

Before discussing the details of our experience with the Birge and Juday data set it is worth considering some general problems with historic data sets. The following comments include a discussion of data sources and publication forms, data reliability, and the use of oral histories to recover otherwise undocumented information about historic data.

Data Sources

Historic data sets can be categorized into three types: 1) planned, 2) opportunistic, and 3) serendipitous. Such data sets range considerably in size, scope, quality, degree of documentation, internal consistency, and availability.

Planned data sets are collected over several years by a relatively small group of coordinated investigators with the intent of studying phenomena or changes that can be observed only over long periods of time. The long-term records of atmospheric CO_2 from Mauna Loa, Hawaii (Keeling et al., 1976; Keeling et al., 1982) and the records of watershed studies at the Hubbard Brook experimental site (Likens et al., 1977; Likens et al., in press) are good examples of planned data sets.

Opportunistic data sets have been collected over many years, but they differ from planned data sets in that they comprise a coherent collection of data gathered with short-term goals and funding periods. These types of data sets are long-term only because the person(s) responsible for the

data sets stayed with the project or project area for a long
time. For example, in a recent address to LTER researchers
at Lake Itasca, MN in May 1984, Goldman asserted that the
Lake Tahoe and Castle Lake data sets have a long-term element only because he chose to stay at the University of
California at Davis, and turned down several job offers over
the years that would have marked the end of the "data set"
had he chosen to leave.

Opportunistic data sets have long-term value because
they have come to be of historic interest to a broader group
of scientists. Good examples of these data sets, in limnology, include the lake manipulations at the Experimental
Lakes Area (Johnson & Vallentyne, 1971; Schindler & Fee,
1974) and the long-term studies on Lake Tahoe (Goldman,
1981).

Serendipitous data sets could be described as long-term
records originally collected with no long-term scientific
purpose in mind. Examples include weather data collected by
private citizens, fish and wildlife harvests which can be
used as indicators of long-term game populations, paper
industry records (total production by region) used to estimate clear-cutting and forest maturity indices, and coal
consumption by region used to estimate atmospheric loading
of sulfur.

Table 1.   Matrix of data sources for historic data sets.

	Single Investigator, Group, or "Institution"	More Than One Investigator, Multiple Groups or "Institutions"
Continuous Records of Research Data	BEST (Rarest)	(Common)
Discontinuous Records of Research Data	(Relatively Rare)	WORST (Probably Most Common)

The sources and quality of historic data on ecosystems
range considerably, from well planned, systematic collections to data pieced together from numerous unconnected
studies. The range of sources is summarized in Table 1.

Scientific data systematically collected by an individual or group of researchers from an integrated program

provide the best source of long-term data. These well planned and documented data sets are the rarest type of historic data set. The goals of the LTER program include collecting high quality data that will be valuable for evaluation of long-term change in ecosystems; however, the systematic collections of data in this program are in their infancy. At the Northern Lakes LTER site our attempts to make use of historic data sets arose directly from needs generated by long-term research goals.

Data sets comprised of discontinuous records from a single individual or institution are somewhat more common. Cases exist where data collection was resumed by an independent group of scientists following departure of the earlier researcher(s). For example, the long-term research at the Northern Lakes LTER site began 39 years after the work by Birge and Juday was completed.

Collections of data from more than one source can be patched together to give a continuous record of critical ecosystem variables. For many large government databases the data must be pieced together from a variety of collections from diverse investigators and institutions. Unevenness in data quality, documentation, and state of preservation make these collections some of the most difficult historic records to use. Less than 10 percent of all national databases require common standards of quality and documentation (W. Michener, pers. comm.). Although databases comprised of data from diverse sources are useful at some level, responsibility for quality assurance is left to the user, who is faced with the difficult task of sorting high quality data from those of low quality. The larger the scope of these databases the larger the potential for uncertainty in the data. It is ironic that these are the databases most often turned to for government or administrative decisions.

Published Forms of Data

Data useful for long-term ecological studies are readily obtained from a variety of sources. Historic records are available in published, formal, peer-reviewed journals; "gray-literature" publications (governmental in-house reports, agency reports, Environmental Impact Statements, Open-file Reports); unpublished records and manuscripts; and in formalized databases.

Science traditionally has given preference to formal, published data over unpublished forms, however, both have advantages and disadvantages. Published data are peer-reviewed, generally of high quality, and widely available to researchers. Often such studies are connected with short-term goals, and therefore have more limited value to researchers interested in long-term records. Perhaps more important is the close editing for brevity most peer-reviewed, published material undergoes. All too often critical details of methods are missing or data are condensed to illustrate the point of the paper rather than to provide the reader with the complete range of data collected. Thus, demands to keep publication costs down can screen out large data sets useful for long-term studies.

Reports generated for agency "in-house" reports, such as Environmental Impact Statements, are common and of potential value as long-term historic data. However, unpublished records typically lack peer review, which is designed, in principle, to assure high quality data. Therefore, the value of these records is limited. Some of these reports are, nonetheless, of high quality and the lack of journal editorial restrictions on the size of the data set may make them more useful for ecological studies.

Some data of value have been collected but never published. Typically such records remain with the scientist who collected the data, and thus are difficult to obtain. Sometimes such data have been recovered only on the death of the scientist, making documentation of techniques and data quality difficult to verify.

Reliability of Data

Quality control and careful documentation of procedures and instrumentation are paramount to the evaluation of the usefulness of historic (and recent) data sets. Data collected by different investigators may vary widely in quality, and such variation is almost impossible to document. Careful intercalibrations, interlaboratory standard exchanges, and sample replications were much less common in the past and, therefore it is difficult to document the quality of many historic data sets.

During the past 50 years there have been significant changes in methodology for numerous types of environmental measurements. These changes generally represent improvements in precision and/or accuracy. Lack of intercomparisons of techniques makes it difficult to tie data obtained with different techniques to a common reference. An excellent example is found in the changes in pH measurements over the past 50 years, which are summarized in the discussion of the Birge and Juday data later in this paper.

The taxonomy of many organisms has evolved over the years, and even today different investigators may use different names for the same organism. Lack of uniformity in the descriptive terminology of various taxa has caused considerable difficulty for investigators in ecology (Kratz et al., in prep.). Preservation and archiving of organisms is a potential solution to the problem of changing taxonomy, but for certain types of organisms (e.g., bacteria and algae) preservation techniques are not always reliable and some techniques may lead to selective preservation of organisms (Crumpton & Wetzel, 1981).

Oral Histories

Loss of valuable data through a researcher's death is also a concern, especially for data collected more than a few decades ago. For example, Birge's death in 1950 meant the potential loss of many years of data on northern Wisconsin lakes. Through the foresight of Dr. John Neess (Zoology Department, University of Wisconsin) a significant amount of the original data and some instruments were kept from being thrown out.

Personal conversations with researchers can provide valuable insights into otherwise undocumented, but critical, details about historic sampling and analysis. For example, two years ago the Center for Limnology at the University of Wisconsin invited many of the staff, technicians, and students who worked in the Trout Lake area with Birge and Juday to a conference held to obtain more information on the research conducted during that period. Discussion with conference participants revealed much about the sampling methods and analyses that was never recorded in the published records.

Natural Records

Although natural research records are not data sets in the same sense as described previously, these records are nonetheless important sources of information on climatic and ecological change. These valuable records cannot be ignored by those interested in historic changes in ecosystems. Most natural records have "integrating" processes that average out short-term change in the system, such as sediment bio-turbation (Jones & Bowser, 1978). However, for the study of long-term trends in natural systems it may be advantageous to have such short-term integrators (Allen et al., 1984). Tree ring data provide a record of growth rate that can be used to reconstruct climatic change in time periods from tens to hundreds of years (Fritts, 1966). Lake sediments, especially varved sediments, also preserve many useful indicators of environmental changes. Other examples of natural records include chiton, to evaluate zooplankton population variation (Kitchell & Kitchell, 1980; Kitchell & Carpenter, in press); phaeopigments and hydrocarbons, to measure auto-chthanous/allocthanous carbon variations in lakes (Gorham & Sanger, 1976; Züllig, 1981); pollen, to measure terrestrial floral changes (Maher, 1982); diatom assemblages, to evaluate lake chemistry through time (Davis & Norton, 1978); and carbonized wood fragments, to chronicle pre-historic forest fire frequency (Swain, 1973, 1978). Because such records allow ecologists to examine environmental conditions before the effects of man, they have potentially great value.

HISTORICAL DATA FROM TROUT LAKE

The Birge and Juday Data Set

The work of Birge and Juday spanned seven decades and comprised more than 400 publications, (Juday & Hasler, 1946). Their work extended to many lakes, principally in the state of Wisconsin, and the literature is rich with references to pioneering work they performed on the biology, chemistry, and physics of lakes (Hutchinson, 1975; Wetzel, 1983). Their collaborative studies with V.W. Meloche (chemistry); W.H. Twenhofel and V. McKelvey (geology); A.D. Hasler, D. Frey, and R. Pennak (zoology); and E.B. Fred and E. McCoy (bacteriology) contributed greatly to limnology (Frey, 1963).

Historic data on lakes in northern Wisconsin dates from before the turn of the century when E.A. Birge was studying Cladocera (Birge, 1893), but it wasn't until 1925 that Birge and Juday initiated extensive studies of the lakes (Frey, 1963). For 15 years, Birge, Juday, and co-workers collected biological, chemical, and physical data on more than 500 lakes in the region. Their survey ended in 1941, and following World War II, the tradition of limnology at the University of Wisconsin was carried on by A.D. Hasler and his students with a more experimental or manipulative approach (Frey, 1963).

In 1979, research groups working with the Northern Lakes LTER project, the Wisconsin Department of Natural Resources, and the Environmental Protection Agency began studies on the pH and alkalinity of northern lakes to evaluate changes that may have occurred since the lake surveys by Juday, Birge, and Meloche. Our experience with these comparative measurements led us directly to some of the problems associated with archiving, documentation, and quality assurance of historic data sets.

Two major problems with the Birge and Juday data set were the frequency of measurements over the 15-year period and the limited sampling period within each year. Although the data were collected from more than 500 lakes in the region, nearly all lakes were represented by only one or two analyzed samples. Only 53 lakes were represented by more than two measurements (Bowser et al., in prep). For even fewer lakes were there measurements even approximating yearly intervals over the 15-year survey.

Seasonal changes in lakes, particularly productive lakes, were recognized as early as 1924 by Juday on Lake Mendota (Juday et al., 1924). However, lack of year-round facilities at the Trout Lake Station and teaching responsibilities in Madison restricted Birge and Juday's work in northern Wisconsin to the summer months. Therefore more recent comparative measurements of long-term trends in lake chemistry undertaken by Bowser, Lehner, Magnuson, and Rasmussen were restricted to measurements made over the months of July through September.

Meloche, Birge, and Juday measured dissolved oxygen, pH, chloride, sulfate, phosphorus, alkalinity, calcium, magnesium, sodium, potassium, iron, and manganese. Their analytical methods were described in several articles (i.e., Juday et al., 1935, 1938). Several problems arose in attempting to replicate their techniques; most of these resulted from the incomplete descriptions of instruments or methodological details in their published works. Moreover, a substantial amount of their data remained unpublished. On Birge's death in 1950 his office materials were boxed and placed in the hallway to be thrown away! Only through the foresight and quick attention of Dr. John Neess (Zoology Department) were such records saved. They were declared archival material and are available today through the Zoology Department Museum. Original chemistry data sheets, which proved essential for summarizing pH and alkalinity measurements between 1926-1941 (Lehner, 1980), were recovered from the museum archives at Wisconsin.

Terminology has also changed over the intervening years. The terms "bound" and "free" CO_2, for example, correspond to the modern terms "alkalinity" and "CO_2-acidity," respectively (Juday et al., 1935). Considerable time was spent reviewing the literature of the period to assure that the older terms were clearly defined.

pH Measurements

Measurements of pH provide a good example of problems encountered when using historic data. Before the mid-1930's it was common practice to measure pH with colorimetric indicators, since then the glass electrode has been used almost exclusively (Juday et al., 1935; Bates, 1954). After the introduction of the glass electrode, the design of pH and reference electrodes has changed considerably, reference buffers for electrode calibration have changed (Bates, 1954), and pH meters been modified significantly. Measurements taken with the most modern instruments and electrodes are still subject to considerable discussion regarding proper pH measurements and interpretation of results (Kramer & Tessier, 1982). Recently Herczig and Hesslein (1984) and Herczig et al. (in press) discussed some of the problems with electrode pH measurements, and suggested that pH can be better determined by calculation from measurements of total carbon dioxide and P_{CO_2}.[2]

Juday et al. (1935) collected pH measurements from 1926 to 1941 using both colorimetric and electrometric techniques. Two-thirds of the measurements were made using colorimetric indicators (pre-1932). Kramer and Tessier

---

[2]An interesting footnote to problems of pH measurement is found in Rae and Meloche (1942). The article was published after 15 years of experience measuring pH using colorimetric and electrode techniques. The following quote is from the description of methods for the determination of pH by photometric techniques.

"Colorimeter. The Evelyn Photoelectric Colorimeter equipped with filter No. 540 was especially selected for this work. It was chosen for the high degree of standardization of its components permitting the interchange of calibration curves and constants from one colorimeter to any other colorimeter of the same make. Thus it has been possible to offer calibration of data of universal applicability for the determination of pH."

Several points are worth noting. First, the Evelyn Photoelectric Colorimeter is no longer manufactured nor is the company that manufactured the instrument still in business. Second, describing the spectral filter as a "No. 540 filter" is not particularly useful to anyone unfamiliar with the means of designating filters at that time. It clearly would have been more useful to describe the filter in terms of its wavelength/ absorption characteristics. But more important, the technique for measuring of pH is no longer used. To try to recalibrate measurements published using this technique would be a difficult task indeed! It is interesting to speculate what troubles researchers 50 years from now will have using data generated even today.

(1982) have discussed errors associated with such techniques, especially for low conductivity waters where acidity of the indicator can contribute to the pH of the solution. Juday, Birge, and Meloche evidently recognized this problem and in 1932 converted to electrometric measurements using the quinhydrone electrode and acids of various strengths to calibrate the electrodes. In the first year of electrode measurements they conducted 124 intercalibration experiments on 24 lakes (Juday et al., 1935). More recently I have conducted similar experiments and found measurements to average 0.1 pH units lower for colorimetric measurements, similar to the difference suggested by Kramer and Tessier (1982).

More interesting, however, were comparisons of pH measurements made in the field with measurements made in the laboratory within two to four hours of sample collection. We measured pH on site using a field pH meter with an accuracy of ± 0.1 pH unit. We later measured pH on samples collected in polyethylene bottles, filled to exclude air bubbles, and kept cold and in the dark until analyzed two to four hours later with a laboratory meter accurate to ± 0.05

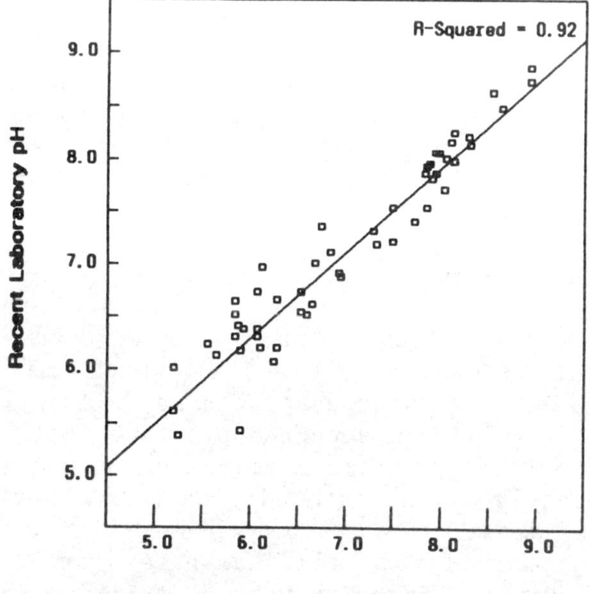

Fig. 1.   Comparison of recent field and laboratory pH measurements. Regression line is plotted. Lab pH is higher than field pH over the lower pH range, and lower over the higher pH range.

pH units. The results show a difference in pH measured with the two techniques (Fig. 1). The laboratory pH was slightly higher for low pH samples and slightly lower for high pH values. These results were consistent with slight gains and losses of CO_2 during sample storage. The problem was then to decide whether to compare laboratory or field measurements with the Juday et al. (1983) data.

Did they determine their pH's in the field directly on collection of the samples or did they return the samples to the laboratory for measurement? Nowhere in their published articles did they mention the time the pH measurements were made. Given the ease of colorimetric measurements in the field, it would be assumed that the pre-1932 measurements were made in the field. The solution to the question came during a personal discussion with Robert Pennak who was directly involved in the field sampling from 1935-1938. He informed us that samples were collected in large containers (in effect having enough thermal mass to minimize sample cooling during sample collection) and returned to the laboratory within two to four hours for laboratory pH determinations. Examination of Figure 2, which shows the relationship between reported Birge and Juday values and our laboratory measurements, reveals a 1:1 slope, thus giving us further confidence in our selection of values to compare. This experience alone was important in our decision to hold an oral history conference with various investigators and camp staff active in the Trout Lake area in the 1920's and 1930's.

Alkalinity Measurements

Historic alkalinity measurements were made using the Methyl Orange technique (M.O.A.) in which titrations were made to a fixed end point described as "a faint pink color" (Juday et al., 1935), an accepted practice that was included in Standard Methods for the Examination of Water and Wastewater as recently as the 14th edition (APHA, 1975). Several difficulties arose with using this published technique; foremost was that "faint pink" was never formally described in terms of pH units, even though, at the time, instrumentation was available to measure it. Our experiments with pH measurements of the M.O.A. endpoint and those reported by Kramer and Tessier (1982) suggest using an endpoint pH of 4.1-4.3. More precise modern titrations using the Gran procedure (Gran, 1952; Edmond, 1970) use a pH of about 4.5 for the endpoint of low ionic strength waters.

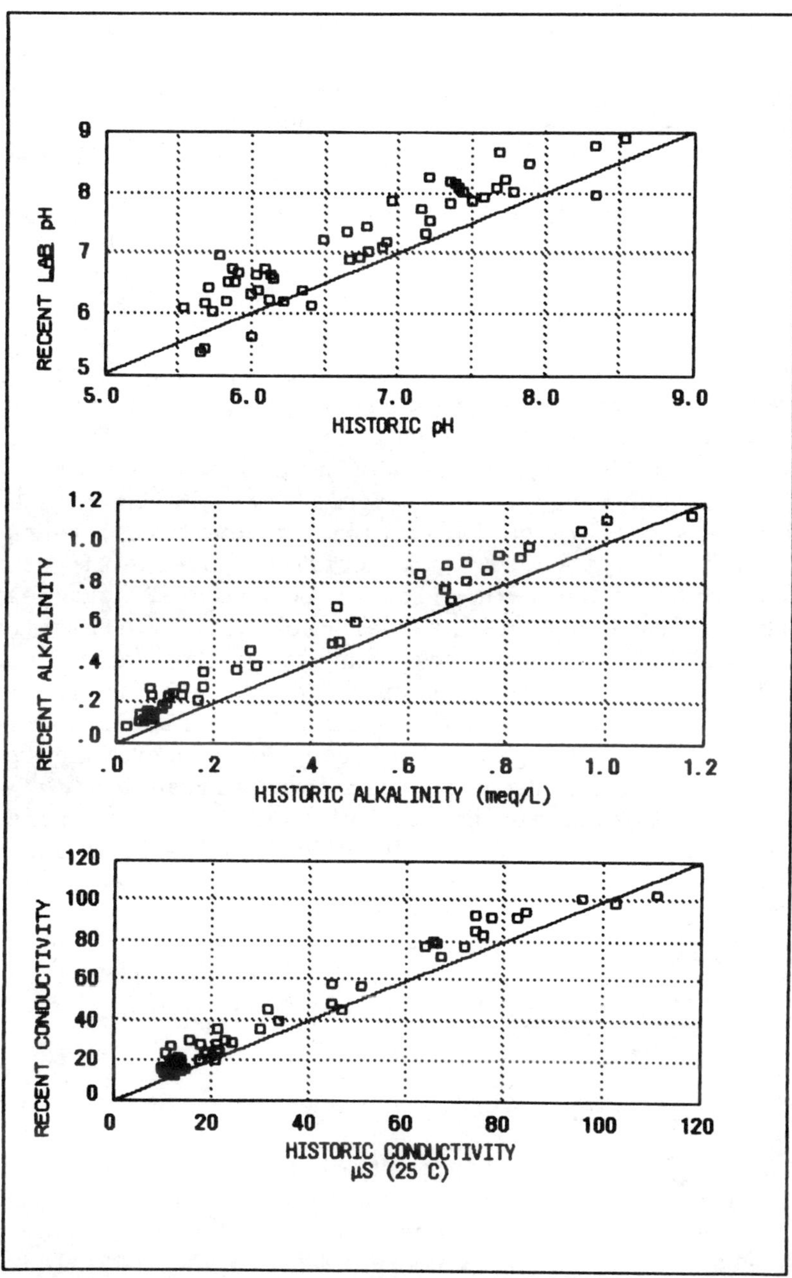

Fig. 2.   Comparison of historic and recent pH, alkalinity, and conductivity measurements on 53 lakes. Plotted lines are 1:1 lines, and not the regression lines.

Clearly the Juday <u>et al</u>. (1935) measurements overestimated alkalinity. Thus, to insure comparability of our measurements with historic values we chose to duplicate their techniques instead of trying, as did Kramer and Tessier (1982), to correct historic values to present day values determined by Gran titration. A comparison of results for 53 northern Wisconsin lakes is shown in Figure 2 with the data for pH and conductivity.

<u>Conductivity Measurements</u>

Juday and Birge (1933) measured conductivity with a "Dionic Water Tester" manufactured by Evershed and Vignoles in London, England. The meter is no longer manufactured. Fortunately a working model was archived along with other Birge and Juday data after Birge's data. Although the instrument was calibrated for a larger conductivity range than that used for the low conductivity waters of northern Wisconsin, the meter did allow us to calibrate our measurements and theirs against standard solutions, with favorable results. Historic standard conductivities were reported corrected to 20°C. Today accepted practice is to report the results corrected to 25°C. To normalize historic and recent values, we used the generally accepted temperature correction factor of 2.5 percent per degree centigrade.

The principles of conductivity measurements were reasonably well understood in the 1920's and 1930's and thus, among the variables in this historic data, we are most confident of these measurements. The archiving of a working model of the conductivity meter used by Birge and Juday proved crucial to our comparisons. Without access to this working instrument, confidence in their measurements would be much lower.

<u>Meteorological Temperature Measurements</u>

Although not directly part of the Birge and Juday data set, recent examination of meteorological data from the Trout Lake area by researchers at the Northern Lakes LTER site (John Magnuson and Dale Robertson) has revealed additional information of interest to those concerned with calibration and interpretation of historic data. At several meteorological stations within 15 miles of the Trout Lake research station temperatures were recorded daily with maximum-minimimum thermometers. These measurements were averaged to give mean daily temperatures. A recent study on

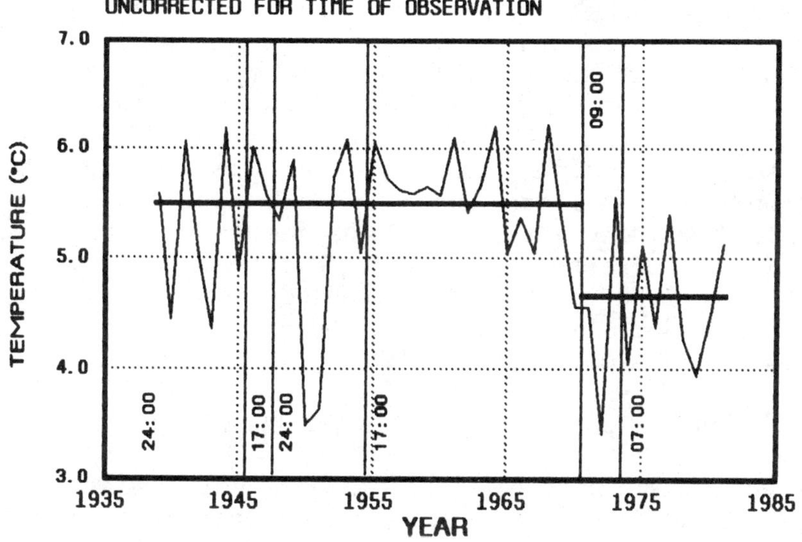

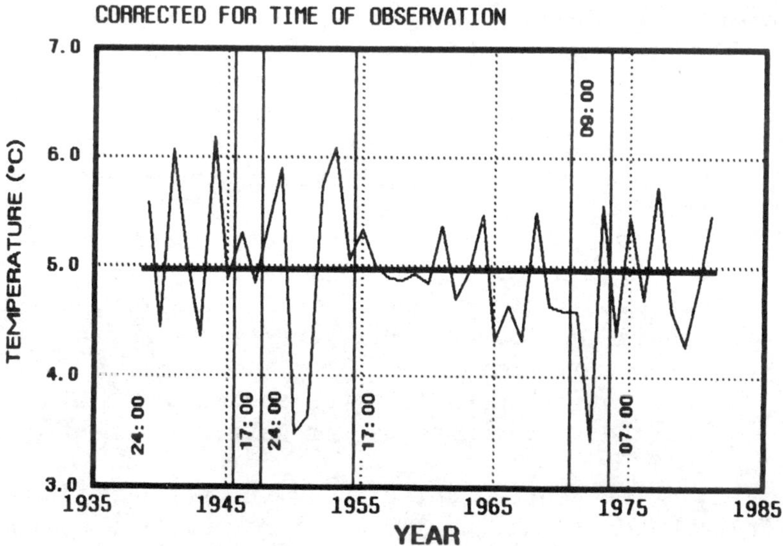

Fig. 3. Average annual temperatures calculated from max-
min thermometer data (Rhinelander, WI). Plots are
shown for raw data and corrected for the time of
observation (Baker, 1975). Times of observation
(24 h basis) are shown vertically on the plots.
Solid horizontal lines represent mean values calculated for 1939-1982 (uncorrected) and the 1939-
1969 and 1970-1981 periods (corrected).

the effect of observation times on the calculated mean daily
temperature showed that the time of measurement on max-min
thermometers can affect the mean temperature calculated
(Baker, 1975). Corrections for time of day measurements
were given by Baker, and calculated mean annual temperatures
from such data were shown to vary as much as 1.8°F.
Magnuson and Robertson (pers. comm.) applied corrections
to data from a station about 30 miles from Trout Lake.
Their results are shown in Figure 3. Times of observations
ranged from 7:00 am to 5:00 pm. Uncorrected values show an
apparent decline of mean annual temperatures after 1968,
whereas corrected values suggest no significant change.

LESSONS FROM THE PAST, LESSONS FOR THE FUTURE

Facing the problems involved in using historic data
sets has increased our concern for quality control and our
awareness of the need for proper documentation and archiving
to insure that our present data collections will be valuable
to future researchers. All of us who are interested in
long-term changes in ecosystems must be as concerned for the
future value of data as we are for our present research
needs. Important questions must be considered:
* What could be made of our data collection and
    database systems if certain key personnel were to
    leave our projects tomorrow?
* What would happen to data sets if one of us were
    to lose funding sources within the year?
* What would happen to records describing our tech-
    niques, quality control, and/or documentation if a
    key person were to leave for another job?
* Whan would happen with the sudden death of key re-
    searchers, technicians, or data management person-
    nel?

Rarely do we concern ourselves with such questions, yet
the survival of a quality data set of long-term research
value depends to a large extent on our ability to plan for
these events.

The following comments are an attempt to state some of
the more critical lessons we have learned from dealing with
historic data sets at the Northern Lakes LTER site. These
lessons, which should guide us in planning long-term research programs, can be divided into two major categories:
1) proper collection, and 2) efficient management of data
for maximum benefit to both current and future research.

Data Collection, Analysis of Samples, and Sample Preservation

Commitment to the stability of data collection is a foremost concern. Typically that means a commitment to long-term stability of personnel, both in project directorship and at the technical and research level. This commitment to personnel makes it possible to develop procedures for data collection, recording, and quality control that are uniformly applied over the long-term.

Standard procedures for data collection, data recording, and quality control should be formalized with research groups. Numerous resources are available to those not familiar with such procedures (APHA, 1975, 1981). The need for proper documentation of such procedures is equally as important. Because the pressures of publication costs and the need to keep article length to a minimum precludes the complete documentation of many data collection and sample analysis procedures in the published literature, other forms of documentation may be necessary.

These procedures may be handled best at each site by having technique manuals, which are updated regularly and which contain information on sample collection and preservation, analytical techniques (referenced to the available literature), data recording protocols, sample labeling protocols, and documentation. Our site is paying careful attention to documentation of the time spans over which specific techniques were used. We use a "current methods" manual and a "prior methods" manual that document the methods and the period of time during which the methods were used. It is useful to share these procedural manuals with interested researchers at other sites and to update them continually in response to accepted practice.

When new methods are introduced or the need to change methods arises, it is essential to use the two techniques on the same samples for whatever period of time necessary to determine that no systematic errors arise from shifts in techniques or, if such biases exist, they can be corrected in future studies. The example of how average temperature data collected with max-min thermometers can give apparent changes in mean annual temperatures, and the example illustrating how the shift from methyl orange alkalinity measurements to electrometric measurements using the Gran procedure produced differing alkalinity values illustrate how such "discontinuities" can appear in the literature when careful attention is not paid to intercalibration of techniques.

In a recent example, Shapiro and Swain (1983) reported on dissolved silica trends in Lake Michigan waters and showed that evident historical changes in silica were due instead to changes in analytical techniques.

It is important that interlaboratory comparison of techniques on reference samples be performed regularly. Such procedures are common with agencies such as the EPA, and it is not difficult to participate in these interlaboratory comparisons. Research sites could initiate such sample exchanges with other appropriate research groups as well.

Documentation of analytical procedures (especially instrumentation) and plant and animal taxa is important for the future use of research data. The changing taxonomy of various plant and animal groups and disagreement over nomenclature for specific organisms has led to much confusion over the years. Photographic documentation could alleviate some of the confusion.

Preservation of reference organisms and historic samples has proven valuable for several studies, but much remains to be done for some areas of systematic ecology. Phytoplankton samples are particularly difficult to preserve in a state requiring little or no sample maintenance, although recent work by Crumpton and Wetzel (1981) offers hope that long-term preservation of such samples will be possible soon. Unfortunately, no such procedures exist for long-term storage of water samples for chemical analysis. Some sample solutes are reasonably stable for periods up to several months, but long-term storage is virtually impossible. Water loss through container walls, irreversible solute adsorption on container walls, and biologic degradation of certain key elements (e.g., nutrients) have frustrated attempts to preserve samples for long-term chemical studies.

Data Management

Carefully collected, analyzed, and preserved samples have no value if the results of such studies are not properly organized and made available to the scientific community at large. It is here that data management plays a most important role.

Participants at earlier LTER data manager meetings agreed that data management involves the documentation and systematization of data collected by project researchers, and should be designed to aid the researcher in using the data (i.e., providing routine reports of data or aiding in the analysis and modeling of such data [Sinclair, 1982]).

In addition, the data manager plays a key role in making data available to future researchers. Data management is such an important part of the LTER concept that most sites have expended a considerable amount of time and resources to insure a successful data management program.

Although data management is an essential activity for the LTER program as well as for others who have long-term research goals, there is concern for the proper balance of resources between the "science" and its "management." Clearly, the research is paramount and is the fundamental reason why a database management system is needed in the first place.

Data managers should consider the development of minimum maintenance database systems so that a proper proportion of scientific "costs" goes into the science and not into the administration of long-term research data. There is, however, strong appreciation for the interdependence of the scientific investigations and the data management. It is often difficult to separate the roles of the scientist and the data manager (Marzolf & Dyer, this volume). Because researchers may feel that data management is taking resources from their own research, some might question the amount of support required for maintenance of a good data management system. Their views are short-sighted. Scientists and data managers (in cases where such a separation of duties is even possible) have a mutual concern for the availability and long-term stability of data. Strong ties between the data management process and the research are essential to the success of long-term research projects.

Rapidly changing technology is a significant concern in the area of modern data management. The "computer revolution" has forced changes in the way we collect, store, and use research data. Data management systems that have been devised to fit the needs of one site may not be appropriate for other sites. Groups whose systems have evolved from older systems have much investment in equipment and software and may be using obsolescent approaches. For example, not many years ago computer input was in batch mode with IBM cards and standard batch output was generally on 132 column computer paper. Several LTER sites initially developed data management systems during the period when batch input was commonplace. Some of those systems have evolved considerably to take advantage of microcomputers, terminal input, and graphic output, others may not have.

It is likely that many current methods will be at least obsolescent within a few years. Thus, a major concern for researchers and data managers today is to maintain flexibility in their data management systems. Data managers will always face questions of upgrading current systems. Ways must be found to balance the added costs of such changes with the improvements the upgrades are likely to provide. The only way to handle the problem is to maintain flexibility in one's approach and to avoid commitment to such expensive database systems that there is little room left to change when necessary.

Publication in peer-reviewed journals is still, and perhaps always will be, the best means of ensuring long-term and widespread availability of research data. In addition, peer-review remains one of the best ways to ensure the quality of research and its underlying database. Consequently, publications of such data should be foremost. Much more careful attention must be paid by both authors and journal editors to precise documentation of techniques for sampling, analysis, quality control, sample preservation, and the precision/accuracy of methods. The availability of "data" has generally not been the problem with historic data, it is the unequivocal documentation of techniques that has concerned us most.

The data manager role is central in cases in which research data and documentation do not lend themselves readily to full publication in refereed journals. Archiving such data in readily available form, together with documentation of data quality is essential. Researchers should consider means of leaving replicate copies of essential data and procedural manuals with institution libraries, historical societies, or federal agencies committed to data preservation.

Other Suggestions

The need for long-term stability of personnel is coupled with the obvious need for long-term stability in the support of such research groups. Such needs are central to the concerns of the Long-Term Ecological Research program. Institutional support for such enterprises is essential. The necessity of continuous, long-term support, translates into the need for international, federal, state, and local agencies to recognize the value of long-term data sets. Also, program administrators must be willing to commit to institutional arrangements guaranteeing long-term stability to critical research efforts.

The initiation of the LTER program is a milestone in the ecological sciences, but not necessarily the final solution to long-term ecological research. Funding for such a program is arranged so that research sites can enjoy the relative benefits of five-year funding blocks instead of the more common one- to two-year terms. However, as in all strong scientific programs, the research must be subject continuously to scrutiny by peer researchers and be subject to loss of funding if minimum levels of performance are not met. Additional protections against loss in continuity of data must involve innovative arrangements among more than one agency or institution that will provide funding arrangements to help "buffer" critical research sites against short-term losses of key researchers, key technical personnel, or minimum project funding, while maintaining the standards of quality necessary for long-term investigations.

Oral histories have provided valuable information for our research group about critical details of procedures and techniques used in the past. Many research groups have access to people who were involved in collecting data in a now valuable historic data set, but who may be well along in years. Information obtained from oral histories may not provide all the missing details about past research questions and older techniques, but we found that our group gained a fuller understanding of the historical perspectives of our own research program.

SUMMARY

Interest in historic data sets arises not only from the potential value of long-term research to future generations, but also from the valuable perspectives that historic data sets provide for those who are committed to long-term research in the ecological sciences. Experience with these data sets has provided valuable insights into problems one should expect when collecting data.

The scope, degree of documentation, quality, and availability of different historic data sets varies widely. Both published and unpublished data sets have strengths and weaknesses. Data discontinuity, whether from single or multiple sources, makes data calibration difficult. Quality control is uneven, at best, and is often undocumented. Instrumentation changes have been rapid and intercalibration with new techniques is not practiced as commonly as would be hoped.

Despite these problems there is a real need to use historic records to the greatest extent possible. Certain environmental issues can only be resolved by using historical data sets.

In offering experiences in dealing with historic data sets I have illustrated some of the problems encountered. There are still those who would argue whether one should make any use of such data sets. However, the benefits far outweigh the risks when it is possible to learn from previous successes (and failures). High quality ecological research requires careful and systematic data collection, sample analysis, sample preservation, data documentation, data archiving, and database management. Such data sets will likely grow in both size and value to researchers in years to come. In addition to designing research projects that deal with our own scientific curiousity and the need for solutions to modern problems, we should be prepared to commit a portion of our research time and energies to data sets and database systems that will literally outlive us.

ACKNOWLEDGMENTS

Little of this paper would be possible were it not for the legacy left by Birge, Juday, and co-workers during the 1920's and 30's. Birge and Juday's data were a key factor in the generation of ideas for long-term ecological research at our Northern Lakes LTER site. My thanks go especially to John Magnuson, Tim Kratz, and Barbara Benson with whom I had numerous opportunities to discuss the various benefits, risks, and problems in dealing with historic data. Special thanks go to Ms. Anna Marie Beckel who helped in the final editing of the manuscript. And finally my thanks to all the LTER data managers who, albeit unwillingly, contributed to my ideas based on our frequent discussions on the data management problems connected with each of our individual sites.

LITERATURE CITED

Allen, T.F.H., R.V. O'Neill, and T.W. Hoekstra. 1984. Interlevel Relations in Ecological Research and Management: Some Working Principles from Hierarchy Theory. USDA Forest Service, General Technical Report RM-110. 11 pp.
American Public Health Association. 1975. Standard Methods

for the Examination of Water and Wastewater, 14th Ed. Amer. Publ. Health Assn., Inc., NY. 1193 pp.

American Public Health Association. 1981. Standard Methods for the Examination of Water and Wastewater, 15th Ed. Amer. Publ. Health Assn., Washington, DC. 1134 pp.

Baker, D.G. 1975. Effect of observation time on mean temperature estimation. J. Appl. Meteorol. 14: 471-476.

Bates, R.G. 1954. Determination of pH: Theory and Practice. Wiley & Sons, Inc., NY. 435 pp.

Birge, E.A. 1893. Notes on Cladocera, III. Trans. Wis. Acad. Sci. Arts Lett. 9: 275-317.

Crumpton, W.B. and R.G. Wetzel. 1981. A method for preparing permanent mounts of phytoplankton for critical microscopy and cell counting. Limnol. Oceanogr. 26: 976-980.

Davis, R.B. and S.A. Norton. 1978. Paleolimnologic studies of human impact on lakes in the United States with emphasis on recent research in New England. Pol. Arch. Hydrobiol. 25: 99-115.

Edmond, J.M. 1970. High precision determination of titration alkalinity and total carbon dioxide content of sea water by potentiometric titration. Deep-Sea Res. 17: 737-750.

Frey, D.G. 1963. Wisconsin: The Birge-Juday era. pp. 3-54. In: Limnology in North America. D.G. Frey (ed.). University of Wisconsin Press, Madison.

Fritts, H.C. 1966. Growth-rings of trees: Their correlation with climate. Science 154: 973-979.

Goldman, C. 1981. Lake Tahoe: Two decades of change in a nitrogen deficient oligotrophic lake. Verh. Int. Ver. Limnol. 21: 45-70.

Gorham, E. and J.E. Sanger. 1976. Fossilized pigments as stratigraphic indicators of cultural eutrophication in Shagawa Lake, northern Minnesota. Geol. Soc. Amer. Bull. 87: 1638-1642.

Gran, G. 1952. Determination of the equivalence point in potentiometric titrations: Part II. Analyst 77: 661-671.

Herczig, A.L. and R.H. Hesslein. 1984. Determination of hydrogen ion concentration in softwater lakes using carbon dioxide equilibria. Geochim. Cosmochim. Acta 48: 837-845.

Herczig, A.L., W.S. Broecker, R.F. Anderson, and S.L. Schiff. In press. A new method for monitoring temporal trends in the acidity of freshwaters. Nature.

Hutchinson, G.E. 1975. A Treatise on Limnology, Vol. 1,

Part 2 - Chemistry of Lakes. John Wiley and Sons, Inc., NY. 1015 pp.

Johnson, W.E. and J.R. Vallentyne. 1971. Rationale, background and development of experimental lake studies in northwestern Ontario. J. Fish. Res. Board Can. 28: 123-128.

Jones, B.F. and C.J. Bowser. 1978. The mineralogy and related chemistry of lake sediments. pp. 179-236. In: Lakes: Chemistry, Geology, Physics. A. Lerman (ed.). Springer-Verlag, NY.

Juday, C. and E.A. Birge. 1933. The transparency, the color, and the specific conductance of the lake waters of northeastern Wisconsin. Trans. Wis. Acad. Sci. Arts Lett. 28: 205-259.

Juday, C., E.A. Birge, and V.W. Meloche. 1935. The carbon dioxide and hydrogen ion content of the lake waters of northeastern Wisconsin. Trans. Wis. Acad. Sci. Arts Lett. 29: 1-82.

Juday, C., E.A. Birge, and V.W. Meloche. 1938. Mineral content of the lake waters of northeastern Wisconsin. Trans. Wis. Acad. Sci. Arts Lett. 31: 223-276.

Juday, C., E.B. Fred, and F.C. Wilson. 1924. The hydrogen ion concentration of certain Wisconsin lake waters. Trans. Amer. Microscop. Soc. 43: 177-190.

Juday, C. and A.D. Hasler. 1946. List of publications dealing with Wisconsin limnology 1871-1945. Trans. Wis. Acad. Sci. Arts Lett. 36: 469-490.

Keeling, C.D., R.B. Bacastow, A.E. Bainbridge, C.A. Ekdahl, Jr., P.R. Gunther, L.S. Waterman, and J.F.S. Chin. 1976. Atmospheric carbon dioxide variations at Mauna Loa Observatory, Hawaii. Tellus 28: 538-551.

Keeling, C.D., R.B. Bacastow, and T.P. Whorf. 1982. Measurements of the concentration of carbon dioxide at Mauna Loa Observatory, Hawaii. pp. 377-385. In: Carbon Dioxide Review. Wm.C. Clark (ed.). Oxford Univ. Press, NY.

Kitchell, J. and S. Carpenter. In press. Plankton community structure and limnetic primary production. Amer. Nat.

Kitchell, J.A. and J.F. Kitchell. 1980. Size selective predation, light transmission, and oxygen stratification: Evidence from the recent sediments of manipulated lakes. Limnol. Oceanogr. 25: 389-402.

Kramer, J. and A. Tessier. 1982. Acidification of aquatic systems: A critique of chemical approaches. Environ. Sci. Technol. 16: 606A-615A.

Lehner, C. 1980. Report on the Inventory and Future Use of Historical Water Chemistry and Plankton Data of Birge, Juday and Associates. Center for Limnology, University of Wisconsin, Madison.

Likens, G.E., F.H. Borman, R.S. Pierce, J.S. Eaton, and N.M. Johnson. 1977. Biogeochemistry of a Forested Ecosystem. Springer-Verlag, NY. 146 pp.

Likens, G.E., F.H. Borman, R.S. Pierce, and J.S. Eaton. In press. The Hubbard Brook Valley. In: An Ecosystem Approach to Aquatic Ecology: Mirror Lake and Its Environment. G.E. Likens (ed.). Springer-Verlag, NY.

Maher, Louis J., Jr. 1982. The Palynology of Devils Lake, Sauk County, Wisconsin. pp. 119-135. In: Quaternary History of the Driftless Area. Wisconsin Geological and Natural History Field Survey Field Trip Guide Book No. 5. J.C. Knox, L. Clayton, and D.M. Mickelson (eds.).

Marzolf, G.R. and Dyer, M.I. 1985. Future direction for research data management in ecology. pp. 409-417. In: Research Data Management in the Ecological Sciences. W. Michener (ed.). Belle W. Baruch Library in Marine Science, No. 16. University of South Carolina Press, Columbia.

Rae, J. and V.W. Meloche. 1942. A photoelectric method for determination of pH. Trans. Wis. Acad. Sci. 34: 195-212.

Schindler, D.W. and E.J. Fee. 1974. Experimental lakes area: Whole-lake experiments in eutrophication. J. Fish Res. Board Can. 31: 937-953.

Shapiro, J. and E.B. Swain. 1983. Lessons from the silica "decline" in Lake Michigan. Science 221: 457-459.

Sinclair, R.A. 1982. Long-Term Ecological Research Data Management Workshop. Urbana-Champaign, IL, Nov. 22-23, 1982.

Swain, A.M. 1973. A history of fire and vegetation in northeastern Minnesota as recorded in lake sediments. Quat. Res. 3: 383-396.

Swain, A.M. 1978. Environmental changes during the past 2,000 years in north central Wisconsin: Analysis of pollen, charcoal, and seeds from varved lake sediments. Quat. Res. 10: 55-68.

Wetzel, R.G. 1983. Limnology, 2nd Ed. W.B. Saunders Co., NY. 767 pp.

Zullig, H. 1981. On the use of caratenoid stratigraphy in lake sediments for detecting past developments of phytoplankton. Limnol. Oceanogr. 26: 970-976.

MAINTAINING DATA INTEGRITY
IN RESEARCH DATABASES

Donald A. Holzworth and Steven K. Seilkop

ABSTRACT

When data are stored in large research databases, there is a strong potential for a lack of integrity in the data resource from which inferences and/or regulatory decisions must be drawn. This paper discusses the problems associated with scientific data collection, storage, and retrieval, and their impact on research investigations. Examples of the data integrity dilemma are taken from the Environmental Protection Agency's Love Canal monitoring efforts and the National Toxicology program's lifetime toxicity testing studies. Suggestions are offered to improve data integrity through specific design goals for large databases.

INTRODUCTION

In today's research environment, the use of computer technology to collect and manage experimental data is the rule rather than the exception. The management and subsequent analysis of a large, complex database generated by many scientific research efforts would be virtually impossible to complete in a reasonable amount of time without the aid of computer technology. Environmental monitoring and health effects experimentation are typical of such current data-intensive research activities. Not only are the data from these investigations voluminous, but they also are often complex and are collected from diverse sources over long periods of time. In many large databases, the data were collected from perishable samples and opportunity for irreversible error exists in the data collection process.

Data management procedures are therefore critical to assuring the integrity of these data resources which are used in scientific evaluations and regulatory decision-making.

Both data processors and researchers are well aware of the potential for error in the scientific data collection and computerization process. Though these problems are recognized, too often their importance is lost in the process of building a computerized data resource with an interface suitable for end user access and analysis. This is particularly true in a distributed data collection environment where experimental results are begin drawn from multiple on-site locations and transmitted to a centralized database.

The need for scientific data quality indicators (tags) in a database has been recognized for several years by prominent industry groups and governmental concerns. A workshop on Data Quality Indicators (Bowman et al., 1982), sponsored jointly by government and industry, concluded that data quality indicators should contain information about four distinct elements: the method used to obtain the data, the extent to which the data have been evaluated, the source of the data, and an indication of the accuracy of the data.

These general conclusions served to further raise the issue of data integrity (the quality or suitability of data for their intended purposes) in many of the commercially available data resources being used today. This paper attempts to address some of the issues of concern to data managers and researchers who are building and using databases for original scientific investigations. If the integrity of these data resources cannot be assured, then no broader attempt to "tag" data as good or bad is going to enhance the quality of commercially available databases.

DISCUSSION

A lack of data integrity in large databases results from three main sources: data collection and entry errors, database design errors, and observational and measurement biases. Data entry errors are generally a function of the transcription process. They result from keying errors, redundant data entry, and data omissions. The design of a collection system for a large database can act to minimize such errors through simplification of the transcription process and on-line data checks for consistency, redundancy, and completeness of data. However, the same collection system can, through inadequate design, perform erroneous transmissions to the database which result in redundancies and omissions that look very much like data entry errors.

Data collection and database design errors are to some degree controllable and can be guarded against through proper design of data management and collection systems (see Gurtz and Stafford et al., this volume). That is not to say that designing such systems is a trivial matter, but that every effort should be made to create databases which are free of those kinds of errors.

Although we can improve data quality through database design and data procedures, we have little control over the actual measurement and observational processes and their effect on data integrity. These sources of error are certainly not unique to large, complex databases but are often overlooked in the analysis of large volumes of data. The diversity of data sources associated with large databases contributes heavily to experimental variability but is often buried beneath all the other potential problems in analyzing large data sets.

Large databases can be used for identifying and examining the consequences of these error sources. Part of the problem with this type of evaluation as well as other ad hoc analyses using large databases is the inaccessibility of the data. Using a data collection, storage, and management system does not guarantee that data can be easily retrieved for analysis. Databases are often designed for data collection and storage with little regard for how the data will be used later.

Two large government programs provide the basis for our discussion of the data integrity dilemma in large databases: the National Toxicology Program and EPA's Love Canal Monitoring Study.

These particular efforts to collect, store, and analyze high quality data from multiple sources represent two broad data type categories: environmental and health effects. Each category has its own attributes which makes it difficult to do repeated sampling and expect the same result.

The National Toxicology Program

The National Toxicology Program (NTP) is an interagency program involved in long-term in vivo testing. These tests involve long-term exposure of animals to known or potential carcinogens in order to assess human health risks from these chemicals. The resultant data are voluminous due to the number of animals, the variety of desirable information collected, the number of chemicals being tested, and the number of geographically dispersed laboratories involved in the testing. The NTP is typical of large-scale health research and has the following attributes: fixed study objectives, complex data structures, lifetime (2-year rodent studies) generation of data, diverse data sources, various interpretation methods used in the data collection process (e.g., diagnoses of tumors, degree of severity in non-neoplastic lesions), and the need to use the database for historical (baseline) comparisons.

The National Toxicology Program has developed several manual and computerized data collection systems over the last decade. The NTP-developed system for collecting and reporting data is called the Carcinogenesis Bioassay Data System (CBDS). This system is based on manual data collection forms that are completed by respective responsible parties and forwarded to a data processing support contractor. The data processing support group validates the forms, encodes where necessary, and performs a batch update of the mainframe resident master data files. The data collected by CBDS fall into three major categories: limited test protocol information, gross live animal data, and microscopic pathology findings. CBDS, which is based on manual forms shipped from diverse geographical locations and processed by various people, exhibits a less than desirable error detection/correction cycle time, and likewise, the response time is less than desirable for data reporting back to the originating laboratory.

Computerized data collection systems were first developed for NTP by the National Center for Toxicological Research (NCTR). The earlier NCTR-developed system for collecting and reporting of data was named the NCTR Integrated Data System. The NCTR System is based on source data automation. This capability is provided by placing data collection terminals as close to the source of the data as possible. The data are collected during the day and transmitted to the mainframe for batch update and reporting at night. This system suffered from three major problems:

    1.   The experimental design used in the original system's construction had changed radically in terms of sample size. At the same time it became much

less predictable as to how the experimental protocol would be conducted.

2. The data systems were developed as independent
   entities and later linked via a rudimentary query
   system. This led to some duplication of data and
   conflicts of terminology or data format since the
   original pathology data collection system relied
   on optical mark sense forms which require strin-
   gent quality control and are very inflexible.

3. The advent of the Good Laboratory Practices (GLPs)
   required that extensive manual documentation of
   data changes be kept on some of the systems and
   total redevelopment of some of the systems to provide the audit trail information required by the
   GLP.

Since both NCTR and NTP desired to redevelop their systems of data collection, an interagency agreement was
formed to develop a real-time data collection system known
as the Toxicology Data Management System (TDMS). The TDMS
was developed and is now in use both at NCTR and at many NTP
contractor laboratory sites.

It was the intent of the TDMS system designers to construct a modular integrated data system which would be flexible enough to provide support for testing programs of the
scale and diversity anticipated at NTP and NCTR as well as
encompass many future systems with minimal change and no
impact to those applications in production.

To meet these goals, a distributed system architecture
was designed. A central facility would contain all master
databases. Information governing the collection and validation of data would be extracted from the central database
and utilized by data collection systems located at the various remote laboratories. Information collected at these
remote sources would flow back into the central databases
for archival storage and reporting.

The central databases would be based on application-
independent tools centered around a common data dictionary.
Since data systems of this size tend to require that certain
processes be performed in a similar fashion across a variety
of data groups, it was desirable that, wherever possible,
application generators would be developed. These application generators would remove the programmers from the chore
of producing many programs providing the same function
varying only by the data elements processed. The data

dictionary containing pertinent information as to data structure, data type, validation criteria and other documentation would provide the common source of input to the various generators. Three areas were targeted for development of application generators: validation of data transactions, on-line mainframe screens for data input/output, and database data extraction modules.

The remote laboratory data collection systems to be used in high volume or complex data areas would be designed specifically for each application. This would allow the systems to be fine-tuned to the environment in which they were targeted to operate.

Over the years NCTR and NTP data integrity has frequently been compromised due to the following factors: the scientific state-of-the-art protocols for toxicology and carcinogenicity testing have changed more rapidly than any of the data collection systems' ability to adequately collect and store the data in a consistent fashion.

The rapid introduction and changes of microcomputer technology in the data collection process has prevented the continued use of a stable real-time, source data collection system for more than several months.

Early attempts at designing a "home grown" DBMS failed to meet the later requirements of the Good Laboratory Practices (GLP) guidelines. A recent move toward using a commercial DBMS has raised further data transcription questions as the statistician is now faced with an old versus new database.

Data are collected and verified under GLP's by individual contractor laboratories and then transmitted to NCTR where data transcription errors can occur when building the master database. Few, if any, database-to-data generator verification checks are made.

EPA's Love Canal Monitoring Study

EPA's Love Canal Monitoring Study was conducted in the summer and fall of 1980 to assess environmental contamination in the vicinity of the Love Canal landfill. The study included the collection of almost 6500 field samples of air, water, soil, sediment, and biota which were chemically analyzed for a large number of known or suspected contaminants. The analyses resulted in the compilation of approximately 150,000 individual measures of environmental contamination. In addition, the Quality Assurance/Quality Control program consisted of the analyses of 5700 samples whose results served as documentation and standards for the precision and accuracy of the analyses of field samples. Analytical data were supplied by 13 different laboratories with requirements for a strict chain of custody procedures for the collection, transportation, and reporting of individual samples.

The research was characterized by severe time constraints imposed by political and legal necessities; changing study objectives; large volumes of data; state-of-the-art, expensive analytical techniques; multidimensional (physical, spatial, temporal) data; diverse analytical methodologies, and diverse data sources.

In contrast to NTP's source data automation in TDMS, EPA's Love Canal monitoring study used a forms-driven data collection system linked to several mainframe databases. The Love Canal database was quickly designed due to political necessity; time limitations precluded sophisticated on-line systems for data entry. Samples were numerically identified and associated with individual geographic sites and sample media. Chemical analyses were performed by subcontracting laboratories. Their results were entered into a raw database by a coordinating group. Listings of the raw database were compared against sample logs from analytical laboratories to assure that all samples were entered into the database. The listings of measurements were sent to the analytical laboratories for verification of individual values. Corrections to the raw data were made and a verified database was created. Both field and quality assurance data were stored in both the raw and verified databases. Quality assurance data were used to validate field samples through accuracy checks on "spiked" samples and quality assurance (QA) samples with known chemical concentrations. Field samples from analytical runs in which QA samples did not meet accuracy criteria were rejected.

Throughout the verification process, range checking was used to tag questionably large values. Since little was known about the chemical profiles of field samples, consistency checks among chemical substances were not possible. In general, the assurance of accuracy in the raw database was the responsibility of each analytical laboratory, and the accuracy of updates to the raw database was the responsibility of the coordinating group. The key to assuring data quality in this system, at least in terms of accurate transcription, was the laboratories and coordinating group's perserverence in detecting errors at the verification step. Documentation of changes and authorization for changes was largely on paper.

Since the error detection/correction cycle was long, the amount of time available for statistical analyses was significantly reduced. In addition, limited statistical software was available on the computer which contained the database, the database structure was complex, and there were no user-friendly query capabilities. The resultant inaccessibility of data made it more difficult to understand the measurement process and adjust for sources of measurement error.

This system is assumed to have controlled data entry error and assured data integrity in the verified database. No formal sampling of the data has been performed for purposes of verifying the accuracy of computerized data relative to that found on the original collection forms. The continual checking of data by each analytical laboratory has been used as a basis for asserting that the computerization process has resulted in a very small number of data transcription errors. This assertion is probably correct, although only spot checks have been available to attest to the accuracy of the transcription process.

SUMMARY

These large, complex databases exemplify the data integrity dilemma, namely that we often have little assurance of the quality of the data that we are analyzing. Large-scale projects pose the following database problems for the researcher:

1.   Data are presumed correct and are analyzed to draw investigative conclusions which could be rejected due to data errors.
2.   Data collected over time may vary in quality and comparability.
3.   Data may be suspected of being incorrect, but the source of error cannot be detected due to the lack of an appropriate audit trail.
4.   Accessibility of the data may be limited, reducing the contact with the data which is necessary to understand and correctly analyze the process under investigation.

It would be sometimes accurate to state that end users are not to blame for the data integrity dilemma, since in large government studies they are usually not in control of the data system development, data collection, and database building stages of a study. What is found, however, is that

end users are caught in the continuing dilemma of having to
make the best of questionable data from well intentioned
scientific programs of major scope.

RECOMMENDATIONS

The following insights are offered as design goals for
high integrity, large, complex data collection and storage
systems based upon the assumption that research data integrity relies on error prevention, detection, and correction.

1. The system must be responsive to user requirements. It must collect and store data in a format
   that the eventual user needs, wants, and can use
   within a reasonable time frame.

2. The system must provide as much verification and
   editing of the input data as is possible at the
   time the data are generated.

3. Data errors must be corrected quickly and easily
   and only by properly authorized individuals. Complete documentation of the correction process must
   be maintained.

4. Data transcription steps must be minimized. This
   cuts response times, error rates, and operational
   costs of the error correction process.

5. Data collection procedures must be well established, documented, and consistent.

6. The system must allow tracking of a sample
   throughout the period of experimentation including
   any data values found to be in error and the corrected values. This results in a complete audit
   trail for supporting the later analyses and interpretation of results.

These goals can be translated into a series of objectives (Table 1) for three major system components. To meet
these objectives, and for a large, complex database to
effectively meet the user's needs, each of the individuals
involved in the development and ongoing maintenance process
must meet a predefined set of responsibilities.

Database Administrators must be responsible for assuring that the database fits the user, is designed according
to the documented specifications, and documentation of the
database structure is accurate. Also, problems must be
tracked and corrected in future updates. Validation plans
should be developed as part of the database design and updated if and when the database design changes.

User Support Personnel should review user documentation for accuracy and usability and aid in the accurate recording and reporting of problems which are encountered with the database. They should also assure that the validation plan reflects the real world and the extremes in which the system must operate.

End User Personnel must be responsible for reviewing end user design criteria and documentation and designing standard interfaces in conjunction with user support personnel for using the system in day-to-day operations. End users must also monitor database performance (Are the data

Table 1.   Objectives of major system components.

I. Data Collection Objectives

- A. Standardization of data elements
- B. Conformance and adherence to desired data entry procedures
- C. Verified sample identification
- D. Data completeness
- E. Data validation
- F. Adherence to error correction procedures
    1. Nondestruction of original entry
    2. Documented reason for change
    3. Time/date and operator performing change
- G. Management and scientific review of data entry

II. Data storage and Reporting Objectives

- A. Consistency of storage format
- B. Organization of support rapid access (multiple indicies, etc.)
- C. Redundancy for backup recovery
- D. Establishment and conformance to reporting standards
- E. Ongoing review of experiment reporting

III. Data Retrieval and Audit Objectives

- A. Users must have ability to retrieve each and every data element upon request.
- B. Users must have ability to interrelate all data elements collected during experimentation
- C. Data must be easily accessed for statistical analysis
- D. Data must be summarized to reflect experimental results

---

consistent and accurate? If not, under what conditions do problems occur?) and communicate problems to support personnel as accurately and in as great detail as possible.

As researchers, the credibility of our work and the ease with which we carry it out rely on good data resources. In large research projects, data management and collection systems directly affect data quality and accessibility. Those with experience in data analysis and applications have much to offer in designing these systems. Unfortunately, we often shy away from this activity, and later pay for our neglect through reduced data quality and cumbersome databases. Participation in the design of databases and data collection systems, if nothing else, draws us closer to the processes by which data are generated, helping us to perform more meaningful and appropriate analyses. It is essential that we fully understand the complexity of the data we are dealing with; involvement in the development of databases and collection systems can better equip us to utilize these data as well as assure their quality.

LITERATURE CITED

Bowman, C.M., G.C. Carter, K.D. Fisher, L.H. Gevantman, S. Siegal, and J.A. Steel. 1982. Workshop on Data Quality Indicators, Summary Report and Recommendations. February 10-12, 1982. Chemical Manufacturers Association, Washington, DC. 61 pp.

Gurtz, M.E. 1985. Development of a research data management system: Factors to consider. pp. 23-38. In: Research Data Management in the Ecological Sciences. W. Michener (ed.). Belle W. Baruch Library in Marine Science, No. 16. University of South Carolina Press, Columbia.

Stafford, S.G., P.B. Alaback, K.L. Waddell, and R.L. Slagle. 1985. Data management procedures in ecological research. pp. 93-113. In: Research Data Management in the Ecological Sciences. W. Michener (ed.). Belle W. Baruch Library in Marine Science, No. 16. University of South Carolina Press, Columbia.

QUALITY CONTROL IN RESEARCH DATABASES:
THE US ENVIRONMENTAL PROTECTION AGENCY
NATIONAL SURFACE WATER SURVEY EXPERIENCE

P. Kanciruk, R.J. Olson, and R.A. McCord

ABSTRACT

The usefulness of a research database is in part determined by the quality of the information it contains. The procedures involved in ensuring the production of a high quality database are collectively termed quality assurance (QA). To ensure that data are as accurate as possible, the research database manager ideally should have the research and data management background to be involved from the point of project design, data verification and validation, through the documentation and dissemination of the final data. This ideal is not often realized.

One exception is Phase I of the Environmental Protection Agency's National Surface Water Survey. Beginning in the fall of 1984, this 3-month field survey will sample 2100 lakes in the northeast, southeast, and upper midwest regions of the United States. Another 900 western lakes will be sampled in the fall of 1985. The survey uses helicopters to take water samples that are chemically analyzed at remote laboratories. The survey will synoptically define the extent of existing surface water acidification.

The information obtained will be used by a variety of researchers and managers concerned with acid deposition questions. The scientific and political importance of this project requires the utmost confidence in the validity of the final database. The QA procedures used for this project include consultation on data forms design; input data screening; double data entry; range checking and relational scanning of data; data verification; and data validation using statistical, thematic, and graphic techniques.

INTRODUCTION

Unfortunately, an often neglected aspect of research projects is the design and implementation of the research database, and in particular, those efforts collectively termed quality assurance (QA) that ensure the integrity of the information stored in the database. We are all aware of data sets whose contents do not reflect the original care used in collecting the information. The need for QA in the design and construction of any database is important, but it is especially necessary for research databases because they are unusual in several aspects: 1) they are usually expensive to create, 2) the direction of additional research and even legislative action often may depend on their content, and 3) they are usually unique databases--and frequent reliance is made on custom software that is prone to error. For these reasons, most data managers would agree that a strong QA program is essential to the success of any major research database project. This paper describes the QA measures used during the design and implementation of a large research database at Oak Ridge National Laboratory (ORNL).

As part of the larger National Acid Precipitation Assessment Program (NAPAP), the Environmental Protection

Fig. 1.   Lake locations for EPA's National Lake Survey. About 3000 lakes are to be sampled in this survey during the fall of 1984 and 1985.

Agency's (EPA's) National Surface Water Survey (NSWS) will sample about 3000 lakes in the northeast, southeast, upper midwest, and western regions of the United States on a one-time, synoptic basis (Phase I) (Fig. 1). A major portion of the lake survey took place in the fall of 1984, with the western region (about 900 lakes) scheduled to be sampled in the fall of 1985. A subset will also be sampled as part of a long-term monitoring project that will provide EPA with accurate data on water quality trends that can be used for NAPAP assessments.

The scientific and political importance of this survey required that the most rigorous QA procedures be employed in the database design and creation. The Environmental Sciences Division of ORNL designed and implemented the NSWS Lake Survey database for EPA.

DATABASE MANAGEMENT

What do we mean when we speak of quality control for research databases? Intuitively, we think of the data as being of 'high quality' or 'accurate.' Does this only mean that there are no typographical or computer-generated errors in the database and that the data manager has ensured adequate data backups and access security? Minimally (and historically), QA for research databases includes those measures that ensure that information entered into the system is not corrupted. Data corruption includes typographical error, incorrect interpretation of data forms, adulteration of information due to incorrect data definitions or data transformations ($\underline{e}.\underline{g}.$, unit conversions), poor design of the database ($\underline{e}.\underline{g}.$, data loss during file merging), and accidental erasure or unauthorized alteration of data.

The research database manager bridges the sometimes significant technical gap between the research scientists and the computer scientist. The research database manager ideally should have both the data management and scientific backgrounds to work with the researchers involved in data collection in order to develop unique QA measures that ensure a database of the highest quality. The database should be as free of errors as possible, both those created at the data center and those previously introduced into the data through poor experimental design, sampling bias, contamination, equipment malfunction, inconsistent sampling techniques, field/laboratory human error, etc.

DATABASE ERRORS

Although there are innumerable ways to make errors, essentially there are two general types of errors (or data loss): (1) incorrect information, and (2) missing, incomplete, or nonretrievable information. The first type is easily understood as an error of commission. Bad instrument readings and typographical error during data entry are examples (Table 1). The second type of error is the error of omission, that is, not including important information relating to a legitimate data value (Table 2). This is not the same as having a missing value for a database variable (when in truth a value does exist)--that is an example of an error of the first type, incorrect information. Here an error of omission is not recording a piece of information directly associated with a variable, a data "qualifier," that affects its correct interpretation. Our qualifiers are somewhat similar to the remark codes in STORET (although many of those codes are labels and not true qualifiers). Qualifiers that assist the interpretation of data values have also recently been incorporated into the WATSTORE database (Yurewicz, in press).

Table 1.   Examples of errors of commission. These errors occur when misinformation is entered into the database.

---

1.   inaccurate selection of subject (sampled wrong location)

2.   use of improper technique (wrong preservative)

3.   inaccurate instrument calibration (wrong scale)

4.   inaccurate reading of an instrument (perceptual error)

5.   inaccurate recording of a value onto the data sheet (human error)

6.   error in transcribing a value from a data sheet to a summary sheet (bad penmanship)

7.   errors during inputting (typographical errors, omitting a value)

8.   improper storage, transformation, or merging/subsetting of variables in the database proper (data management error)

Table 2.   Examples of data qualifiers.   The use of data qualifiers prevents errors of omission.

1.   oil slick on lake surface, sample may be contaminated

2.   instrument slow to calibrate, value recorded anyway

3.   instrument unstable, value recorded anyway

4.   air bubble in sampling syringe, might affect reading

5.   label on vial smudged, sample identity uncertain

6.   re-analysis, first value discarded

7.   lake stratified, sampled anyway

Data qualifiers are not variables to be analyzed, but they help determine how data should be analyzed. Despite their importance and high frequency of occurrence in most research projects, data qualifiers are usually not incorporated into research databases. It is important, for example, to record the correct pH reading of a lake water sample; it may be equally important to indicate that the pH meter was unstable or took extra time to equilibrate. That information might influence the interpretation, or even use, of that value. Research database managers often plan for perfect data. Qualifying information is not anticipated, not allowed, or is usually buried in some general, unstructured 'COMMENTS' field where it is difficult to retrieve, and its relationship with a particular data value is lost.

It might be suggested at this point that the data qualifier be made a true data variable (e.g., check off pH meter unstable, yes or no) and included in the database, negating the need for a data qualifier. This would be a solution if research databases were simple and data qualifiers were limited in type and predictable in nature. However, research databases have large numbers of variables and many potential qualifiers. The working lake survey database contained over 3000 variables; the inclusion of six predefined data qualifiers as legitimate variables for each variable would quickly render the database unwieldy.

A research database that does not utilize data qualifiers has reduced quality and is of less use than the database that can accommodate data qualification. The National Lake Survey database incorporates the ability to handle data qualifiers for all its quantifiable variables.

QUALITY ASSURANCE AND THE NATIONAL LAKE SURVEY

The National Lake Survey is a complex project with hundreds of field and laboratory variables to be entered into the database. Dozens of personnel are involved, with information gathered from multiple sites under varying conditions in a short period of time. Quality assurance programs have been developed for the Lake Survey in three general areas: (1) experimental design, (2) analytical procedures, and (3) database management. All of these function together to ensure high quality data. Those relating to research database management are discussed here.

DATA FLOW

The data flow for the Lake Survey is summarized in Figure 2. Water samples are collected in the field using small helicopters with pontoons that land on the lake surface. A 4-L Van Dorn sample is taken as well as _in situ_ measurements on pH, conductivity, depth, and Secchi disk visibility. Watershed characteristics are also noted. Helicopter sampling takes only 10 to 20 min per lake, allowing a large number of lakes to be quickly sampled. Two helicopters and a mobile laboratory are assigned to each of the five base sites used in the Lake Survey. Each helicopter brings the water samples back to the base site, where additional water chemistry information is obtained. The samples are prepared and sent to a number of analytical laboratories, which are contracted to perform the majority of the water chemistry analyses. Field forms with the information obtained at the lakes and the mobile laboratories are sent to the data center at ORNL; copies are sent to the EPA site control center.

Both the field data forms and the contract laboratory data forms are double entered directly into data sets, scanned for typographic errors and suspicious values, added to the growing raw data file, printed out as "verification" reports, and sent out for review by those in charge of the analytical laboratory QA. The reviewed data are returned to the data center on magnetic tape and any corrections or flags (see below) are automatically entered into the verified data sets. The old data values are maintained in raw data sets as historical backup.

The verified data for each geographic region are accumulated until all the data for a given region are obtained.

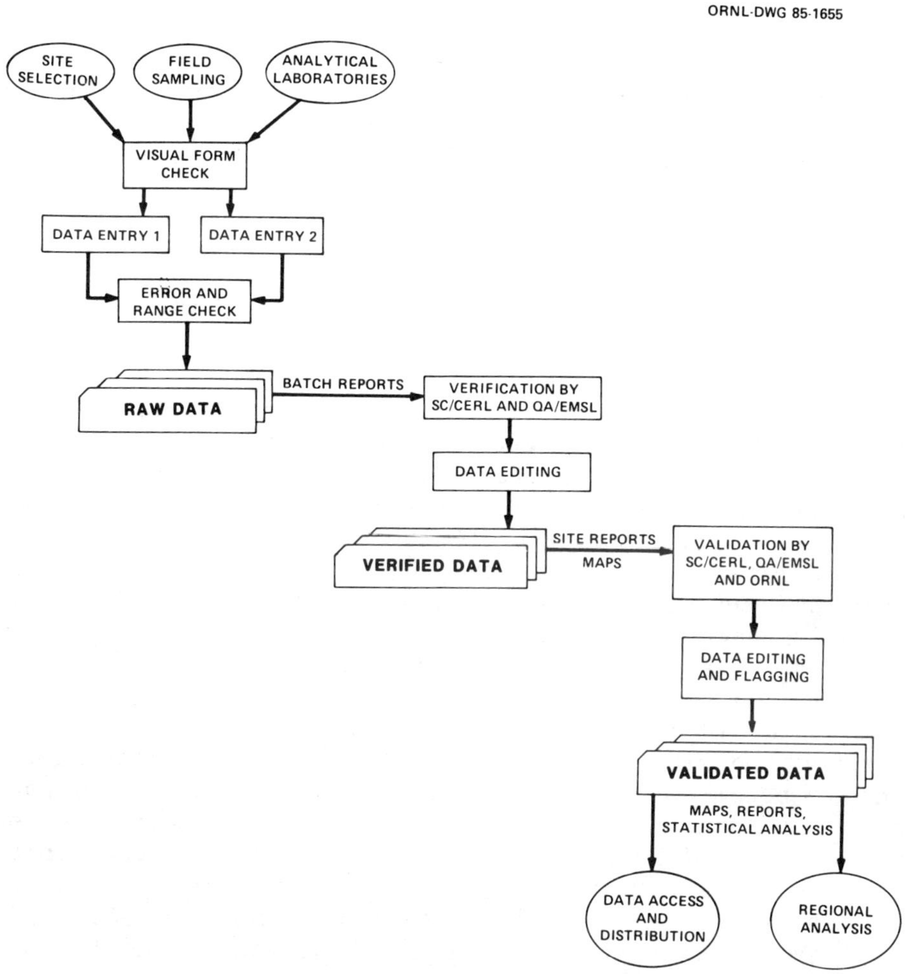

Fig. 2.   Data flow. Data from the field and laboratory flow into the ORNL Database Management Center where numerous QA checks and reviews are a part of the database system. The site control (SC) group is responsible for tracking daily survey progress in the field.

Then the data for each region are subjected to 'validation, using selected graphical and statistical techniques that examine the data from a regional/ecological perspective. Suspect values are highlighted in validation reports that are distributed for review by the Lake Survey management team. Any corrections deemed necessary to data values or data flags are entered, and the database then becomes available for scientific analysis.

FORM DESIGN

Interaction between the database manager and the research scientists started early with the design of the field and laboratory data forms. This was important for a number of reasons. The database software (SAS*) chosen to implement the Lake Survey database is a relational system. It was important that each data form be a single or repeated record in a relational sense to allow easy translation of each data form into a separate SAS data set. For simplicity and easy conceptualization of the database by the research scientists, a one-to-one relationship between data forms and data sets was maintained. The final survey database will be a combination of many separate data sets incorporated into one database. Some of these will not have originated with the Lake Survey, such as the concurrent National Oceanic and Atmospheric Administration climatic information.

Important aspects of cooperation between the data manager and research scientists include unit definition, and defining data qualifiers and comment fields. Because the final database will use international metric conventions, the form designers initially assumed that the field personnel would enter data in metric form. This was not always directly possible (e.g., helicopter cockpit instruments), and it became apparent that conversions in the field would be necessary. The database manager realized that such conversions would be prone to error and that nonmetric units, where necessary, would be better entered "as is" and converted within the database with less error. The inclusion of data qualifiers (Table 3) in the database and the preprinted data forms were important QA measures.

Table 3.   The data qualifiers used in the Lake Survey for the field form.

A -- instrument unstable;
B -- redone, first reading not acceptable;
C -- instruments, sampling gear not vertical in the water column;
D -- slow stabilization;
E -- hydrolab cable too short;
X -- other (explain) ___________________________.

*Statistical Analysis System, SAS Institute, Inc., Box 8000, Cary, North Carolina 27511.

DATA QUALIFIERS

It was evident that important qualifying information could not be discarded, and equally clear that anticipating every possible qualifying remark that could be made about the hundreds of variables in the survey would be impossible. The compromise was to allow persons entering data to 'tag' data values directly on the data sheets with predefined qualifiers (Fig. 3). Tags are one letter characters that directly qualify the data value. For example, if a pH reading of 6.6 is recorded only after re-analysis, it is recorded as '6.6R' on the data sheet. The number enters the data set as a pH value, "PH" of 6.6, and a pH tag, "PH_T" of R. Up to six tags can be strung together to modify a data value (although multi-qualified data were rare), and user-defined tags are allowed where the person entering the data can include an unanticipated qualifier.

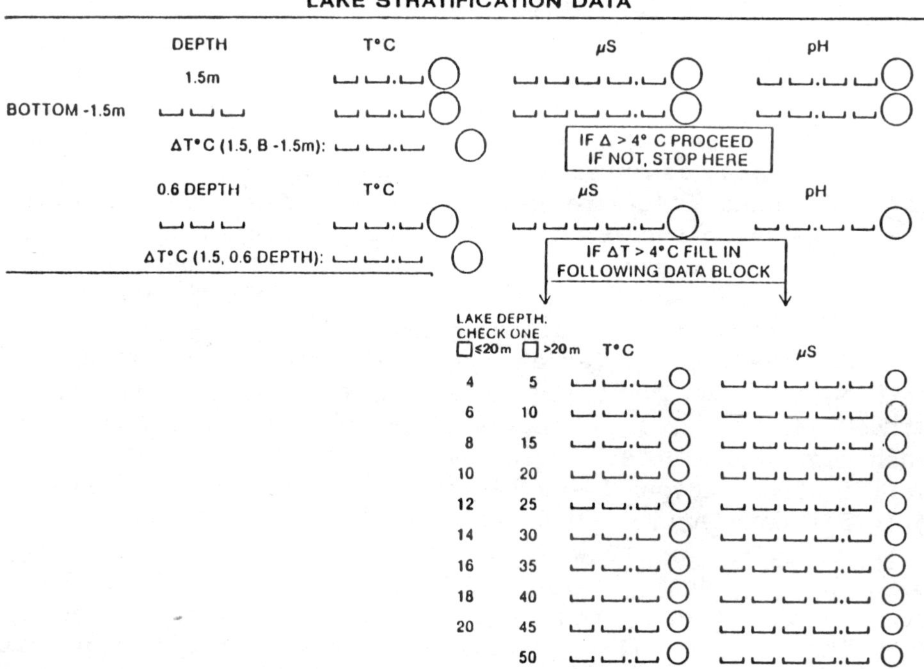

Fig. 3.   Tags. Each quantitative variable in the survey can be qualified using data tags that are predefined or user-defined single alphabetic characters that, in this case, are placed in the circles next to the variable being modified.

Calibration values (taken before and after sample measurements, recorded on the data sheets, and incorporated into the data set) also can be thought of as data qualifiers because they can be used to interpret or modify the actual data values observed. Every quantifiable data value in the Lake Survey database can have a qualifier, and although this effectively doubles the size of the database, the benefit of having complete documentation of every data value is well worth the programming and storage costs.

DATA SCANNING

Data forms, replete with data values and tags, are entered directly into SAS data sets. There are three QA benefits to direct data entry:
1. Data screens are designed to look almost exactly like the data forms. Variable names are displayed and maximum lengths defined, helping reduce keyboard errors (Fig. 4).
2. Data are computer-scanned as they are being entered for correct type and range. A pH value of 74 or 7A instead of 7.4 will be immediately rejected and data entry halted until an acceptable value is entered into the data set.
3. When data are entered onto the screen, they are automatically a part of a SAS-defined data set. Errors incorporated into data sets when they are appended from card-image files into database systems are automatically avoided.

For the Lake Survey database, all data forms are double entered into two separate data sets by two different operators. Double data entry was standard practice in the past when data were entered via keypunch machines and separate verification keypunch machines and operators checked the original data entry. With data entry via terminals, double data entry is not always practiced. For the Lake Survey database, a custom program (COMPARE) was developed in SAS to compare the two data sets and identify any inconsistencies.

The use of double data entry to reduce typographical errors is easily demonstrated. If a keyboard operator has an error rate of 1 in 200 and the data are entered only once, the overall database error rate will be 1 in 200 or 0.005. If two operators are used and the two resulting data sets are compared, all errors that are not identical (same error in the same position) will be detected and corrected.

```
                         FSBROWSE Screen                     : SCREEN  2
COMMAND ===>                                                 :----------
                                                             :
------------------------------------------------------------------------
                      FSBROWSE Screen Modification           : SCREEN
   Secchi Depth Disappear: ____ (tag): ______


********************** LAKE STRATIFICATION DATA **************************

       DEPTH            TEMPERATURE        CONDUCTIVITY           pH

TP: ____ tag: ____ T: ____ tag: ____ C: ______ tag: ____ pH: _____ tag: ____
BT: ____ tag: ____ T: ____ tag: ____ C: ______ tag: ____ pH: _____ tag: ____

Temperature difference (top-bottom): _____   (tag): ______

0.6 DEPTH

 D: ____ tag: ____ T: ____ tag: ____ C: ______ tag: ____ pH: _____ tag: ____

Temperature difference (Top-0.6*Depth): _____   (tag): ______
```

Fig. 4.   Input screen design. The input screens developed
           for the database are designed to mimic the actual
           field and laboratory forms as much as possible.
           Variable names and permissible variable lengths
           visible on the screen help reduce keyboarding
           errors.

The overall database error rate will be the probability that
two operators will make the same error in the same position,
and can be calculated

$$error\ rate = key1 \times key2 \times prob\ same$$

where error rate is the probability of two operators making
the same error in the same position (and therefore being
undetectable by the comparison program). This is the overall database typographical error rate. Key1 is the probability that any given keystroke will be an error for the
first operator. Key2 is the probability that any given keystroke will be an error for the second operator (both rates
assumed to be 1 in 200 or 0.005). Prob same is the probability, given both operators have made a mistake in the same
position, that both have made the same keystroke error
(assumed conservatively as 1 in 4 or 0.25). Under our
assumptions

$$error\ rate = 0.005 \times 0.005 \times 0.25$$
$$= 0.000625$$
$$or\ 1\ in\ 160,000$$

Single keyboarding of data yields an overall error rate of 1
in 200. Double keyboarding yields an overall error rate of

1 in 160,000. Doubling the keyboarding effort is rewarded with an 800-fold decrease in typographical error rate.

DATA FLAGS

After each package of data forms is keyboarded and checked for errors, each incremental data set is appended to the growing database. This is accomplished with an 'ADDER' program, which appends the new data, scans the data, and sets flags for those that are suspect for some reason. This scanning often uses narrower acceptable value ranges than the input screening program or uses combinations of values in evaluating data.

For example, the data entry screen would not allow a pH value of 24 to be entered, but would allow a value of 11.5 to be input. The scanning procedure would flag this value as being a legitimate, but suspect, pH value for a natural lake. As another example, a pH value of 4.9 would be accepted as reasonable for a lake system in this study, but would be flagged if the same lake had a high alkalinity measurement (e.g., over 500 µeq), because this is suspicious based on lake water chemistry. As a last example of flagging, all the values of a given lake might look reasonable in isolation, or in simple pairs, but the overall anion/cation balance might be askew. In this case, all the values used in the balance calculation would be flagged. Thus, our data set now contains, for each quantifiable variable, its value, a tag, and a flag field. Both tags and flags can be multiple characters.

The value of this scanning and flagging process is substantial. Many, if not all, of the errors so detected will have their origin external to the database management center. Their identification and rectification depend on the ingenuity of the research database manager and the researcher in devising the appropriate scanning algorithms.

DATA VERIFICATION

Data verification is the logical next step after data flagging. Verification reports are created that summarize all the data and highlight the flagged data as needing special consideration. These reports are sent to appropriate reviewers, especially those who have information that allows decisions to be made to accept, modify, additionally flag, or reject these values. In the Lake Survey, many of these determinations are made by the analytical laboratory QA personnel who reviewed the laboratory QA blank and audit sample data which were entered into the database at ORNL from laboratory forms.

DATA VALIDATION

Validation is an extension of verification, but validation data are scanned from a larger perspective. In the case of the Lake Survey, our validation process is concerned with reviewing all the lakes contained within a region or subregion. Values that appear reasonable in isolation or when reviewed against other values for the same lake may be distinct outliers when compared with values from similar lakes in the region, based on elevation, size, watershed characteristics, etc. The validation process looks at lake water chemistry values on a regional basis and flags outliers or suspect values for review. For example, we calculated the distance from each lake in the upper northeast to the nearest ocean shoreline and plotted this against measured chloride levels in each lake (Fig. 5). One would expect the concentration of chloride to drop off exponentially with distance from the ocean, and it did with few exceptions. These outliers were identified and other measurements checked (e.g., anion/cation balances) to determine if, perhaps, road salt contamination was the cause of the aberration. Validation is the last step of quality control. When data entry, verification, and validation are completed, the database is ready for analysis.

MISCELLANEOUS

The previously mentioned measures describe the formal QA procedures used in the database management of the Lake Survey. Numerous ad hoc QA measures were developed as the survey progressed. These measures were not anticipated, but were recognized as useful. For example, lakes were selected using a set of criteria and were marked on United States Geological Survey topographic maps. Their characteristics, along with their latitude and longitude, were transcribed and entered into our database. These latitudes and longitudes were to be entered into the helicopter LORAN navigation systems and had to be accurate. Simple computer plotting of the selected lake locations on regional maps showed a few that were obviously in error (located in nonsampled states or in the Atlantic). However, any miscoded locations

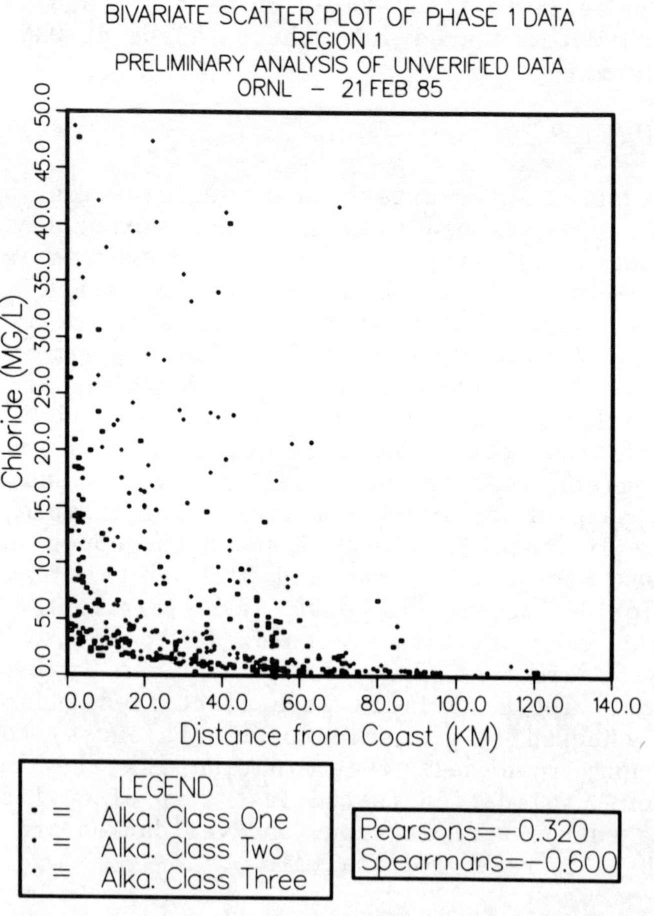

Fig. 5.   Plotting chloride ion concentrations against distance from the ocean isolates those lakes that
show abnormally high chloride concentrations.
Further investigation can determine if, for example, these lakes received road salt contamination.
This is an example of regional-based data validation.

that were not grossly in error could not be identified this
way. Our solution was to plot these lake locations on
transparency paper using the exact projection and scale of
the original topographic maps. They became overlays for the
original maps used for lake selection. Any lake location on
our transparency that was not on the original map (or vice
versa) was an error in the database. This proved to be an
easy and useful method for checking lake locations.

CONCLUSIONS

The methods of QA described in this paper are not, in themselves, unusual. They are based on sound data management practices and common sense. What is, perhaps, unusual is the importance placed on QA for this research database and the totality of the approach. Quality assurance, unfortunately, is often a neglected aspect of research database management. Often funds and time allocated to database management are tacked on to research projects as an afterthought and are inadequate to properly manage the data. The temptation is to concentrate on the collection and analysis of the data, and the production of results and conclusions. However, the effort and expense invested in QA measures before analysis will repay the researcher many times over.

ACKNOWLEDGMENTS

This research is sponsored by the US Environmental Protection Agency under Interagency Agreement No. 40-1441-84 with the US Department of Energy under Contract No. DE-AC05-840R21400 with Martin Marietta Energy Systems, Inc. Publication No. 250, ESD/ORNL. This research has been funded as part of the National Acid Precipitation Assessment program of the US Environmental Protection Agency. The research described in this report has not been subjected to EPA's or NAPAP's required peer and policy review and therefore does not necessarily reflect the views of these organizations and no official endorsement should be inferred.

LITERATURE CITED

Yurewicz, M.C. (In press). Quality assurance of WATSTORE water-quality filed data. pp. 27-29. In: Proceedings of the 1983 Integrated Data Users Workshop. R.J. Olson and N.T. Millemann (eds.). CONF-831117. National Technical Information Services, Springfield, Virginia.

AN INFORMATION MANAGEMENT SYSTEM
FOR ECOLOGICAL IMPACT ASSESSMENT

Craig Brandt, Charles Comiskey, and Terrell Farmer

ABSTRACT

This paper describes a Scientific Information Management and Analysis System (SIMAS) for ecological impact assessment programs. The SIMAS is part of an overall strategy which integrates information management, sampling design, data analysis, and ecological modeling into an efficient method for quantitatively addressing scientific questions related to environmental pollution. The central component of the SIMAS is the Project Database (PDB), which is an integrated repository for all information relevant to the study. Associated with the PDB is a library of programs for entering, retrieving, analyzing, and displaying the information in the PDB. The role of the SIMAS in conducting an ecological impact assessment is also discussed.

INTRODUCTION

Ecological impact assessment programs require planning and coordination of a wide variety of scientific activities in related, but distinct, technical disciplines. During the last several years, Science Applications International Corporation (SAIC) has developed a flexible and cost-effective approach to designing, organizing, and managing such programs. A primary objective in our approach has been the flexibility to modify the design of the study during the course of the program. Such modifications may result from new insights into the ecosystem under study or the discovery of additional factors affecting the ecosystem, which were not considered in the original design.

In order to achieve this objective, an information management system is required that permits the rapid computerization, retrieval, and cross-referencing of the data collected throughout the program. We have designed the Scientific Information Management and Analysis System (SIMAS) to fulfill the information management needs of all program participants. This paper describes the SIMAS and its role in conducting an impact assessment program. Although the examples are drawn from our work in marine impact assessment, the SIMAS is also applicable to other types of ecological investigations.

ORGANIZATION OF THE SIMAS

Figure 1 illustrates the organization of the SIMAS. Central to the SIMAS is the Project Database (PDB) which is an integrated repository for both scientific data and management-related information. Linked to the PDB are five application modules for information entry, retrieval, analysis, display, and program management. The SIMAS uses the Statistical Analysis System (SAS) (SAS, 1982) for most of the data management, analysis, and display activities. SAS is a sophisticated software system capable of managing and analyzing large volumes of data. It contains extensive and powerful tools for editing, subsetting, concatenating, merging, and updating data sets, together with a comprehensive set of statistical procedures. SAS output capabilities permit display of information in both tabular and graphical form.

Project Database

The PDB is based on the relational data model (Codd, 1970) in which each file is represented as a two-dimensional table. Columns in the table denote fields and the rows are specific values of the fields. Different tables can be joined or merged on the basis of common fields. The PDB includes the five basic types of files shown in Figure 1; dashed lines represent the linkages available between the file types.

The inventory file describes where and when each sample was collected and is the central directory for all historical and project-generated samples. Each entry in this file corresponds to a specific data type from a particular sample and typically consists of a sample identification code, a data type code, the collection date and time, and the location of the sampling station.

The data files contain the project-generated and historical data. Different types of data such as physical, geological, chemical, and biological are stored in separate files. The basic unit of organization within a data file is the sample. Each sample is identified by a unique code that is used to cross-reference the data files with each other and with the inventory file.

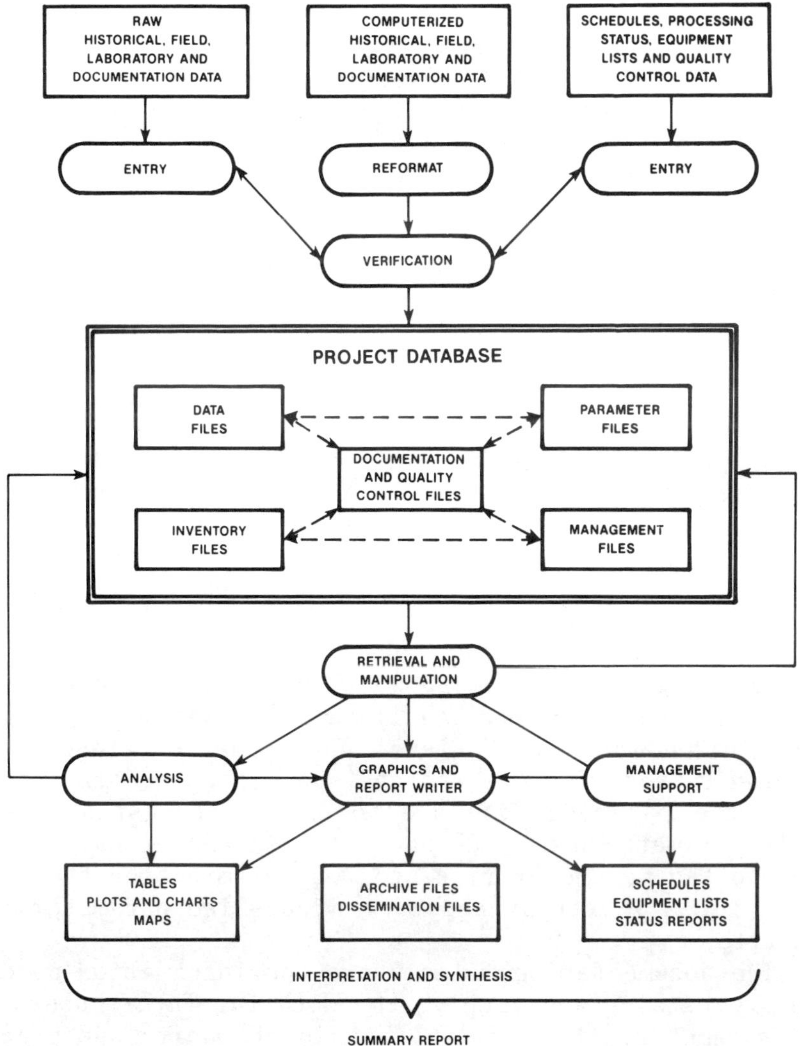

Fig. 1. Organization of the Scientific Information Management and Analysis System (SIMAS).

There are two basic forms of data storage in the PDB. For data with a limited and fixed number of parameters, each sample is stored as a single record with the requisite number of fields needed to store all of the sample measurements. For example, a sediment grain-size sample consists of the sample identifier and the percent of sediment in various size classes, and each record in the sediment grain-size data file contains the data from a single sample. In contrast, taxonomic identification and abundance data are characterized by a large number of sporadically occurring taxa. The taxa in any one sample often represent less than 10 percent of the total number of organisms collected in the survey. The problem is especially severe if the total number of taxa collected is high and the commonality of each taxon is low. If the approach used for fixed parameter data was applied to taxonomic measurements, each record in the taxonomic data file would include a field for every taxon collected in the survey, and many of these fields would contain zeros. In the PDB, such data are stored as pairs of parameter codes and measurements. A unique parameter code identifies each taxon and only the non-zero measurements are stored in the data file. This results in a considerable savings of space compared to the storage format used for fixed parameter data.

A parameter file is used to identify the parameter codes used in the data files. An entry in a parameter file includes the parameter code, the parameter name, and other descriptive information. For example, a taxonomic parameter file for a marine study contains the parameter code that identifies the organism, the scientific name of the organism, the National Oceanographic Data Center hierarchical taxonomic code (NODC, 1984a), the literature reference describing the organism, and linkages permitting aggregation of the taxon into higher level phylogenetic categories. In addition to reducing the storage space required for the data files, the parameter files also ensure consistent spelling of the parameter name. Changes in a parameter name are confined to the parameter file, thereby eliminating the need to search for all occurrences of a misspelled parameter in the data file.

The documentation files store the information needed to properly catalog and archive the data for future users. For each parameter, these files contain the units and precision of the measurements, formats, missing value codes, range of values, and source of the data. Equipment descriptions and summaries of methods are also stored in the documentation files. Instrument calibration data, error summaries, and measurement audits are contained in the quality control files. This information is useful to primary and secondary users of the data in assessing the quality of the information in the PDB.

Information Entry and Verification

A primary application of the SIMAS is the timely and accurate incorporation of information into the PDB. This information may be computerized or non-computerized and includes project-generated data, data from past and ongoing projects in the same area, taxa lists, sample schedules, and data documentation. Figure 2 is a flowchart of the data entry, verification, and error-correction procedures used in the SIMAS.

The bulk of the information entered into the PDB originates as handwritten data sheets, including field and laboratory data, sample manifests, cruise schedules, sample inventories, and some essential historical information. To expedite computerization, handwritten data are recorded on forms which are designed to facilitate data entry. These forms are also used to record comments and other types of qualitative information for entry into the documentation files of the PDB.

Handwritten data are computerized using the EasyEntry (Applied Information Systems, Inc., 1982) software package. EasyEntry is a key-to-disk entry system employing an interactive computer program to display a blank form on the CRT terminal. As each part of the form is completed, the computer prompts for the next item of information. When the form is completed, the information is written to a computer file and a blank form is again displayed on the terminal. If necessary, a series of different forms can be linked together for data that span several types of forms. There are several advantages in using an interactive, full-screen program for data entry. First, since the program formats the data, errors due to misaligned numbers are virtually eliminated. Second, the program can automatically duplicate the contents of fields which are identical across several records. Third, the program checks the range and internal consistency of the data entered and prompts for verification of suspect values.

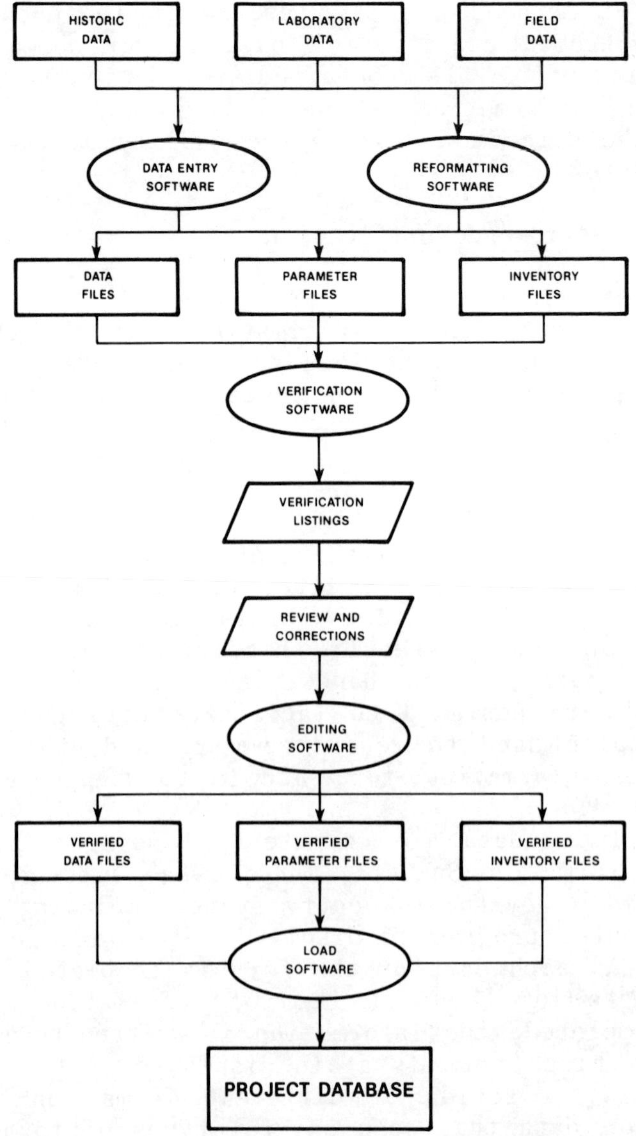

Fig. 2. Flowchart of data entry, verification, and error correction procedures.

Computerized data require a different set of procedures in order to be properly integrated into the PDB. The first step is to make a reconnaissance pass through the data set to verify the file structure and generate tables that show the spatial and temporal extent of the data. The tables are

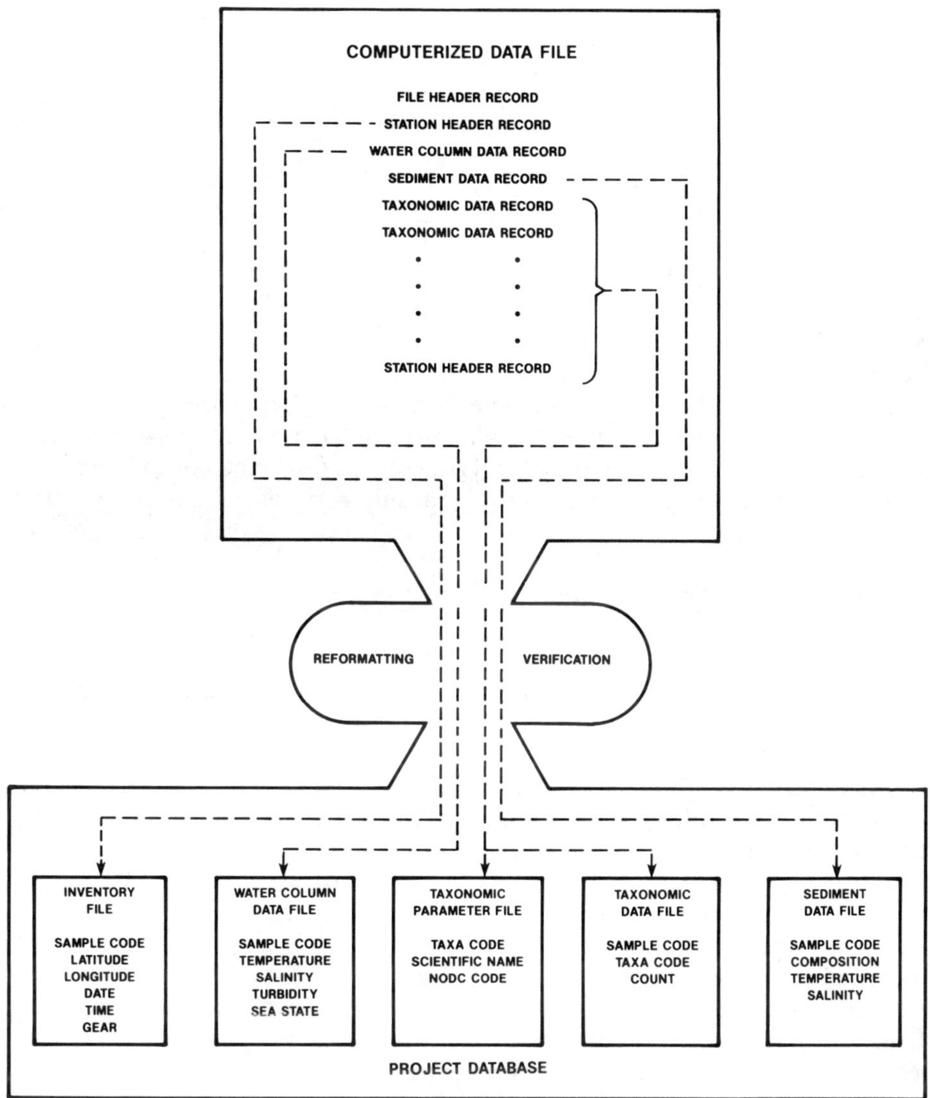

Fig. 3.   Incorporation of computerized historical data into
the Project Database.

compared with laboratory reports to ensure that the correct
number and types of samples are present. The tables also
help identify problems in the sample identifiers, such as
erroneous dates and station codes, and eventually form the
basis of cross-referencing files that link the original
sample identifiers with those utilized in the PDB. The
final step is reformatting the computerized data and loading

it into the PDB (Fig. 3). Reformatting involves separating
the data into inventory, data, and parameter files, and
accessing the cross-reference tables to determine the new
sample identifiers and variable names.

After entry or reformatting, the data are subjected to
a series of quality control procedures designed to minimize
errors. Verification listings are generated and compared
with the original data sheets or laboratory reports. Lists
of sample identifiers and parameter names are generated and
reviewed for accuracy and completeness. Any errors identified during the quality control check are recorded on a data
correction form. The data are corrected using the EasyEntry
data entry software described above. In this case, each
record requiring modification is displayed on the terminal
screen with the appropriate form. After the operator completes the corrections, the updated record is inserted back
into the data file.

Information Retrieval and Mainpulation

Once in the PDB, the data retrieval and manipulation
module provides the tools for generating necessary subsets
of information. Queries to the PDB can be either interactive or by batch program. The interactive mode is generally used for ad hoc and exploratory queries. In this mode,
a user can modify the search criteria during the query.
Query via batch program is generally used when large and
well-defined subsets of the PDB are being generated.

Three basic types of retrieval are available in the
SIMAS: spatial/temporal, parameter, and observational. In
a spatial-temporal retrieval, a subset of the inventory file
is created which contains entries matching a set of date,
time, and location criteria. The inventory subset is then
merged with the appropriate data files to generate a data
subset. In parameter retrievals, the user creates a data
subset based on the presence of certain parameters (from the
appropriate parameter file) and merges this subset with the
data file to create the data subset. Parameter subsetting
is particularly important with taxonomic data where a user
may wish to conduct an analysis on a particular assemblage
of organisms. The third type of retrieval is used to create
data subsets based on the values of specific parameters.
For example, a user may wish to select all bottom sediment
samples containing greater than 20 percent coarse material.
The retrieval software allows varying levels of generality
in the search criteria. For example, a user may want to

retrieve all data collected at one location without regard to time. Alternatively, the user may wish to retrieve the data collected at a specific station on a specific date.

Complex queries of the PDB are possible by combining several simple queries with the logical operators AND, OR, and NOT. Another common query is to create a subset of one data file and then cross-reference this subset with another data file to create a merged data subset. For instance, a user may first create a subset of the sediment grain-size file which contains those samples with greater than 10 percent clay. This subset can then be merged with the taxonomic data to create a subset consisting of taxonomic abundance data at stations characterized by fine-textured substrate.

The SIMAS also permits the user to perform mathematical manipulations of information within the PDB. Standard mathematical operators and statistical functions are available as SAS functions. New variables which are combinations of existing variables can also be easily created. Mathematical and statistical manipulation of the PDB is an important capability in dealing with biological data, which often require summarization or transformation prior to analysis.

Management Support

The SIMAS is designed to aid program administrators as well as researchers. The management support module provides program administrators with the information required to manage and monitor the study from initial planning to final report preparation. This module aids the administrator in several ways. First, the module helps in planning and scheduling the field collections and laboratory analyses. Lists of samples to be collected are entered into the PDB, and then used to generate sampling and processing schedules. Second, the module provides an accurate accounting of the location and processing time of each sample. With this information, management can monitor laboratory sample turnaround and identify sample and data processing bottlenecks before they become critical. Finally, the module contains background and logistic information useful in planning complex field collections. In a marine program, cruise plans, sampling schedules, and shipboard duty rosters are stored in this module.

Data Analysis

Analysis of data from impact assessment programs requires sophisticated statistical and mathematical software. Any of the PDB files or subsets generated from the PDB can be analyzed with the SAS statistical procedures. In addition, many of the SAS statistical procedures allow the user to generate SAS files containing the analysis results, which can easily be added to the PDB as new data files.

However, SAS does not include all of the data analysis tools necessary in impact assessment. For example, the Cornell Ecology Programs (Gauch, 1982, 1984) and the Numerical Taxonomy System (Rohlf et al., 1974) contain several multivariate techniques that have been developed exclusively for environmental analysis. This software can be used, but it requires that the data first be output from the PDB into a format that can be processed by the software. Several SAS programs are available in the SIMAS for outputting the data from the PDB.

Graphics and Report Writing

The report writer and graphics module is used to display information in the PDB, present analysis results, generate input files for specialized analysis programs, and create files for dissemination to other sites. SAS contains several procedures for displaying data in both tabular and graphic form. A user can also create customized tables and plots within SAS to display information in unusual formats.

ROLE OF THE SIMAS IN ECOLOGICAL IMPACT ASSESSMENT

The SIMAS is part of an integrated, three phase approach which we routinely use in conducting ecological impact assessment programs. The SIMAS plays a critical role in our approach since this system serves as a centralized information resource for all phases.

Phase I focuses on the formulation of a conceptual model of the ecosystem under study. This model identifies the important inputs, output, processes, and regulators of the ecosystem, and provides a tool for identifying the direct and indirect effects of human activities on the ecosystem. For example, in a marine dredge material disposal assessment, the conceptual model would describe the physical, biological, and geochemical components of the ecosystem together with the impacts of dredge disposal on the various components of the ecosystem. The conceptual model also provides the framework in which the existing historical information can be organized, data gaps identified, and research priorities established. As new data are acquired, computerized, analyzed, and synthesized, the model is refined and quantified. In the process, new insights into ecosystem structure and functioning are achieved, possibly leading to changes in the program design.

Phase II is a reconnaissance sampling effort in which spatially and temporally extensive, unreplicated data are collected. These data are used to identify and define the components of the ecosystem and to identify the major sources of spatial and temporal variability in the study area. Replicated data at carefully selected locations may also be collected in order to determine the statistical nature of the data. The magnitude of efforts required during Phase II will depend on the adequacy of the conceptual model. If the study area has been well studied in the past and good historical information exists, then the level of effort for Phase II will be reduced. However, if reliable historical data are not available for the study area, then the Phase II effort will need to be increased accordingly. The preliminary analysis of data collected in Phase II identifies specific hypotheses to focus on in the third and final phase.

Building upon the results of Phases I and II, Phase III seeks to test specific hypotheses regarding impacts in the study area. Data collection in Phase III is oriented towards replicated sampling at specific locations. The goal here is to test specific a priori scientific hypotheses concerning possible environmental impacts. This requires replicated sampling to permit hypothesis testing, using analysis of variance to determine if there are any significant differences between control stations and impacted stations. The outcome of Phase III represents the culmination of our approach. At this point, the practical importance of detected significant differences are evaluated in terms of alternative management strategies.

Identification and Acquisition of Historical Data

Historical data constitute an important source of information for formulating the conceptual model and designing the field surveys for Phases II and III. Historical data

can come from a variety of sources including published articles and reports, original data sheets, and archives of computerized data. Identifying relevant historical data can be
a time-consuming and expensive task. Fortunately, there are
several services available which help to simplify this task.
For example, the National Environmental Data Referral
Service (NEDRES, 1984; see also Freeman, this volume) and
the National Oceanographic Data Center (NODC, 1984b) maintain computer-searchable indices which contain information
on sources, availability, and format of various data sets
(also, see Olson & Kanciruk, this volume).

Design of the Project Database

Using the conceptual model as a framework, the study
team designs the sampling plan for Phases II and III. Once
the plan is finalized, the data manager compiles the detailed specifications for the PDB, including the types of
files, the field types, and the field names. The data manager also ensures that the necessary fields are present in
each file to permit cross referencing.

The data manager devises an overall coding system which
uniquely identifies each sample and compiles a master list
of all samples to be collected. The list is then entered
into the SIMAS, forming the basis of the collection scheduling, sample tracking, and data tracking systems. A taxonomic parameter file is also compiled for each biological
data type to be collected. The file contains the parameter
code, the scientific name of the organism, the National
Oceanographic Data Center hierarchical taxonomic code (NODC,
1984a), and other information. A similar list is also prepared for any chemical parameters which will be measured
during the study. These lists are the basis for coding the
taxonomic and chemical data sheets. Finally the data
manager prepares the data recording forms and sample labels
to be used in each field survey and by each laboratory. At
this point, the interactive data entry programs are coded
and tested.

Sample Collection

Before each survey, the participants meet to plan the
field effort and review the methods and equipment to be
used. A draft cruise plan is prepared which describes the
sampling sites and procedures, equipment needs, sample preservation and storage techniques, cruise tracks, and quality

assurance and control requirements (e.g., blanks and dupli-
cates). The SIMAS is used to generate lists of sample identifiers, tentative sample locations, cruise schedules, and
equipment lists for the cruise plan. Following review of
the draft, a final cruise plan is prepared that serves as
the basic work document for the field effort.

In the field, the cruise leader designates one person
in each sampling crew to be the data recorder. The recorder
uses the cruise plan as a checklist of the number and type
of samples required at each sampling station. The recorder
is responsible for ensuring that the appropriate information
is recorded in the field log, and that the samples are properly labeled. For each successful sampling effort, the
recorder notes the time and location of the sample and other
pertinent information such as weather and sea conditions.
If a sampling attempt fails, the reason is noted in the
field log. Before departing a station, the recorder checks
that all samples have been collected. If a different number
of samples is collected than planned, the recorder notes the
reason in the field log. After the cruise, copies of the
field collection logs are forwarded to the data manager, who
enters this information into the SIMAS inventory file.

Laboratory Processing

Upon completion of the survey, the cruise leader is
responsible for ensuring that the samples are transferred to
the appropriate laboratories for processing. A manifest
specifying contents of the shipment is sent with the samples to the laboratory. Upon receipt, the laboratory data
coordinator compares the samples with the manifest to ensure
that all samples are present and correctly identified. Dis-
crepancies between the shipment contents and the manifest
are noted on the form, and the cruise leader is contacted
immediately to resolve the discrepancy. After the transfer
is verified, the laboratory data coordinator signs and dates
the manifest. One copy is retained by the laboratory, one
copy is returned to the cruise leader, and a third copy is
sent to the data manager for entry into the management
files.
When a sample is withdrawn from storage for analysis,
the laboratory technician copies the sample identification
information from the label to the appropriate data sheet.
Transfers of subsamples to secondary laboratories are accompanied by a manifest and the appropriate data sheets. A
duplicate copy of each completed manifest is sent to the

data manger for entry into the management files of the PDB. When the sample processing is completed, the technician initials and dates the data sheet. Completed data sheets are reviewed by a laboratory data coordinator. Periodically, the coordinator photocopies the completed data sheets and forwards the copy and a manifest to the data manager.

For biological samples, a reference (or voucher) collection with identification cards is established. These cards are critical to the data management system, and are the basis for the taxonomic parameter file. For each taxon, the laboratory coordinator assigns a reference number that is entered on the voucher specimen identification card and on the voucher specimen labels in the reference collection. As new organisms are identified in the laboratory, the complete classification (lowest practical identification level) is entered on the identification card. New or unknown species are denoted by alphabetical epithets (e.g., *Mediomastus* sp. A). Such species are usually sent to taxonomic experts for collaborative identification. In this case, two cards are produced: the first card is completed by the laboratory with the lowest level of identification indicated, and the second card is completed by the outside taxonomic expert. Copies of all voucher cards are periodically submitted to the data manager for incorporation into the PDB. The laboratory retains the original cards with the voucher collection.

Incorporation of Information into the Project Database

The data manager is responsible for incorporating information into the PDB. The information may be computerized or handwritten and may come from archives of historical data, field surveys, or laboratories. One of the most demanding tasks in integrating data from numerous sources is standardizing sample identifiers, parameter names, and measurement units. This is especially true for taxonomic data because there may be inconsistencies among laboratories and between studies.

Upon receipt of handwritten field and laboratory data, the data manager reviews the forms for completeness and checks that data are present for all samples on the manifest form. Data sheets requiring the assignment of parameter codes are forwarded to the data coding clerk (e.g., taxonomic data), who consults an alphabetically sorted master parameter list and notes the appropriate parameter code on the data sheets. New parameters that are not in the master list are assigned temporary codes which are entered on the data sheet. A list of the temporary codes is sent to the data manager, who assigns permanent codes and adds these codes to the appropriate PDB parameter file. A cross-reference file is created to convert the temporary codes to the permanent parameter codes.

Once the data are in the PDB, parameter lists are created and sent to experts for review. These lists are especially useful for taxonomic data where misspellings and synonymies are likely to occur. Tables showing the spatial and temporal distribution of samples are also generated and these tables are used to check that the proper number and type of samples are present in the PDB.

Data Analysis and Synthesis

There are a variety of numerical and statistical tools available, which, when collectively used in a structured framework, can aid in reducing the multidimensionality of complex ecological data sets to fewer, more interpretable dimensions. We have incorporated many of these methodologies into a hierarchical analytical approach for the identification and testing of patterns within and between abiotic and biotic data sets (Comiskey & Brandt, 1982). The SAS statistical procedures are used for most data analyses, although additional software has been integrated into the SIMAS.

The stages in the analysis scheme are designed to provide a progression of evaluations, with the results of each stage guiding the activities in subsequent stages. The stages in the analysis system are meant to correspond closely to the phases of the program design. Thus in Phase II of the program emphasis is placed on the exploratory and classification analyses, while in the third phase of the program hypothesis testing is stressed.

It is important to note that each stage of the data analysis system provides information to the next stage and is also a decision point regarding the direction of the next level of analysis. Depending on the phase of the assessment program, the results of data analyses are used either to establish the design for reconnaissance sampling (Phase II) or to test hypotheses regarding impacts (Phase III). Results of data analysis, especially in the reconnaissance stage of an impact assessment, feed back into the formulation of goals and hypotheses and can therefore alter the experimental design. The design of the impact assessment sampling may also be modified for optimum allocation of resources during the course of the impact assessment sampling.

SUMMARY

Although the SIMAS has been designed to fulfill the information management needs of a consulting firm involved in applied research, we feel that the system is also applicable to basic ecological research programs. Both applied and basic research programs share many of the same types of data and needs for processing this data. We hope that this paper provides some insight for other researchers involved in ecological research.

LITERATURE CITED

Applied Information Systems, Inc. 1982. EasyEntry Data Entry System. Applied Information Systems, Inc., Chapel Hill, NC. 75 pp.

Codd, E.F. 1970. A relational model for large shared data banks. Comm. ACM 13(6): 377-387.

Comiskey, C.E. and C.C. Brandt. 1982. Quantitative impact assessment. pp. A1-A80. In: Marine Ecosystem Monitoring. J.D. Allen (ed.). Marine Ecosystem Monitoring Task Group, U.S. Environmental Protection Agency. Washington, DC.

Freeman, R.R. 1985. The National Environmental Data Referal Service: A publicly available on-line data catalog. pp. 143-154. In: Research Data Management in the Ecological Sciences. W. Michener (ed.). Belle W. Baruch Library in Marine Science, No. 16. University of South Carolina Press, Columbia.

Gauch, H.G. 1982. Multivariate Analysis in Community Ecology. Cambridge University Press, NY. 298 pp.

Gauch, H.G. 1984. Catalog of the Cornell Ecology Program Series. Ecology and Systematics. Cornell University, Ithaca, NY. 10 pp.

NEDRES. 1983. NEDRES User's Guide. National Oceanic and Atmospheric Administration, Washington, DC. 115 pp.

NODC. 1984a. NODC Taxonomic Code, 4th Edition. National Oceanographic Data Center, Washington, DC. 554 pp.

NODC. 1984b. NODC Users Guide. National Oceanographic Data Center, Washington, DC. 238 pp.

Olson, R.J. and P. Kanciruk. 1985. Locating machine-readable ecological data. pp. 125-141. In: Research

Data Management in the Ecological Sciences. W. Michener (ed.). Belle W. Baruch Institute in Marine Science, No. 16. University of South Carolina Press, Columbia.

Rohlf, F.J., J. Kishpaugh, and D. Kirk. 1974. NT-SYS, Numerical Taxonomy System of Multivariate Statistical Programs. State University of New York, Stony Brook, NY.

SAS. 1982. SAS User's Guide: Basics. 1982 Edition. SAS Institute, Inc., Cary, NC. 921 pp.

STATISTICAL INTERPOLATION PROCEDURES FOR MAPPING ECOLOGICAL DATA

Steven K. Seilkop

ABSTRACT

Contour and three-dimensional maps are powerful tools for depicting the spatial variability of ecological data. They enhance the researcher's ability to detect spatial trends and patterns, and are being used with increasing frequency due to the availability of graphics software. Most computer programs for producing these maps rely on deterministic interpolation methods, which are often sensitive to outliers and can produce maps which do not accurately characterize the spatial variability of the process under investigation. This paper discusses the advantages of a statistical interpolation method known as "kriging." Among its desirable characteristics is the unique ability to provide estimates of interpolation error. This feature allows the researcher to assess the significance of observed patterns relative to the estimation error associated with the interpolation process. An application of kriging and methods to use its interpolation error estimates are demonstrated with data from EPA's Acid Deposition System database.

INTRODUCTION

The spatial distribution of organisms and the geographic variability of their environment is fundamental to many ecological concepts. Although the ability to characterize and describe spatial processes is critical to both the confirmation and generation of ecological hypotheses, developing such representations from sample data can be a formidable task. Contour and three-dimensional maps are often used by researchers in the analysis of spatial data.

Through interpolation and smoothing techniques, these maps can help to identify systematic spatial trends against a background of random variation.

Recently, the wide availability of computer technology has resulted in a number of "automatic" contouring software packages that offer a broad spectrum of interpolation algorithms (e.g., distance weighting, polynomial interpolation, least squares, spline fits). The user of these programs is often presented with a large number of interpolation methods from which to choose, and very little theoretical support to justify a choice in a particular algorithm. Many of these procedures can result in extremely irregular surfaces that do not reflect the physical reality of the process being analyzed. In addition, surfaces generated by some interpolation routines are extremely sensitive to outlying points. Most importantly, nearly all contouring software is based on deterministic (as opposed to stochastic) approaches that are incapable of providing an indication of the precision of the results.

BACKGROUND

Examples of three mapping techniques commonly used to represent spatial data are shown in Figures 1-3. The first is a simple plot of data values on a geographic background. It has the advantage of explicitly showing the magnitude of data points and portraying spatial variability. However, it requires a good deal of thought and imagination on the user's part to visualize and identify spatial trends in the data. The other two figures improve upon this deficiency with increased pictorial impact, but reduced quantification detail. For example, Figure 2 is a good representation of the spatial distribution of samples, but gives only a rough idea of the magnitude of the data. In Figure 3, a bivariate spline interpolation method was used to estimate non-sampled points in order to produce a smoothed surface of the data shown in Figure 1. The major features of this surface are easily grasped, but it is difficult to obtain much quantitative information about the response variable. Contour maps are often used as an alternative to these graphs due to their ability to depict the magnitude of trends in specific regions of the geographic background. However, like three-dimensional smoothed surfaces, contour maps provide characterizations that are sensitive to the interpolation and smoothing algorithms which create them.

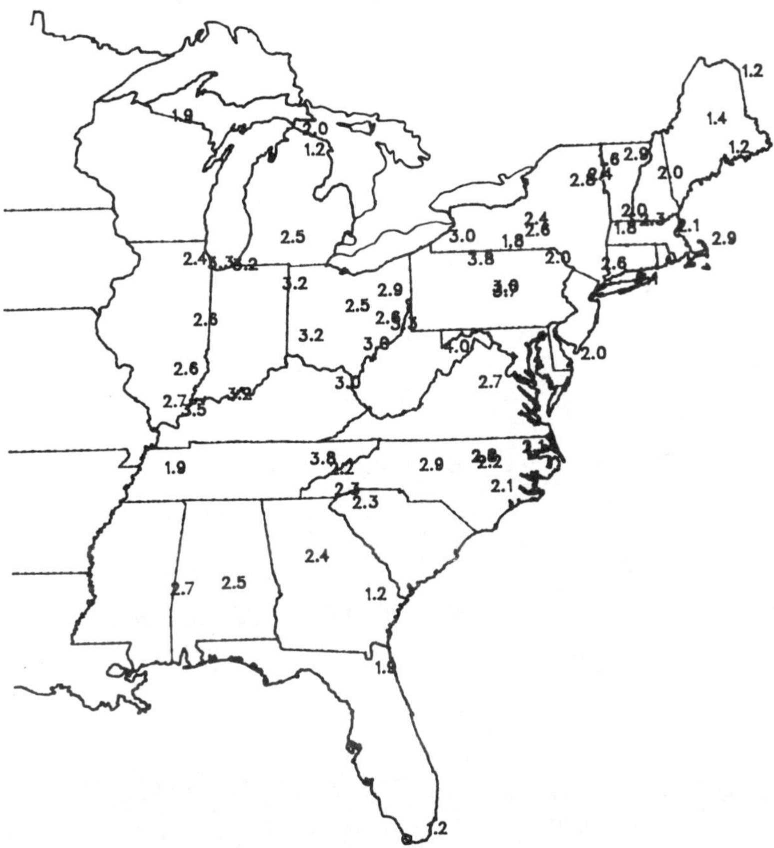

Fig. 1.   Geographic distribution of sulfate deposition (g/m^2) in the Eastern US.

To appreciate the problems that can arise in using software for contouring, consider the two computer-generated contour maps in Figures 4a and b. As a result of changing the smoothing "parameters" used as inputs to the program that generated the plots, one obtains two extremely different portrayals of the data depicted in Figure 1. Looking at the raw data, it is clear that Figure 4a has oversmoothed the surface, and that Figure 4b is a better representation. Moreover, aside from the investigator's intuition, there is no way to assess how well these maps represent the data.

The documentation for most contouring programs gives little or no instruction concerning the alteration of smoothing parameters or interpolation methods; it is up to the user either to accept the default values or experiment

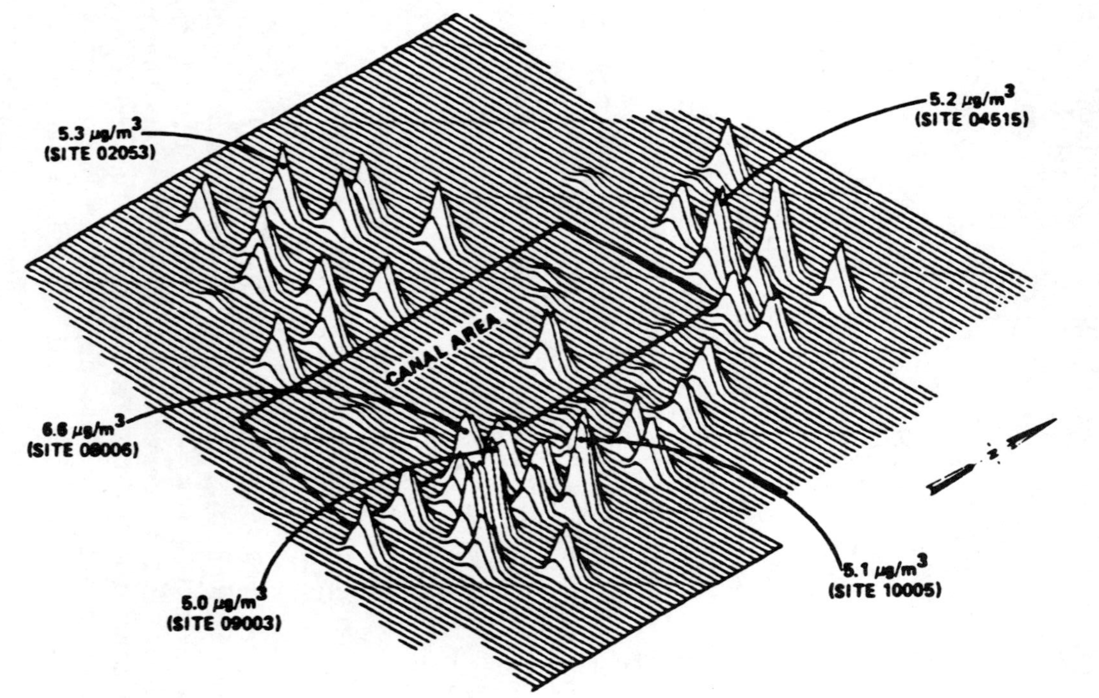

Fig. 2.   Three-dimensional scatter plot depicting concentration of 1,1,2,2-tetra-chloroethylene in living areas of homes at Love Canal (EPA, 1982).

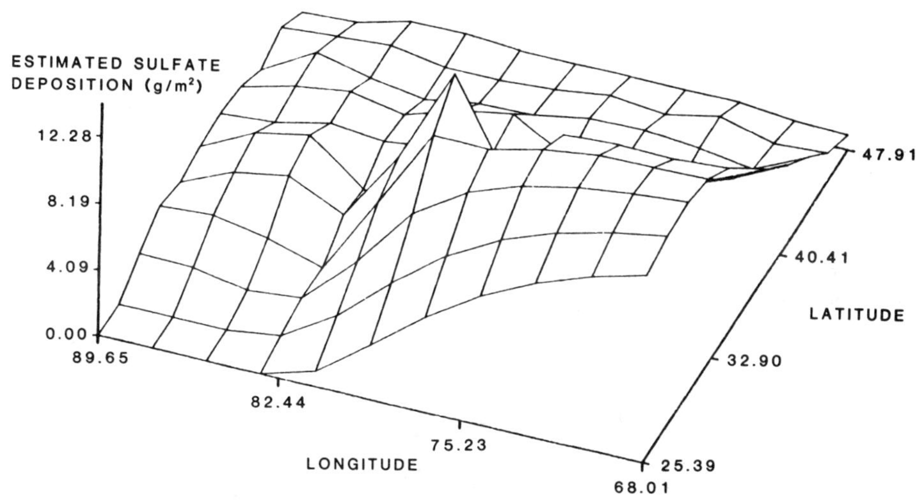

Fig. 3.   Spline interpolation of sulfate deposition data in
          Figure 1.

until a "good" contour map is produced. This procedure for
selecting a contour map is largely subjective and also
fails to utilize the variability in the data to determine
the appropriate interpolation scheme. Most importantly,
this process provides no way to get a feel for the effect of
sampling error on the estimated surface and its associated
contours.

When analyzing and graphing two-dimensional data, scientists have grown comfortable with confidence bands on regression lines and various representations of uncertainty
associated with sample means (standard deviations, standard
errors, confidence limits, etc.). It seems both appropriate
and desirable to expect an analogous measure of uncertainty
associated with interpolated surfaces or contour maps.
Recognizing the inability of deterministic methods to provide such measures, an important contribution to interpolation techniques was made by the Soviet school of hydromete-
orology. Drozdov and Shepelevskii (1946), Kagan (1967), and
Gandin (1970) proposed methods for utilizing spatial variability in interpolation weighting schemes and in estimating
interpolation error. A similar, but more comprehensive,
"geostatistical" approach to these problems was developed by
G. Matheron (1965) in France. Matheron named his estimation

procedure "kriging" in honor of D.H. Krige, a South African mining engineer specializing in ore estimation. This technique has been used most extensively in the mining industry, but has also been applied in a wide variety of earth sciences research (see Barnes, 1980, for descriptions). Only recently has this methodology been utilized in analyzing ecological and environmental data.

Kriging is based on Matheron's theory of regionalized variables (Matheron, 1965, 1973), which was designed to mathematically describe phenomena distributed in space and/or time and provide an appropriate means for solving spatial estimation problems. The theory is ideally suited to characterizations of geographically oriented (regionalized) variables (such as species density and diversity) as well as spatially variable environmental quantities. As with most interpolation methods, the estimation of unknown values in these processes is accomplished using a weighted moving average of data values. In such procedures, a moving "window" brackets sample locations to be used in the interpolation of an unknown value at another location. The window is not necessarily a fixed area, but rather a predetermined number of points to be used in the interpolation. Normally, samples nearest the location of the unknown value are chosen, although directional constraints are sometimes applied. A weighted average is then used to estimate the unknown value. Interpolation procedures differ based on the way that the weights are chosen. The simplest approach is to assign equal weights of 1/n to each point to obtain an average of the surrounding data values. This weighting scheme is not typically used since we expect some distance-related correlation in the data; locations that are separated by small distances are likely to have data values which are more similar than those separated by greater distances. Inverse-squared distance is a popular weighting scheme used to reflect distance-related correlation. As its name implies, weights on interpolating points decrease with the square of their distance from the point to be estimated. The problem, with this and other deterministic weighting procedures, is that the degree of variability in the data dictates their utility. For example, in one data set inverse-squared distance may provide a reasonable map, but for a more variable data set, a simple reciprocal distance weighting scheme may be more appropriate. The investigator must subjectively decide upon the most reasonable scheme.

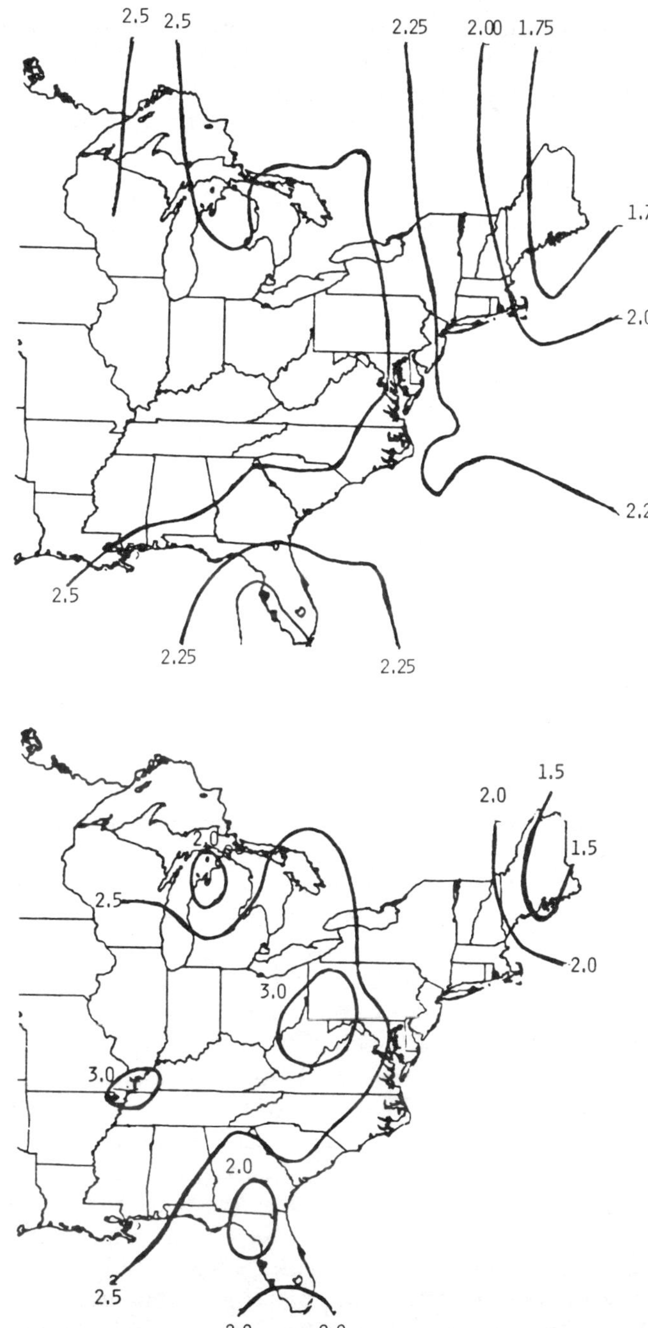

Fig. 4a & b.   Contour maps generated by a computer contouring
package using differing smoothing algorithms.

Kriging departs from traditional deterministic interpolation methods in that the determination of interpolation weights depends on the covariance structure of the regionalized variable. A basic and critical part of kriging is the estimation of the model of this structure, which is known as the semi-variogram. The relative geometry of the data and the estimated point is used in combination with the semi-variogram to obtain optimal interpolation weights. The estimator that results from this procedure has several appealing characteristics. First, it is an exact interpolator; if a known point is interpolated, the exact value of the point will be reproduced. This is not true of all interpolation procedures (e.g., least squares). In statistical terms, the kriging estimator is also the best linear unbiased estimator. This implies that the estimator gives no systematic error (over or underestimation) and has the smallest estimation variance among unbiased estimators that are linear functions of the data. In addition to these features, kriging is unique among interpolation schemes because it provides an estimate of the variance between the true and estimated values. This "kriging error" can be used to place error or confidence bands on the location of contours used to depict spatial patterns. The variance is also extremely useful in selecting the optimal location of a new sampling point or in assessing the performance of sampling designs. For a general discussion of kriging, see Delhomme (1978) and Clark (1979).

In addition to individual point interpolation, kriging can also provide estimates of average values over blocks or fields. This feature can be utilized to represent spatial trends through a partitioning of the geographic area of interest, and subsequent estimation of averages in each of the partitions. This technique has been used recently by Bilonick (1983) in investigating spatial patterns of hydrogen ion concentration in precipitation. As with the kriging technique's point estimates, its block averages have variances associated with them, thus allowing the statistical comparisons of spatial or temporal means. There is some question as to the distribution of the kriging block averages that estimate these means, especially in environmental applications where the underlying data are often not normally distributed. It has been suggested (Rendu, 1979) that often the parent distribution is lognormal, and logarithmic transformation of the data prior to kriging allows the use of standard normal theory in comparison of areal averages.

However, transformation of the data can result in biased estimates in the original scale unless a bias correction factor is applied in transforming back from the logarithmic scale. Unfortunately, an accurate estimate of the correction factor is dependent on an unbiased estimator of kriging variance, which may be difficult to obtain.

Kriging is not a single method, but rather a family of interpolation techniques. "Simple" kriging is dependent upon the assumption that the underlying process is viewed as stationary, with a constant or locally stable mean over the entire spatial or temporal range. However, if the variable contains a nonconstant deterministic component, a trend which is referred to as drift, simple kriging can lead to a biased estimate of variance. Some authors (Journel & Huijbregts, 1978) contend that in general use serious estimation bias will not be encountered if simple kriging is used in situations of moderate drift. It is also argued that since variances estimated from a nonstationary process will be overestimated, these error estimates will lead to conservative statistical comparisons, reducing the chance of making false inferences. However, inferential conservatism is not always desirable, especially in environmental analyses where the volume of data is often small and statistical power is at an undesirably low level.

To redress the nonstationarity problem and to account for drift (trends), "universal" kriging has been developed. The method involves the introduction of polynomial trend terms into the system of equations used to determine optimal interpolation weights. Unfortunately, this technique relies on correctly predefining the polynomial form of drift and using a semi-variogram that does not include the drift component in its estimation. The estimation of drift-free semi-variograms has been addressed (Huijbregts, 1970; Sabourin, 1975), but has not been satisfactorily resolved; it requires complex computing algorithms and equidistant data points. The necessity of having to preselect drift forms is also a definite drawback that has not received much attention in the literature. The most common approach to drift form selection is based on kriging known values, using all combinations of first- and second-order terms. The trend model with the smallest mean square error is selected. Other methods for accounting for drift have been suggested. One approach is to initially use least squares to fit surfaces to the data, and perform simple kriging on the residuals. The approach has been used by Volpi and Gambolati

(1978), who attempted to demonstrate its superiority to universal kriging. Criticism of this methodology centers on the fact that the estimates are biased, although the seriousness of the bias is not clear.

The problems associated with universal kriging have resulted in its present stage of development known as the "generalized convariance" approach. The ideas for this technique are drawn from those used in the analysis of non-stationary time series. The basic concept is to replace the first order differencing used in simple and universal kriging with higher order finite differences. These increments of orders two, three, four, etc., filter polynomial drift of degrees one, two, three, etc. Due to the complexity of calculations, this methodology is totally reliant on sophisticated computer software for its application. The current direction in software design for the technique has been to provide "automatic" identification of the appropriate order of the generalized covariance, and subsequent estimation of the variable of interest. Several programs which employ this philosophy are available; two well known packages are BLUEPACK (Delfiner, 1975) and POLYPAK (Davis & David, 1978). Unfortunately, these programs have turned the generalized covariance approach to kriging into a "black box," which operates under assumptions that the user may neither appreciate nor understand. Although the technique is used widely, little research has been done to assess its estimation characteristics. There is some evidence to suggest that the least squares methods used to estimate the parameters of the generalized covariance function are hampered by the nature of the distribution of the dependent variable and the multicollinearity of the independent variables (Starks & Fang, 1982). A misrepresentation of this function could severely impact estimates of variance generated by the technique, leading to inappropriate inferences. There is also some question concerning the sampling variability of the estimated generalized covariance.

Even with these deficiencies, the generalized covariance approach to kriging represents an estimation technique that can potentially enable scientists to better understand the spatial and temporal dynamics of acid precipitation. There is some question as to whether it is significantly superior to the other forms of kriging; further research is needed to weigh the relative merits of these forms. Much of this research effort will probably be directed towards the cornerstone of all kriging procedures, the estimated covariance structure, regardless of whether that structure is represented by the generalized covariance or the semi-variogram.

EXAMPLE OF KRIGING

The data depicted in Figure 1 will be used to demonstrate the use of kriging and some of its features. They represent total sulfate deposition values (g/m^2) for 1982, which are summarized in the US EPA's Acid Deposition System (ADS) database. This database was developed and is being administered for EPA by the Battelle Northwest Laboratories. It consolidates data from seven nationwide US and Canadian precipitation chemistry monitoring networks.

The goal of the example is to demonstrate the use of kriging in characterizing the spatial distribution of sulfate deposition in the Eastern US. This is essentially an estimation problem in which unknown data values are interpolated from those shown in Figure 1. The surface, which is produced by kriging, and a contour map representing it will be used to identify spatial trends in the example data. An estimate of the semi-variogram for these data will be developed and discussed, and confidence limits on the interpolated surface will be presented. The efficacy of this approach in characterizing the variability of such data was demonstrated by Finkelstein and Seilkop (1981). Other statistical techniques for the spatial/temporal analysis of rainfall data have been suggested (Rodriquez-Iturbe & Mejia, 1974; Lenton & Rodriquez-Iturbe, 1977), but kriging appears to be gaining acceptance among meteorologists. Analysts of acid precipitation data have found the technique particularly useful. Recent investigations using kriging include an analysis of spatial trends in hydrogen ion concentration in New York State (Bilonick, 1983), an evaluation of the performance of kriging in interpolating acid precipitation parameters in the Eastern US (Finkelstein, 1984), and a study of temporal/spatial patterns of pH and rainfall (Eynon & Switzer, 1983).

As indicated above, the basis of kriging is the estimation of the spatial covariance structure. The main assumption used by the technique is that the covariance between two points is a function of only the distance between them. The semi-variogram is an easily estimated representation of the structure, which is convenient for

deriving interpolation weights. For this analysis, simple
kriging will be employed and a semi-variogram will be estimated. It is defined as the difference between the underlying variance of the spatial process and the covariance
between values at given distance. The semi-variogram is
given by

$$G(h) = V - Cov\ (Z_i,\ Z_j)$$

where Z_i and Z are measured at points that are separated by
a distance of h units; V = variance of the regionalized
variable; Cov = covariance; and h = distance.

The semi-variogram is estimated from the sum of the
squared differences of all sample data values. The estimated variogram is given by

$$G_h = \sum_{i>j} \frac{(Z_i - Z_j)^2}{N}$$

where h is now a distance category dictated by the range of
data; and N is the number of points in the distance category.

Distance categories are normally chosen so that an adequate number of pairs of points is available to estimate the
semi-variogram's value. A set of forty pairs is often taken
as a lower limit. In this example, distance categories of
160 km (100 miles) were chosen to fulfill this minimal requirement. Thus the intervals were 0-80 km, 81-240 km,
241-400 km, etc. The calculated semi-variogram values and a
fitted straight line are shown in Figure 5. Note that the
semi-variogram is an increasing function that depicts positive spatial correlation in the data; as distance increases,
the correlation between the data decreases, and the variogram approaches the underlying variance of the spatial process. If there had been no spatial correlation between the
data, calculated semi-variogram values would show no
distance-related trend.

Traditionally, a limited range of functional forms have
been used as semi-variogram models in geostatistical analysis. Linear, spherical, exponential, and gaussion models
are most commonly used. Any function is acceptable provided
that it is positive definite (i.e., it results in positive
estimates of the kriging error). Monotonically increasing
functions of distance satisfy this requirement. Semi-
variograms which asymptote to a constant value (called a
sill) are intuitively pleasing because they represent

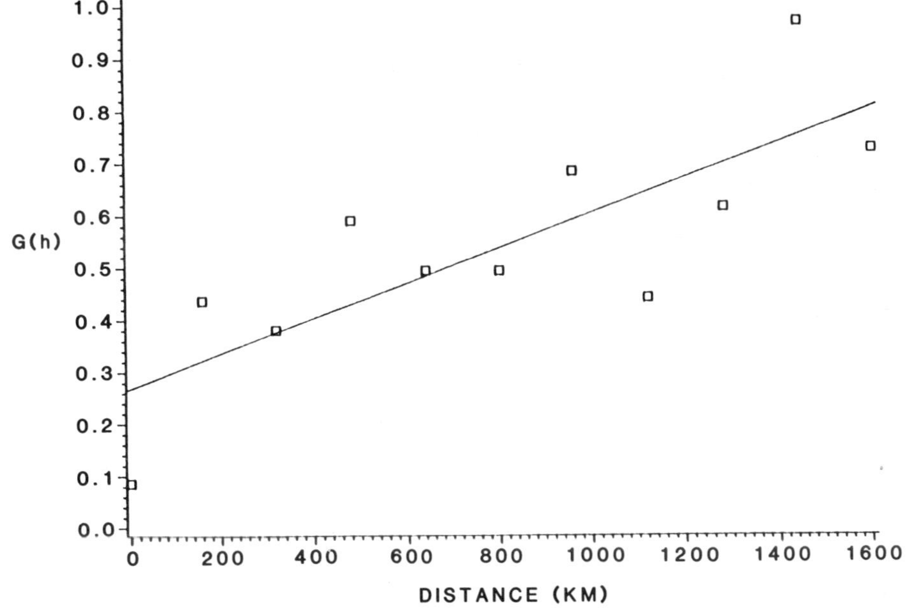

Fig. 5.   Semi-variogram for sulfate depostion in Figure 1.

processes in which the covariance limits to the overall
variance among points. A semi-variogram lacking correla-
tional structure is said to have a "pure nugget" effect.
It is possible that the covariance structure is direction-
ally oriented, a situation known as anisotropy, requiring
that directional semi-variograms be estimated. When
anisotropies are encountered, the estimation process becomes more complex, since the interpolation weights assigned
to data points are functions not only of distance but also
direction. For a general discussion of variograms see
Delhomme (1978).

In Figure 5, the estimated semi-variogram does not pass
through the origin. Given that the semi-variogram represents the difference between the variance and the distance-
related covariance, one might expect it to have value zero
at distance zero. In actuality, at distance zero this is
true. However, points that are separated by infinitesimal
distance have a nonzero semi-variogram. Mathematically one
can interpret this as the semi-variogram's value limiting to
a constant as the distance approaches zero. A nonzero
variogram intercept is usually taken as an indication that
measurement variability is embedded in the spatial process.

With the example data, sulfate deposition, this is an entirely reasonable interpretation since co-located samples (which are actually located a small distance apart) exhibit measurement variation of the same magnitude as that predicted by this variogram (Finkelstein, pers. comm.).

Using a straight line model for the semi-variogram, a set of kriging equations (see Delhomme, 1978, for details) was solved for each of 624 geographic locations in the Eastern US. Values from the eight nearest sample locations were used to interpolate each of these known points to produce the surface in Figures 6a and b.

In this surface one can see evidence of high values in the Ohio Valley (southwestern Illinois, Indiana, Ohio, and eastern Pennsylvania). There is also a clear trend of decreasing deposition to the north, south, and east of this region. In viewing the raw data, this appears to be a plausible representation of the underlying spatial process. Of course, without knowning the true process, it is impossible to judge whether this is a better characterization than that produced by a deterministic contouring package. However, kriging provides theoretically superior interpolation estimates. In addition to providing a data-based weighting scheme, it allows for the estimation of the error in the interpolated surface. In this example, for each interpolated point an error variance was estimated. The surface representing the error variance of interpolated points is given in Figures 7a and b.

These plots have several noteworthy features. First, we see that the largest interpolation variances are on the edge of the plot. Intuitively, one might expect this since interpolated points on the edge of the plot have fewer sample points near them than those in the center of the plot. The greater the distance from the interpolating points to the point being estimated, the less certainty we have in the estimate. The converse of this can be seen in the central portion of the plot. Regions with the heaviest sampling density have the smallest interpolation variances. This is intuitively pleasing since we expect to have the highest degree of estimation precision in these areas. However, in the example data, there is no region in which we are extremely confident of the interpolation estimates. The coefficient of variation (standard deviation divided by the estimated value) ranges between 20 and 50% over the estimation region. With this high degree of interpolation error, the interpolation estimate is expected to be inexact. For

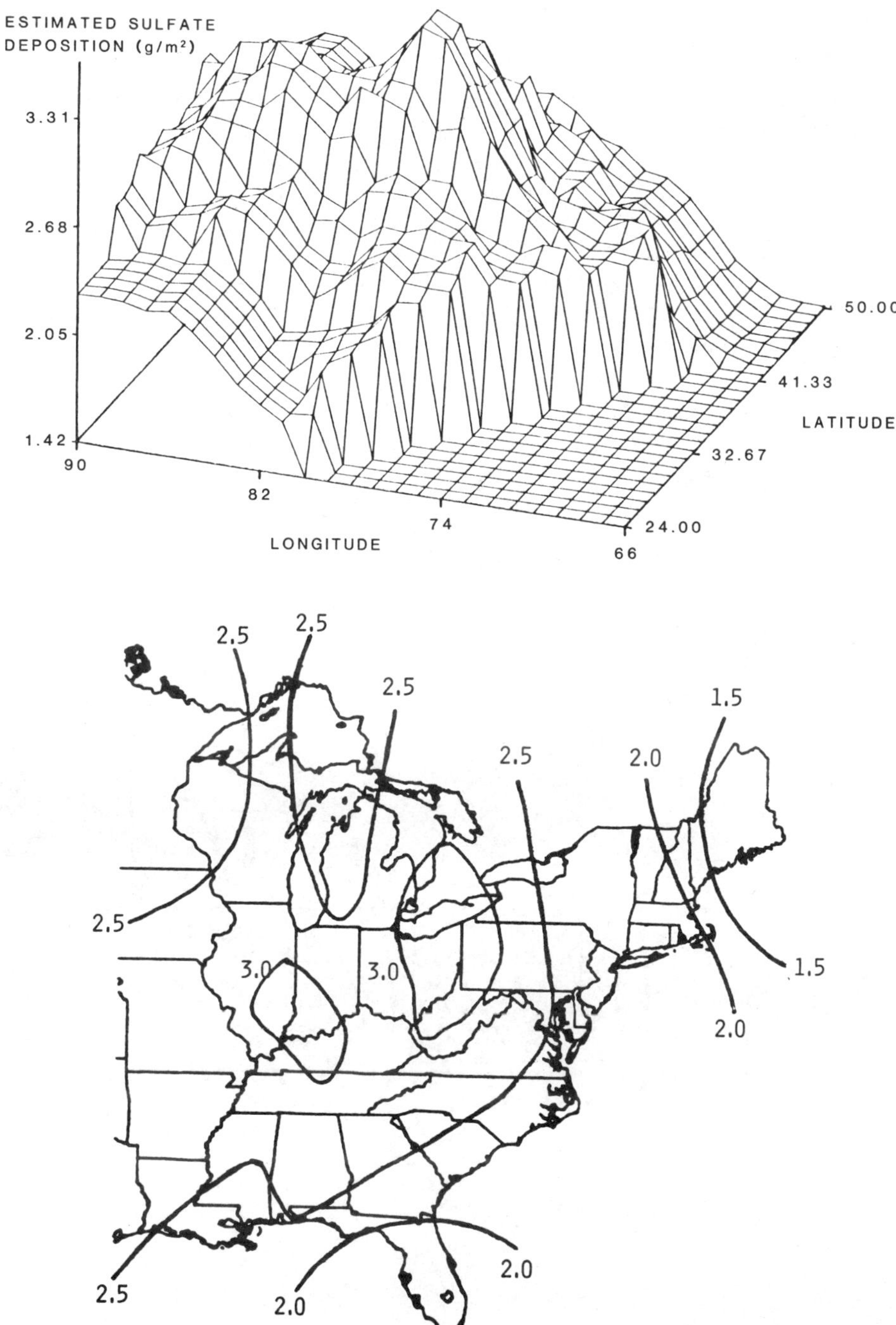

Fig. 6a & b.   Annual sulfate deposition (g/m²) estimated by kriging.

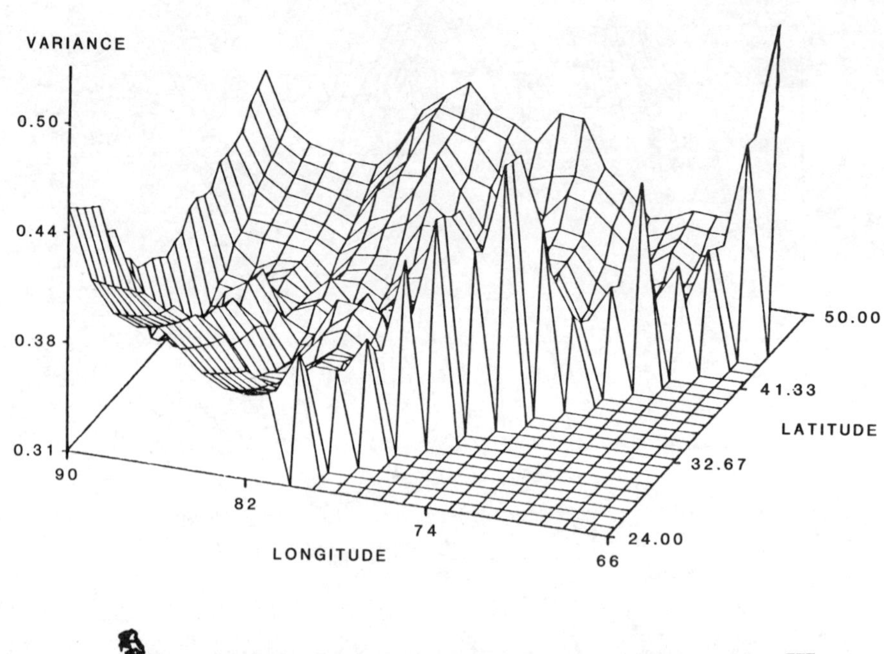

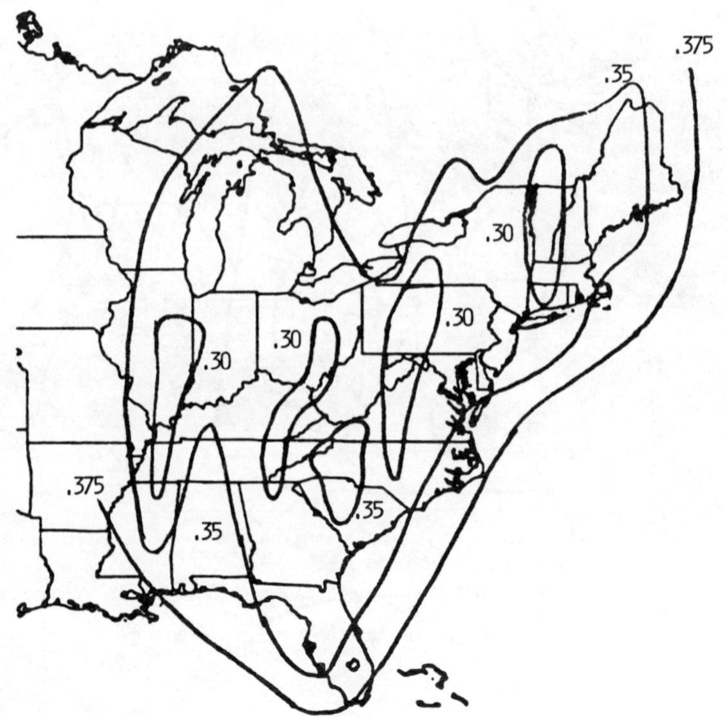

Fig. 7a & b.   Variance of interpolation estimates of sulfate deposition in Figures 6a and b.

example, using standard normal theory, a 95% confidence interval for a point in Central Ohio is approximately

$$2.5 \pm 1.96 \ (.325)^{\frac{1}{2}} = 2.5 \pm 1.12$$

With this amount of error in estimated points, the contours are also imprecisely estimated and could easily change as a function of sampling variation. A different sampling network than that used in the example would be likely to produce a much different set of contours than those observed in Figure 7b. One can again make a normality assumption to produce a set of confidence contours as seen in Figure 8. This plot depicts the upper 80% asymmetric confidence contours for the example data. To get a feel for the effect of sampling variation on the estimation of individual contours, plots like Figure 9 can be used. Here, the 3.0 g/m^2 contour generated by kriging is shown with its upper 80% confidence contour. Clearly, there is a great deal of variability associated with the estimated location of this contour.

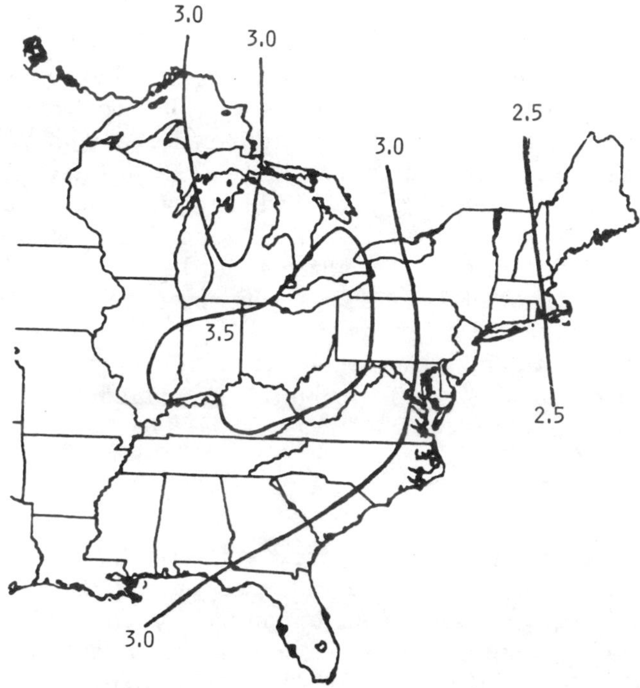

Fig. 8.   Approximate upper 80% asymmetric confidence contours for sulfate deposition data (g/m^2) in Figure 1.

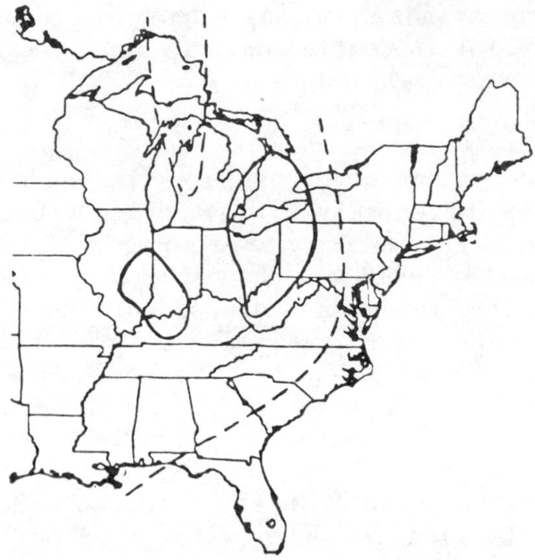

Fig. 9.   3.0 g/m^2 sulfate deposition contour (solid line)
and upper 80% confidence contour (dashed line).

One can also use the confidence contours to obtain further information about the spatial distribution of the data.
The upper contour for a region can be interpreted as that
value which is greater than approximately 80% of the region's data. For example, nearly all of Ohio is within the
3.5 contour on the 80% probability map. One can interpret
this as meaning that 80% of all deposition samples in Ohio
fall below 3.5 g/m^2. In this example, 80% confidence contours were selected to reflect the uncertainty in the interpolated surface in Figures 6a and b. Contours for any such
probability can be generated at the analyst's discretion.
In addition, a group of probability contours can be generated for a specified value.

SUMMARY AND CONCLUSIONS

The example has demonstrated some of the characteristics of the family of statistical interpolation procedures
known as kriging. The maps produced by this methodology can
give us much more insight into spatial processes than characterizations obtained from deterministic contouring
methods. Unlike deterministic approaches, the subjectivity
involved in this methodology relates to the modeling of the
spatial covariance structure of the data rather than the

contouring itself. Due to this statistical basis, interpolation errors can be estimated and the effect of the data's variability on estimated contours can be assessed. Using this information we are far better equipped to analyze spatial data and develop and test spatial hypotheses.

LITERATURE CITED

Barnes, M.G. 1980. The use of kriging in estimating the spatial distribution of radionuclides and other spatial phenomena. TRANSTAT (Statistics for Environmental Studies). #13. Battelle Memorial Institute, Richland, WA. 22 pp.

Bilonick, R. 1983. Risk qualified maps of hydrogen ion concentration for the New York State area for 1966-1978. Atmos. Environ. 17(12): 2513-2524.

Clark, I. 1979. Practical Geostatistics. Applied Science Publishers, London. 129 pp.

Davis, D.W. and M. David. 1978. Automatic kriging and contouring in the presence of trends (Universal kriging made simple). J. Can. Petrol. Technol. January-March. pp. 90-99.

Delfiner, P. 1975. Linear estimation of non-stationary spatial phenomena. pp. 49-68. In: Proc. of NATO ASI Geostat 75. Rome, Italy. Reidel Publishing Corp., Dordrecht, Netherlands.

Delhomme, J.P. 1978. Kriging in the hydrosciences. Adv. Water Res. 1(5): 251-266.

Drozdov, O.A. and A.A. Shepelevskii. 1946. The theory of interpolation in a stochastic field of meteorological elements and its application to meteorological maps and network rationalization problems. Trudy NIV GUGMS, Series I, No. 13. (In Russian).

Eynon, B. and P. Switzer. 1983. The variability of rainfall acidity. Can. J. Stat. 11(1): 11-24.

Finkelstein, P.L. 1984. The spatial analysis of acid precipitation data. J. Clim. Appl. Meteorol. 23(1): 52-62.

Finkelstein, P.O. and S.K. Seilkop. 1981. Interpolation error and the spatial variability of acid precipitation. pp. 206-212. In: Proceedings of the Seventh Conference on Probability and Statistics in Atmospheric Sciences. American Meteorological Society. Boston, MA

Gandin, L.S. 1970. The planning of meteorological station networks. World Meteorological Organization. #265 TP 149. 35 pp.

Huijbregts, C. 1970. Le variogramme des residus. Int. Rep. CMM/N-200. Ecole des Mines, Fountainebleau, France. 22 pp.

Journel, A.G. and C.J. Huijbregts. 1978. Mining geostatistics. Academic Press, NY. 600 pp.

Kagan, R.L. 1967. Some problems relating to the interpretation of rainfall data. Trudy GGO, No. 208, 64-67. (In Russian)

Lenton, R.L. and I. Rodriquez-Iturbe. 1977. Rainfall network systems analysis: The optimal estimation of total aerial storm depth. Water Resour. Res. 13: 825-836.

Matheron, G. 1965. Les variables regionlisees et leur estimation Masson, Paris. 306 pp.

Matheron, G. 1973. The intrinsic random functions and their applications. Adv. Appl. Prob. 5: 439-468.

Rendu, J.M. 1979. Normal and lognormal estimation. Math. Geol. 11(4): 407-423.

Rodruqiez-Iturbe, I. and J.M. Mejia. 1974. The design of rainfall networks in time and space. Water Resour. Res. 10(4): 713-738.

Sabourin, R. 1975. Application of two methods for the interpolation of the underlying variogram. pp. 101-109. In: Proceedings of NATO ASI Geostat 75. Rome, Italy. Reidel Publishing Corp., Dordrecht, Netherlands.

Starks, T.H. and J.H. Fang. 1982. On the estimation of the generalized covariance function. Math. Geol. 14(1): 57-64.

U.S. Environmental Protection Agency. 1982. Environmental monitoring at Love Canal. Volume 1. EPA 600/4-82-030a. 296 pp.

Volpi, G. and G. Gambolati. 1978. On the use of a main trend for the kriging technique in hydrology. Adv. Water Res. 1(12): 345-349.

DATA PROBLEMS IN LARGE-AREA
ECOLOGICAL MODELING

Elgene O. Box

ABSTRACT

Large-area ecological modeling involves geographic and
time-series simulations of ecological processes and structures on continental scale or larger (e.g., global carbon
balance or global hydrologic cycle). Data requirements for
such modeling include coordinated site biological and environmental "hard data" for model development, environmental
data for many other sites (for geographic simulation),
purely cartographic data, and more hard site data plus other
types of data for model validations. Data problems with
such modeling may involve data management but more commonly
concern data availability. Some missing environmental variables can be "reconstructed" if sufficient basic data are
provided. Some gaps in relevant biological data are also
identified. Necessary primary and secondary environmental
variables and carbon-balance variables for terrestrial sites
are suggested. Ecological research sites can standardize
procedures and formats for some basic data without limiting
their own capabilities. This standardization, combined with
improved data quality, documentation, and cooperation, will
result in the development of more geographically useful
databases.

INTRODUCTION

Large-area ecological modeling generally involves simulation, often geographically and/or through time, of ecological processes, structures, or system dynamics over subcon-
tinental or larger areas. Such models attempt to synthesize
many aspects of ecology in order to study the dynamics of

entire systems, including those on a global scale. Geographical relationships are often a major focus of such modeling efforts and require large quantities of various types of data. This is especially true if predictive models are applied geographically to time-series data in order to produce geographic scenarios. Current applications of such large-area modeling include the global hydrologic cycle, the global carbon budget, and other global biosphere-atmosphere interactions (e.g., problems of climate change).

Large-area geographic modeling can be described as involving the following general steps:

1. Collection of coordinated biological and environmental data over the full range of environmental situations possible (e.g., climate and substrate types).

2. Construction of a general model based on the full range of environmental data (curve-fit, system model, etc.).

3. Simulation, using a much larger environmental database, in order to depict the geography of the situation.

4. Quantification and other interpretation of results.

5. Model validation, using independent hard data from many situations.

This approach has been applied to a variety of basic ecological processes including primary production (Lieth, 1975; Box, 1978; Meentemeyer et al., 1982), litter decomposition (Meentemeyer, 1978), phytomass accumulation (Box, 1980), soil pH (Folkoff et al., 1981), soil CO_2 content (Brook et al., 1983), local spatial dynamics of nutrient compounds (Kesner, 1984), and the monthly dynamics of carbon sources and sinks in natural terrestrial vegetation (Box, in prep.; Gillette & Box, in prep.).

This geographic approach, through rigorous, quantitative comparison of data from different situations, seeks insight into environmental control of basic processes. This type of modeling permits synthesis of many factors and processes in order to produce geographic inventories and scenarios. The main limitation is that it must often be rather superficial, attempting to reconcile different types of data and subsystem models at the level of their common denominators. The problems of both data management and model construction in such studies are quite different from those associated with site, taxon, or region-oriented projects.

TYPES OF DATA

The following types of data are usually required for such geoecological modeling:

1. "Hard data:" coordinated measurements of site-specific biological phenomena and expected environmental determinants (e.g., phytomass and production amounts plus site-specific climate and substrate data).

2. Simulation data: climatic and other environmental data for generating the predictive results in different situations (e.g., geographic information systems for geographic simulations).

3. Cartographic data: cartographic and related data for displaying results over the topography of the study region (e.g., computerized base maps and overlays).

4. Validation data: independent data of various types (e.g., other site-specific "hard data," maps, satellite photographs).

The hard data used for both model development and validation must include enough sites to represent the full range of variation and all main situations found in the study area (e.g., all major world climate types). It is also useful to have maps, photographs, ground truth, and other different types of validation data in order to check the spatial resolution as well as accuracy of model predictions. Simulation and cartographic databases are usually much larger than site-specific hard databases. Simulation data may involve a single set of situations (e.g., long-term mean climatic data) or may involve voluminous individual-year, monthly, daily or other time-series data for generating scenarios.

The types of data needed for particular modeling methodologies are summarized in Table 1. One must keep in mind that the large volumes of numerical and spatial "results" from one model may become "data" for other projects (e.g., mapping and map overlaying, input to other models, additions to geographic information systems). An especially important example involves the climatic water-budget results that provide the estimates of actual evapotranspiration, soil water storage, and seasonal water surplus and deficit needed to drive various other models. The problem of error magnification in combined models is of great concern but is not, strictly speaking, a data-management problem.

Databases associated with large-area geographic models

are different from many other ecological databases in the
following ways:

1. The spatial context (relative as well as absolute location, scale and degree of spatial resolution, site density and evenness in relation to topography, etc.).

2. The large number of sites needed for geographic coverage (over a thousand as a bare minimum for representing even the most general world patterns).

3. The necessity of geographic completeness, $i.e.$, that primary databases ("hard data") represent all major environmental situations throughout the study area.

4. The necessity of data coordination, $i.e.$, that hard data for both the target phenomena ($e.g.$, biological processes) and possible environmental determinants represent the same measurement location, measurement period, etc. and be consistent (same units, comparable resolution and measurement period, etc.) with data of the same type for other sites.

5. The necessity of data compatibility, $i.e.$, that the ecological data be compatible with the geographic data structures ($e.g.$, computerized base maps and information systems) used to display the data and model results.

Data compatibility can be especially frustrating since the large volumes of data involved may dictate some of the computer processing methods. These methods may involve technical limitations or other trade-offs.

DATA MANIPULATION

Generally speaking, data manipulation is not a major problem. Nevertheless, there are some data-manipulation problems associated particularly with geographical modeling.

One common problem in geographic modeling involves the adequate, but economic, representation of the outlines and heterogeneity ($e.g.$, topography) of the study region. The size and topographic complexity of the study region greatly determine the spatial resolution attainable and the number of outline and data sites required. As a result, a database that will be used at different scales may have to be stratified in order to activate more sites when needed. Small

Table 1.   Types of databases needed for different types of
models. "Hard data" (i.e., measurements) and
validation data (not used in model development)
are required for all good models. In addition, a
generally much larger number of data sites or
cases is required when models are applied to
describe other situations, with or without mapping
or other explicit geographic representation.
Time-series data represent replicates of the entire suite of data for different times during some
simulation period. Application of a model to such
time-series of data results in a scenario (also a
prediction). Geographic databases may involve not
only situation data for a large number of well
located sites (from which a predictive map will be
produced), but also the purely cartographic databases which may be needed to produce the map.

Modeling Task	Hard Data	Validation Data	Situation Data	Time-Series Data	Geographic Databases
Model development (all model types)	+	+			
(a) Predictive models	+	+	+		
(b) Scenario models	+	+	+	+	
Geographic application of models					
(a) Predictive models	+	+	+		+
(b) Scenario models	+	+	+	+	+

increases in size or complexity generally require large,
geometric increases in site density. Such problems can be
handled best by irregular spacing of sites, but gridded data
structures are sometimes unavoidable, as in global atmospheric circulation models that treat the atmosphere as an
array of contiguous compartments.
     A related problem involves the reduction or amplifica-
tion of databases in order to match the particular spatial
and temporal scales of the study.   The needed data usually

must be obtained from whatever other databases may be available, which have generally been collected for quite different purposes. The available climatic databases, for example, usually involve huge volumes of monthly data from the individual years, which must be read and summarized into long-term averages plus various measures of variability. On the other hand, it is also common that only mean annual climatic values are available (with some qualitative estimate of extremes and their timing). It is often possible to use annual averages and extremes to "reconstruct" the monthly values by curve-fitting procedures. When this is done, error estimates should be included with the estimated variables in the database.

The process of data standardization may involve both computerized and purely mental "data-management" operations. These tasks are tied to philosophical questions of standardization versus site independence and data-gathering flexibility, and will be treated later.

The main problems of data display involve not only graphing and mapping of data and results but also photography or other reproduction, such as the redrawing of computer-maps. Line-printer maps, such as those produced by SYMAP, are easily generated, can be quantified easily, and may be most appropriate for the accuracy of the data and models. Printer maps, however, are relatively difficult to photograph and must sometimes be redrawn. Plotter-produced maps may be easier to photograph but are more expensive, require more special equipment, and may involve pseudo-precision in representing results. Plotter maps also generally cannot show the spectrum of shading levels appropriate for many isoline maps.

Models that must display data frequently, either as geographic feedback during model development or as a time-series simulation, are a special problem. Most readily available mapping packages cannot easily produce series of maps representing time-series simulations. More sophisticated mapping software for diachronic simulations needs to be developed (and made readily available) for geoecological modeling operations.

SITE RECONSTRUCTION

By far the most important data problem in geoecological modeling is the availability of a sufficient number of site measurements. This involves both geographically complete sets of biological sites and the availability of coordinated environmental data for those sites. (Environmental data for simulation represent a different problem and will be treated later.) When site location, substrate and topography, and some annual climatic data are provided, monthly climatic and water-balance data can often be "reconstructed" from the available information. Even when needed values are reported, it is often useful to reconstruct the site's environment in order to understand the values better and to insure that the data have the same meaning at all sites in the database. The data needed and the ability to reconstruct them depend on the particular modeling task.

The key to this estimation of site conditions is the ability to estimate monthly temperature and precipitation accurately, both long-term and for the period of biological measurements. The general procedure consists of the following steps:

1.   Accurate location of the site and description of its topography; location of nearest meteorological station(s).

2.   Estimation of the site's monthly temperature and precipitation values, based on nearby meteorological data, differences in elevation and aspect, etc.

3.   Interpretation of soil type, depth, effective root-zone depth, horizons, texture, and resulting water-holding capacities at field capacity, saturation, and permanent wilting.

4.   Construction of the climatic water budget (Thornthwaite & Mather, 1957), which yields estimates of monthly potential evapotranspiration (PET) and actual evapotranspiration (AET), soil water storage and change, climatic water surplus and deficit, plus annual totals and indices of annual and seasonal humidity and aridity.

Other environmental data that may be needed for the particular study (e.g., microclimate, soil chemistry and nutrient status, or degree of disturbance) can also be estimated in some cases. These are no substitutes for actual measurements, but they may be compared with measured data, where available, to provide more insight into local processes.

In addition to models of annual process rates or average accumulations, the ensemble of environmental data suggested above can be used to estimate monthly scenarios for some aspects of the site's ecosystem dynamics. Table 2 shows a climatic water budget and subsequent estimation of monthly production, respiration, litterfall, lying litter

Table 2. Average climatic water budget and predicted production-respiration balance, litterfall and litter decomposition by months for natural vegetation at Lisbon. The table illustrates the use of readily available macroclimatic data (in combination with climate-based models of individual ecological processes) to provide initial profiles of site water, energy, and carbon budgets. All values are generated from monthly temperature and precipitation, plus soil dimensions, as the only required input data. Potential evapotranspiration (PET) is estimated from air temperature. Actual evapotranspiration (AET) and soil water are estimated by SOLWAT (Box, 1982), a generalized soil water budgeting system designed for both modified and natural soil situations. Monthly values for production, respiration, litterfall and decomposition rate are obtained by partitioning the values from annual models (e.g., Box, 1978; Meentemeyer et al., 1982) based on monthly AET or temperature contributions to the annual total or average.

LISBON 38.7°N, 9.2°W, 37m. Soil Dimensions and water-holding capacities (mm)													
	depth		saturation			field capacity			permanent wilting			available water	
Effective root zone	1000		480			260			110			150	
	Jan.	Feb.	Mar.	Apr.	May	June	July	Aug.	Sept.	Oct.	Nov.	Dec.	Year
Soil water budget													
Temperature	11.0	11.8	13.6	14.6	16.9	20.1	22.7	23.3	21.8	18.6	14.1	11.8	16.7	°C
Precipitation	89	73	131	66	35	22	5	3	29	72	109	134	768	mm
PET	39	44	55	62	78	101	118	118	102	79	53	42	892	mm
P-PET	50	29	76	4	-43	-79	-113	-115	-73	-7	56	92	-124	mm
soil water (root zone)	260	260	260	260	222	175	139	122	116	116	172	260		mm
AET	39	44	55	62	73	69	41	20	35	72	53	42	606	mm
Deficit	0	0	0	0	5	32	77	98	67	7	0	0	286	mm
Surplus	50	29	76	4	0	0	0	0	0	0	0	4	162	mm
Biological Processes													
Gross Production (GPP)	144	162	201	230	270	254	152	74	127	266	196	155	2230	g/m²
Respiration (Rd)	53	56	63	68	79	99	118	123	111	89	65	56	980	g/m²
NPP (= GPP - Rd)	91	106	138	162	190	155	33	-50	16	177	131	99	1249	g/m²
Litterfall	55	37	20	15	19	72	117	75	33	1	19	53	504	g/m²
Rootfall	64	64	45	27	19	26	90	126	67	31	2	25	585	g/m²
Litter decomposition	35	39	45	45	46	50	39	22	37	69	41	34	504	g/m²

pools, and decomposition for Lisbon, Portugal. Such carbon-balance scenarios can be estimated for any site with monthly climatic data and vegetation not deviating too greatly from natural vegetation. Results for many such sites worldwide can then be used to estimate geographic scenarios of bio-sphere-atmosphere exchanges of carbon, water-borne sub-stances, etc., over the course of a year.

NEEDED BIOLOGICAL DATA

The type of modeling described here focuses especially on basic energy and water-budget processes of vegetation, locally and geographically. The required biological data include primary production, decomposition, and other aspects of carbon and other material budgets. The main gaps involve particular environments and regions that are poorly repre-sented (if at all) as well as particular vegetation frac-tions that are harder to measure (e.g., below-ground bio-mass, production in highly variable environments).

It is not possible here to present a detailed catalog of data needs, but some generalizations can be made. Bjorkland (1984) has recently collected a world primary-production database of over 200 terrestrial sites (or annual situations). Poorly represented regions include most of South America and Central America; much of extra-tropical mainland Asia; North Africa and the Middle East; New Zealand; much of Australia; various parts of Africa; and many islands. Less well represented among the earth's main climate-vegetation types (biomes) are the subtropical deserts, the mediterranean regions, the temperate rain-forests of perhumid west-coast climates, the warm-temperate east-coast forest areas (especially in the Western Hemis-phere), and various less typical tropical climates (e.g., drier equatorial climates with bimodal rainfall). Although some data are available for some of these regions, they often express hourly photosynthesis or other values which cannot be translated into monthly or annual amounts (or interannual variability) per unit ground surface area. Vegetation fractions that are especially important but generally poorly inventoried include below-ground biomass, production, rootfall and decomposition; leaf area, biomass and production, plus monthly changes (foliation-defoliation patterns); monthly and other short-term production; respi-ration of different vegetation fractions; and mortality of individual plants.

RECOMMENDATIONS

Ecological research sites in the USA and similar sites elsewhere have sometimes considered the idea of standardization of data collection and formats, in order to facilitate data transfer and storage and to permit more direct comparison of patterns and quantities in different situations. Standardization has often been seen, however, as not recognizing the fundamental differences between different environments (e.g., terrestrial vs. aquatic) and as limiting the freedom of individual sites to do what seems most appropriate in their particular situations. Measurement accuracy and inadequate database documentation are more often seen as limiting the usefulness of data than does the lack of standardization.

One might react by hoping that these goals of flexibility, accuracy, improved documentation, and some degree of standardization adequate for rigorous geographic modeling are not mutually exclusive. Some data, at least for terrestrial sites, are relatively "standard" by nature (e.g., monthly mean temperature and precipitation) and could easily be reported in a standard fashion. Other data are basic and could be standardized relatively easily (e.g., depth of effective root zone, field capacity, standing-biomass fractions) without limiting the freedom of choice of the local investigators. The fact that some of these factors may vary seasonally and spatially does not preclude the development of standard formats, perhaps at various degrees of resolution.

A list of primary environmental data suggested for each terrestrial biological measurement site is shown in Table 3. The most important item is precise location of the site. Description of the site's climate must include both long-term average temperature and precipitation values by months and those during the actual period of biological measurements. Extreme values, diurnal patterns, and measures of variability are also needed. Since water budgets will usually be estimated, it is necessary to have estimates of soil type, dimensions, texture, and water-holding properties. When terrestrial models based solely on climate fail, it is usually because of edaphic constraints such as the extreme infertility of many tropical soils. Soil pH is not an adequate measure of nutrient status, but it is the one variable most commonly measured and should be reported (preferably obtained by several methods at several locations

Table 3.   Suggested primary site environmental data for all well-documented biological sites. These data are generally necessary in order to estimate a climatic water budget for a site and for any reasonable attempt to interpret environmental influences on site ecological processes. Other data may also be required.

Site Location

Site name and geographic unit(s)
Latitude and longitude (minutes or hundredths of degrees)
Elevation (range, average, actual elevation of climatic instruments)
Name of closest meteorological station and location
Map of site location and adjacent terrain

Documentation

Publications: authors, titles, dates, where published (primary sources)

Topography

Type of terrain
Slope and aspect at site (range, average, actual at measurement sites)

Climatic

*Temperature (air): mean monthly and annual; annual and monthly extremes (absolute and long-term average); years averaged; actual values during periods of biological measurements.

*Precipitation: average monthly and annual; annual and monthly extremes (absolute); years averaged; actual values during periods of biological measurements.

Substrate

*Soil type (FAO, USDA, and traditional) and variability (e.g., site soil vs. adjacent general soil type)

*Soil depth, horizons, and parent material

*Depth of effective root zone, depth to less permeable layers and/or groundwater

*Soil texture, porosity, rock content, and field capacity

Soil pH (>1 method), organic content, fertility (at least for A or other uppermost horizon)

Vegetation

Vegetation type: formation type and formation, physiognomy, structure (strata), communities, dominant taxa (in each stratum)

Status: age, successional stage, degree of "naturalness"

Vegetation height, strata, and cover percentage

Seasonality: evergreen vs. deciduous (etc.), foliation/defoliation patterns of stand and of main component plant types

---

*Should include estimates of variation with elevation (with values at other sites) in uneven terrain.

near the site). Complete description of vegetation structure, composition, and seasonal changes is equally important. The variables listed in Table 3 represent a bare

Table 4.   Suggested secondary site environmental data for well-documented biological sites. These data permit more detailed modeling efforts and provide more complete description of site environmental and ecological conditions. The measurements of climatic factors should, of course, coincide with the periods of biological measurements.

Climatic

Evapotranspiration: mean annual and mean monthly potential and actual evapotranspiration (measured; computed may be useful also); monthly extremes; years measured (and monthly temperature and precipitation values for those years).

Solar radiation: mean annual and monthly (measured); measurement device; monthly and annual extremes; years measured.

Diurnal temperature: average and absolute daily maximum and minimum for each month; years averaged.

Frost-free period: average, extremes.

Growing season: how defined, threshold values used (warmth, wetness, phenological, etc).

Substrate

Soil CEC and base saturation (at least for A or other uppermost horizon)

Soil chemical content: N, P, Ca, etc.

Soil water content: monthly means and extremes, vertical distribution (e.g., depths of seasonal wetting/drying)

---

minimum for making a site's biological data useable in comparative geographic studies.

A suggested set of secondary environmental data is shown in Table 4. It may not be possible to collect all of these data at all sites, but all such data collected will be useful and in some cases mandatory for modeling attempts. Measurement of evapotranspiration at biological sites is quite useful since it permits checks on the accuracy of evapotranspiration estimates used in model development and subsequent geographic simulations. Soil cation-exchange capacity (CEC) and base-saturation levels are much better measures of soil fertility than is pH. Soils are patchy, however, and care must be taken to insure that soil data are representative of both the exact site and the surrounding area (up to perhaps 10-100 km^2). If the soil at the exact site differs from the surrounding area (e.g., floodplain site), soil data should be collected for each separately.

Perhaps some probability measures of relative soil-type frequency could also be included. These would be bulky, however, and not always appropriate to the mapping scale employed.

Finally, Table 5 shows what appear to be the most important data relating to the carbon balance at a terrestrial site. Included are amount of living and dead biomass,

Table 5.   Most important carbon-balance data at biological sites. These data permit relatively complete description of the carbon balance of a site. In combination with both long-term concurrent and average climatic data, these carbon-balance data (at enough sites in different climatic situations) also permit development, improvement, and validation of predictive models for site carbon-balance dynamics. Data are needed for the different biomass components, especially for leaves and for the divisions above vs. below-ground and woody vs. non-woody. Data for more than one year are highly desirable. "Mortality" refers to the amounts of biomass (litter) represented by dead individuals as opposed to dead plant parts.

Biomass

Standing biomass: average, monthly changes (if appropriate), year-to-year changes; above and below-ground; by biomass component (boles/stems, branches, leaves, roots, etc.); measurement methods.

Standing dead biomass (similarly, as appropriate).

Lying litter pools (leaf, other traditional "litter," roots, dead individuals, larger branches/trunks, etc.): average monthly amounts, year-to-year changes; lignin content, woody and other fractions; measurement methods.

Biomass Processes

Respiration: monthly and daily amounts (with concurrent climatic and phenologic data); by biomass component; measurement methods.

Net production: monthly and daily amounts (with concurrent climatic and phenologic data); by biomass component; measurement methods; growth vs. production.

Gross production (similarly, if not by combination of respiration and net production).

Litterfall: monthly and daily amounts (with concurrent climatic data); by components (leaves, other "litter," rootfall, larger branches/trunks, dead individuals); measurement methods.

Litter production (similarly, via net production of leaves, etc.).

Mortality: annual and monthly (daily?), via litterfall.

Litter decomposition: annual, monthly, and daily k-values (with concurrent climatic data); by litter components; measurement methods.

---

production and respiration rates, and litterfall and decomposition rates, all by vegetation fraction (leaves, stems, roots, etc.) and by months or appropriate shorter measures of seasonal change. These data are the most difficult to standardize across different environmental situations. The so-called "Woodlands Data-Set" (DeAngelis et al., 1981) represents the best published example of how this standardization must be done (at least for woodlands) and of how much work is involved. It provides a useful model for standardized definition of biomass and production data but is far from including adequate environmental data.

There are also other data problems that have not been mentioned yet. For geographical simulation one needs environmental data (but not the hard biological data) at many sites in the region simulated. Climatic databases such as the world climatic data from the National Climatic Center (Spangler & Jenne, 1979), with voluminous individual-year data, should be reduced to long-term averages and measures of variability in order to provide databases more useable by ecologists. Substrate data should be obtained for these climatic sites and added to the reduced database to provide a standardized world environmental database for 1500-2000 terrestrial sites.

It will soon be possible to combine general ecological models with satellite data in order to simulate patterns more quickly, easily, and accurately than if one had to rely on ground measurements. The interfacing of site-based models and mapping techniques with satellite images (of variables not coinciding exactly with those familiar to ecologists) will cause problems of data management that should be anticipated now.

Another potential problem involves the situation in which data-sites (or model cells) interact significantly, as in watershed hydrology, atmospheric dynamics, and landscape ecology. This may have implications for data management due to the greater number of data-sites that may be required to express the interactions.

The significance of these data problems can be stated fairly concisely. Availability and management of appropriate data often represent the main limiting constraints on ecological modelers and other synthesizers. For the field ecologists who collect the primary biological data, these data problems mean that more environmental data need to be collected and reported in order to ensure that the biological data will be fully useable by others. The Long-Term

Ecological Research (LTER) program has generally agreed to oppose data standardization (various personal communications). Nevertheless, if the goal is to facilitate the availability of data to ecologists, including modelers and others working comparatively at various geographic scales, then problems of scale, data-processing limitations, and legitimate needs for some standardization of some basic data must be appreciated. Ecological research sites, and the LTER network in particular (by example), can help greatly in this effort through their ability to ensure data quality and improved data documentation.

LITERATURE CITED

Bjorkland, R.A.   1984.   Organic-Matter Turnover Rates and Their Relationship with Coimate. M.A. Thesis. University of Georgia, Athens. 233 pp.

Box, E.O.   1978.   Geographic dimensions of terrestrial net and gross primary productivity. Radiation and Envl. Biophysics 15: 305-322.

Box, E.O.   1980.   What determines the amounts of leaf and total standing biomass of climax terrestrial vegetation? Bull. Ecol. Soc. Am. 61: 76 (abstract).

Box, E.O.   1982.   SOLWAT: A Minimal-Input Soil Water Simulation System Applicable to a Full Range of Natural Situations. User's Manual for Version 2.2. Department of Geography, University of Georgia, Athens. 27 pp.

Brook, G.A., M.E. Folkoff, and E.O. Box. 1983. A world model of soil carbon dioxide. Earth Surface Processes 8: 79-88.

DeAngelis, D.L., R.H. Gardner, and H.H. Shugart. 1981. Productivity of forest ecosystems studied during the IBP: The Woodlands Data Set. pp. 567-672. In: Dynamic Properties of Forest Ecosystems. D.E. Reichle (ed.). IBP series, Vol. 23. Cambridge University Press, New York and London.

Folkoff, M.E., V. Meentemeyer, and E.O. Box. 1981. Climatic control of soil acidity. Physical Geography 2: 116-124.

Kesner, B.T. 1984. The Geography of Nitrogen in an Agricultural Watershed: A Technique for the Spatial Accounting of Nutrient Dynamics. M.A. Thesis. University of Georgia, Athens. 121 pp.

Lieth, H. 1975. Primary production of the major vegetation units of the world. pp. 203-216. In: Primary Productivity of the Biosphere. H. Lieth and R.H. Whittaker (eds). Springer-Verlag, New York.

Meentemeyer, V. 1978. Macroclimate and lignin control of litter decomposition rates. Ecology 59: 465-472.

Meentemeyer, V., E.O. Box, and R. Thompson. 1982. World patterns and amounts of terrestrial plant litter production. BioScience 32: 125-128.

Spangler, W.M.L. and R.L. Jenne. 1979. "World Monthly Surface Station Climatology." Tapes and documentation for about 2500 sites (updated annually). National Climatic Center, Asheville, NC.

Thornthwaite, C.W. and J.R. Mather. 1957. Instructions and tables for computing potential evapotranspiration and the water balance. Publications in Climatology (Elmer/ New Jersey) 10: 185-311.

GEOGRAPHIC INFORMATION SYSTEMS

Frank M. Brookfield

ABSTRACT

A geographic information system (GIS) is a specialized database management system used to store and retrieve cartographic data. At the simplest level, a GIS is a method of managing, analyzing, and displaying map information. Computerization of a GIS is a way to automate this process by allowing maps to be digitized, stored, manipulated, and displayed by a group of integrated computer programs. Automation allows greater flexibility and versatility than previous methods in the handling of maps and map information. A computerized GIS requires a computer, digitizer, graphics terminal, and large amounts of disk space to operate. Thus the initial investment for such a system is quite large.

Environmental Systems Research Institute's (ESRI) ARC/INFO is a good example of a computerized GIS system. ESRI has taken a "toolbox" approach by creating a series of general purpose programs that perform specific tasks, thus allowing the users to customize this system to meet their specific needs. There are two subsystems in ARC/INFO. ARC stores coordinate information, while INFO is a relational database management system that stores attribute information. These two subsystems are integrated to create a powerful map-processing system.

This capability of storing and presenting research data in both tabular and graphic form should enhance the value of the research conducted as well as provide a more complete historical account of that research. The capability to store, retrieve, and manipulate cartographic data such as land use, environmental degradation, and changes in biologic distribution patterns will increase the researcher's ability to model these systems.

INTRODUCTION

It was not until the 1700's that maps were used to display information other than locations and landmarks. One of the most famous uses was that by Dr. John Snow, who plotted cholera deaths on a map of London in 1854. By examining the scatter pattern of these deaths, he was able to determine that the Broad Street water pump was the source of the cholera. Once the handle on the pump was removed, the deaths stopped (Gilbert, 1958). Since that time we have used maps to display a variety of information about the world around us. This use of maps makes complicated descriptions much easier and more concise (Robinson et al., 1978).

Throughout history, new inventions have changed the form of maps. Because of increased computer availability and the invention of peripherals that are able to display map information, the computer is making significant changes in the cartographic field. These changes provide for more versatile handling and display of cartographic information, which provides other fields with new and creative possibilities for the usage of mapping systems (Bagrow, 1964).

WHAT IS A GIS?

A geographic information system (GIS) is a specialized database management system used for storing and retrieving cartographic data. At the simplest level, a GIS is a method of managing, analyzing and displaying map information. Computerization of a GIS is a way to automate this process by allowing maps to be digitized, stored, manipulated, and displayed by a group of integrated computer programs. These programs enhance the user's ability to manage this information by providing 1) utility programs designed to manipulate lines, points and arcs used in the drawing of maps, and 2) methods to link attribute information to specific areas which this data describes (Monmonier, 1982).

This form of data management provides for the standardized entry, storage, and display of data. The output from these data is all that the user requires--the system can provide the necessary utilities to manipulate the data. Sorting, searching, and linking data become part of the system rather than a needed programming task that would have to be done for the system. The linking of different data sets is the most powerful of these utilities.

There is a distinction that should be made between geographical and database handling software, and an information system. To be an information system, the system has to inform. This may seem redundant, but there is a considerable difference between information and data. "Information" is processed data that is displayed in a form that fulfulls the requirements of the users. "Data" is the raw form of this information and requires processing before it can be used (Gore & Stubbe, 1979).

A substantial amount of planning is required to develop a complete GIS. The needs of the user and the physical requirements of the system must be addressed. The attainment costs and availability of the required information need to be evaluated. This process is important to both the development of an efficient system and the full utilization of that system (Ullman, 1980).

Equipment requirements for storage are dependent on desired usage of the system and map precision required. In addition to storage capacity, a means for displaying and entering information is needed, including a plotter for hard copies, and a graphics terminal for editing and displaying information, and a digitizer or image analyzer for entering cartographic information. The system also requires a computer with enough processing power and access ports to handle the GIS and all of its users.

There are several companies which work with cartographic information. Some companies offer the software as well as provide a cost analysis of equipment for the system. These companies serve primarily as consultants for the development of a system. Other companies have digitizing services available for specific needs or have digitized areas that can be purchased. Another source of map information is the federal government, which has some very sophisticated operational GIS's and very large data sets of map infomation on specific locations. These sources need to be examined and evaluated for their specific resources and what they could provide to any proposed system.

Actual digitized maps have to be developed either in-house or purchased from outside sources. The sources of the original information have to be found so they can be added to the system. This information can come from aerial photography, satellite photography, or remote sensing of data through specifically defined equipment. The source and quality of information is an important issue (Tufte, 1983); if the data for the maps are not available or in a form that

will not meet specific needs, then the whole system has
questionable value. The system can't supply information
that is not available. There is a cost associated with the
attainment of each additional bit of information. This cost
has to be carefully incorporated into the total cost of the
system and should be evaluated as to the effective value of
the information (Stiner et al., 1972).

A GIS can accommodate a number of levels of complexity,
ranging from a very rough outline of the United State to a
meter grid of an intensive study area. Ganesa Group International, Inc. developed one system that is available on an
IBM Personal Computer with a hard disk drive. This is a
closed system for maps data, because there is no feasible
way for the user to update or add to the geographical presentation. The user is allowed to associate any number of
descriptors to areas as small as census tracts of the United
States (Ganesa, 1984). Businesses interested in the display
of market areas with associated sales dispersions or similar
types of displays form the primary user target for this
software. This orientation has made the system quite coarse
in the display of the areas of the United States, but the
system does provide enough geographic information to fulfill
the needs of this specific user group.

A much more complex system is under development at the
Illinois State Natural History Survey. This system is being
developed on a Prime 750 minicomputer system. It includes
five work stations, each consisting of a digitizer and
graphics display terminal, with direct connections to the
computer. Three plotters are available for use by any of
these stations. The computer has 1200 megabytes of disk
storage and a tape drive. This system is being used by the
five divisions of the Illinois State Department of Energy
and Natural Resources.

This GIS is being developed primarily to service the
Lands Unsuitable for Mining Program. This program's objective is to answer questions from interested parties concerning the effects of mining on the environment in a specific location, the costs associated with mining of that
area, and whether or not the costs justify the gains in coal
production. To fulfill this obligation, a database of
natural resource data for Illinois is being compiled and
associated with geographic locations. An area in question
can be selected and all available information for this location can be displayed for evaluation. Missing information
can be obtained from alternative sources or by site visits.
The final product is a report stating the facts of the par-

ticular case and interpretation by the specialists retained by this project. This report is sent out for a final decision by the State's Department of Mines and Minerals.

A geographical information system can be as simple or complex as the uses require. The cost of such a system can range from $15,000 for a rudimentary system to several hundred thousand dollars. No GIS is cheap; the amount of required data storage, the time involved in digitizing maps, and the costs of the equipment and software are high. Software and machinery prices are decreasing, but the costs for digitizing and data entry are still high. The system has to be completely planned; user needs have to be defined, the resolution of the required maps has to be calculated, and sources of either digitized maps or original map information have to be found. The initial cost is high, but with a well-defined system the return can offset this investment both in terms of improved efficiency as well as in increased analytical capabilities (Marble et al., 1972).

ARC/INFO

Environmental Systems Research Institute (ESRI), a consulting firm, was hired to evaluate the needs and define software and equipment requirements for a system at the Illinois Natural History Survey. ESRI is also providing digitized maps for much of the state-wide database. They have developed ARC for processing the geographical information, and interfaced this system with INFO, a commercial database management system from HENCO (Nenco, 1984), which handles the descriptive data.

ARC/INFO was developed using a "tool box" approach. There are a variety of programs that perform specific tasks. Commands such as "digitize," which is used to record cartographic information; "editplot," which is used to update input; "intersect," which combines two map coverages; or "dropline," which allows the user to eliminate unnecessary lines are examples of such programs (ESRI, 1984). This method allows the user more flexibility in processing map information, but there is also a constraint requiring the user to have an indepth knowledge of the system in order to use it.

In the ARC/INFO system, each point, line, or polygon is identified by a number that the system assigns. Additionally, each of these can be identified by one or more codes assigned by the user. Utilizing these codes, each cartographic unit can be linked to descriptive files (called attribute files) that correspond to these map features. Attributes can be grouped together by codes to form generalized descriptors of areas of interest, or these codes can be linked to files that contain colors or font styles that can be used to identify these positions. In this way, the areas are described on a map and these areas can be identified internally.

The system also provides utilities that can combine areas from one or more coverages. This combining step can be either a union or intersection of attributes. This process makes it possible to create more complex maps structures and provides the ability to ask more in-depth questions about the areas of concern. Another utility ARC/INFO provides is scaling changes in the coverages. Care must be taken when reducing the scale, because some detail is lost by increasing the distance between digitized points. Thus, digitizing a state map and then displaying one county would produce a very rough outline of that county. Conversely, increasing the scale of a map produces large data files with too much detail, thereby increasing the storage costs. It is important to decide on the needed resolution for the required maps and to digitize these maps at an appropriate scale to meet these needs.

The processing of a map in ARC/INFO starts with a base map which is used as a starting point for all the maps produced for a certain area. This is an outline of the main land or other features that will be needed for all information displays. The next step is to digitize overlays that will fit onto the base map. These overlays contain specific features, area outlines, or points of interest to be either displayed or simply recorded. The overlay map is aligned to the base map by at least four standardized points called tic marks. This provides a frame of reference by which the locations can be combined to realistically portray the map features. Many different overlays can be created for one base map and used one at a time or in any combination; providing a very versatile way to store different details for any area.

LARGE RIVER SITE

The Large River Site of the Long-Term Ecological Research project is developing a GIS using ARC/INFO for portions of the Mississippi River. The base maps were created from seven-and-a-half-minute US Geological Survey quadrangle

maps containing the Mississippi River. The outline of the
river, islands within the river, and streams that empty into
the river were digitized. These are the areas of primary
interest for the project and, as such, the basis for the
mapping work.

Pool 19, one of the intensive study areas for this project, is contained on nine quad sheets (Fig. 1). Thus, nine
base maps had to be created so the entire site could be
displayed (Fig. 2). This breakdown of the coverage area
provides enough detail to fulfill the project's objectives.

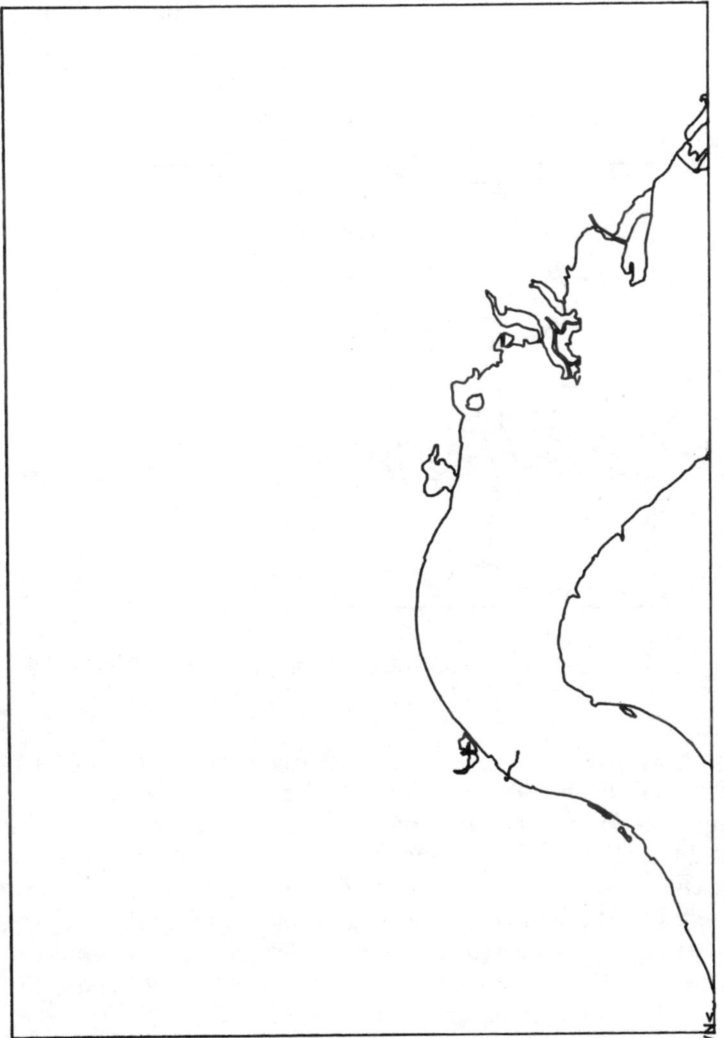

Fig. 1.   Base map for the Nauvoo quad.

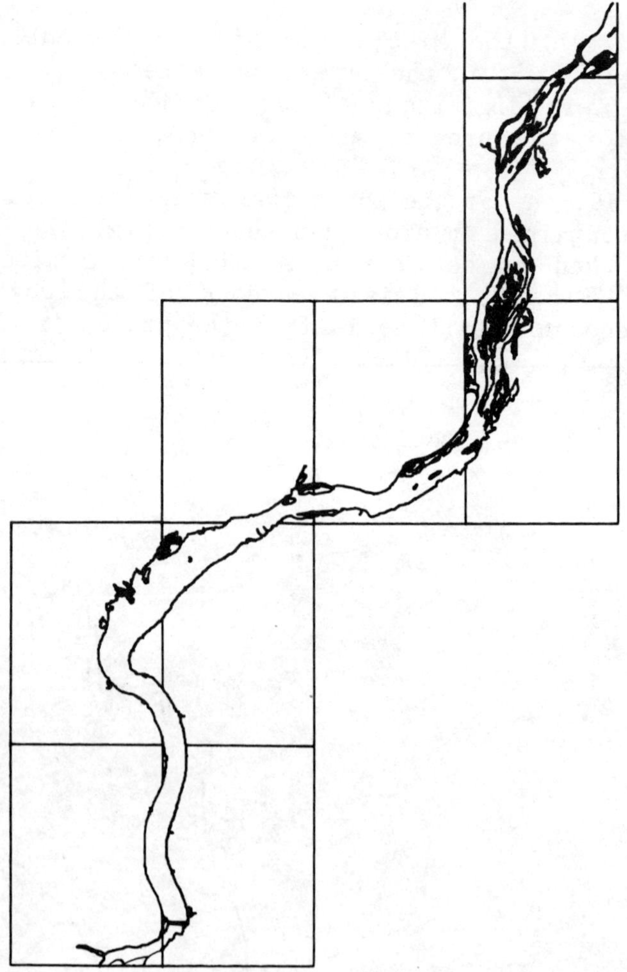

Fig. 2.   Combination of base maps for Pool 19 of the Mississippi River.

The system also allows the combination of the individual pieces so that the whole area can be displayed.

An example of the overlay coverages that have been digitized for the LTER project at the Large River Site are the points sampled by researchers (Fig. 3). Each point represents information ranging from water quality or river-bed materials to biological data such as macrophyte content and type. Each point has an associated code that links that point to the tabular data it represents.

Bed materials of the river have also been digitized

into overlay maps. These are interpolations of the data
collected from river transects, which provided the information on the river bed. Each polygon was identified and
coded to indicate the type of bed materials present. These
codes were then associated with color and line patterns to
display the type of materials represented. This interpolation was done for each quad so that a complete map of the
bed materials for Pool 19 could be displayed.

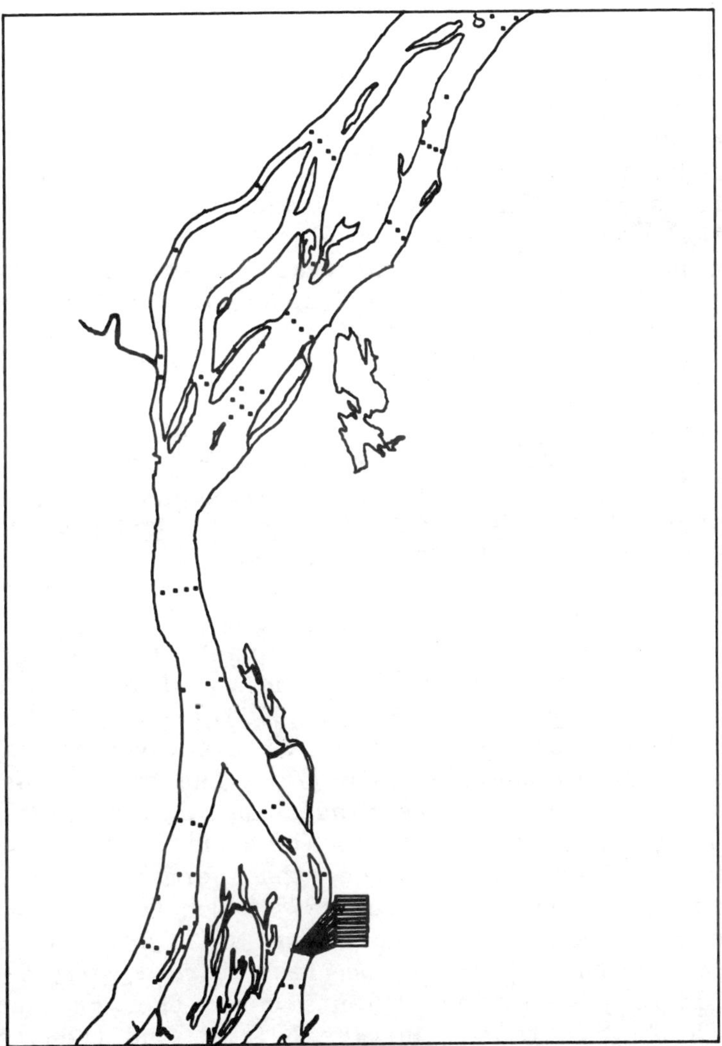

Fig. 3.   Data collection sites for the Burlington quad.

USE OF ARC/INFO GIS

This GIS is new and still in the development stage, but the long-term nature of the research on this system provides a great potential for the future. Currently, the vegetation beds are being mapped from aerial photographs of Pool 19. Old maps have been found that show the river prior to the damming of this section. These maps, when entered into the system and compared to other years, should indicate how past changes have affected the present river structure. Understanding historic changes should allow prediction of what future man-made or natural changes will do to Pool 19 as well as other river reaches.

Other aspects of biology, geology, and hydrology are being added to the database to help develop an understanding of this area. The current standardized collection and recording of this information linked to the sampling sites, gives a physical basis and processing capability for future research. One current use of these data is the comparison of the different sites to define habitat area types for the river. This sectioning of the river is currently being used for our hydrologic and biologic models.

Long-term trends can be recorded and changes in these trends can be evaluated. An example from the Illinois Natural History Survey is the change in the dispersion of the spotfin shiner (*Notropis spilopterus*) and the red shiner (*N. lutrensis*) in Illinois. Figure 4 shows collection sites for the spotfin shiner for both the years prior to 1905 and the years after 1950. The same display for the red shiner is shown in Figure 5. The long-term record indicates that the red shiner is filling the gap left by a receding spotfin shiner population (Smith, 1979).

Short-term trends can also be displayed, such as the effect of the drought in 1976-77 on the plant beds in the Montrose Flats area of the Mississippi River. In Figure 6 the horizontal line represents the mean flow. Notice that in 1976 the discharge was lower than normal, and that there was no flood in the spring of 1977. Due to the lower-than-average rainfall, less runoff and soil erosion occurred. This caused the river to be less turbid, which allowed better light penetration. Plants could then grow at depths in which they could not survive before, and the vegatative beds expanded into deep water (compare Figs. 7 and 8). Since then, the plants have persisted (Fig. 9) despite the return of normal flows and turbidity, because they can grow upward into the well-lighted zone each spring, using stored nutrients for the initial spurt of growth. This change in the physical makeup of the Montrose flats has had far-reaching effects at all levels of the ecosystem in the area. The abundance and kinds of macroinvertebrates have changed at one of our long-term benthic sampling stations (the square on Figs. 7-9), and usage of the area by water fowl has changed.

This information is important both for an historical reference and for future predictions. The long-term effects caused by this natural occurrence can be evaluated and used to predict what consequences similar natural or man-made changes to the river may have.

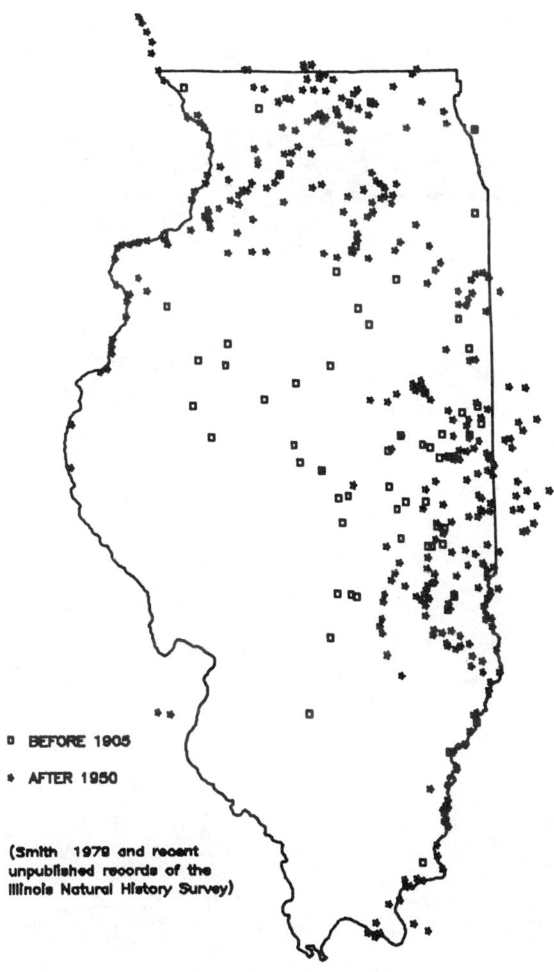

Fig. 4.   Dispersion of the spotfin shiner.

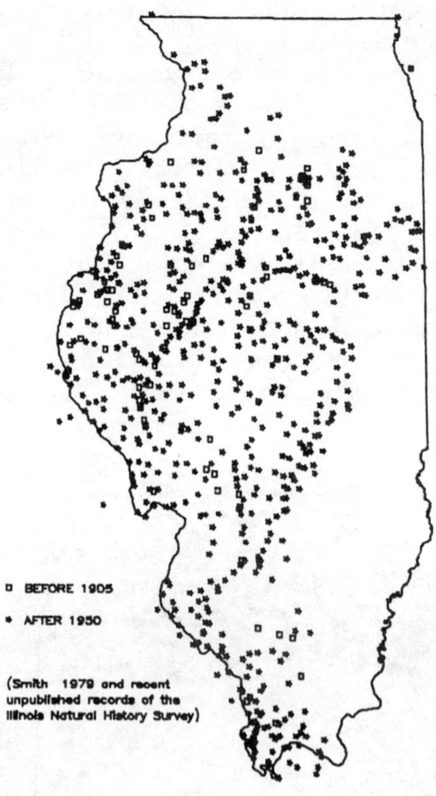

Fig. 5.   Dispersion of the red shiner.

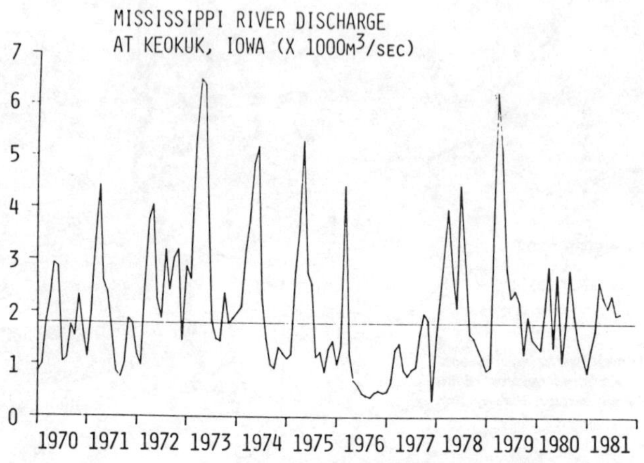

Fig. 6.   Water flow of Pool 19 of the Mississippi River.

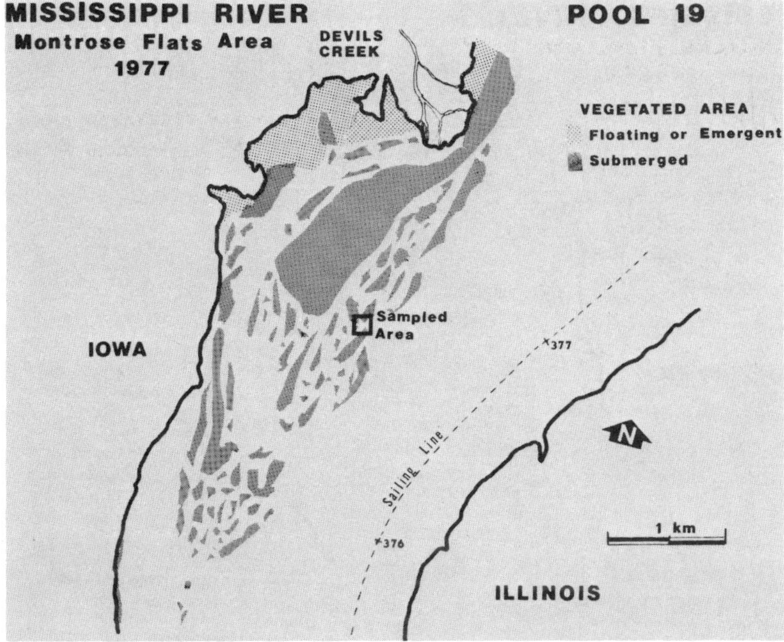

Fig. 7.   Vegetative beds prior to the drought.

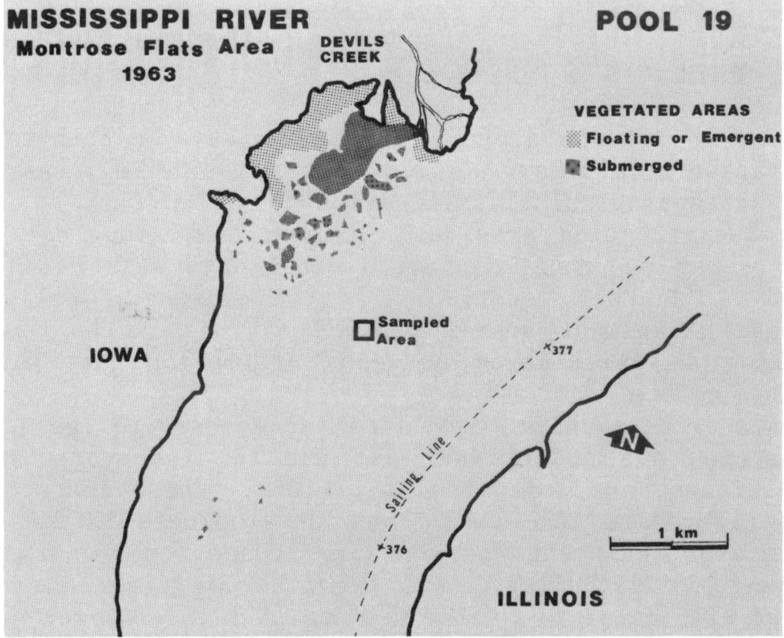

Fig. 8.   Vegetative beds during drought.

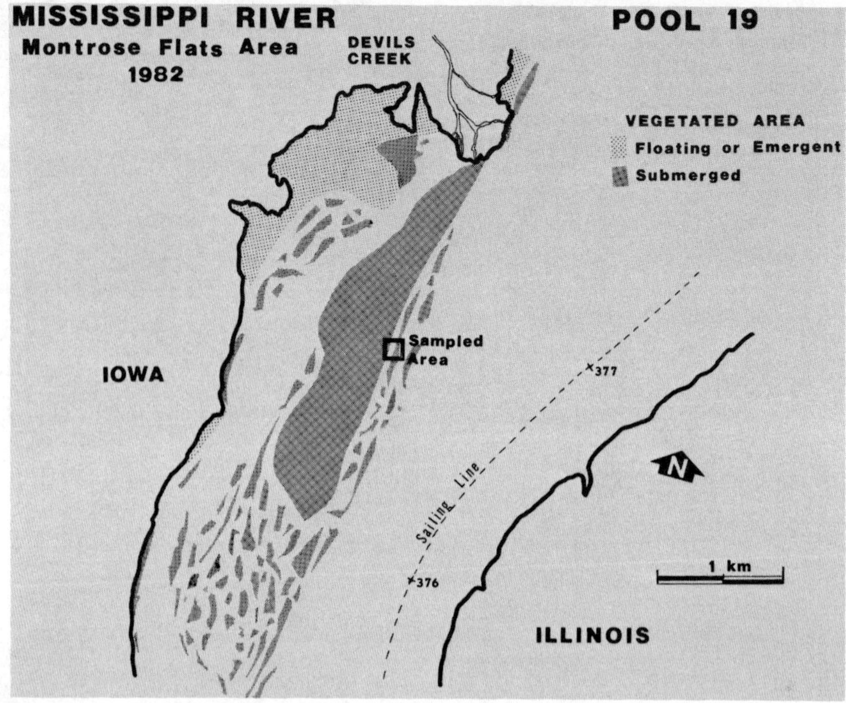

Fig. 9.   Vegetative beds five years after the drought.

FUTURE OF THE GIS IN RESEARCH

The ability to store and retrieve data in both tabular and graphic form allows a greater flexibility in research by making a topological classification possible (Carter, 1980). This classification provides a common link among data sets so they may be associated with other data sets being collected at another location. By the use of a relational database management system these locations can be combined to allow an association between information that was not possible before.

Researchers have had limited access to geographic information systems in the past due to high system costs, and equipment and software limitations. These restrictions are being reduced due to the new developments in the areas of computer equipment and software. Once researchers begin to use this technology, the creative application of such systems will initiate a whole new approach in ecology.

LITERATURE CITED

Bagrow, L. 1964. History of Cartography. Harvard University Press, Cambridge, MA. 312 pp.

Carter, J.R. 1980. Added dimension geographic data. pp. 17-18. In: ACM Proceedings, Annual Conference of the Association for Computing Machinery. Nashville, TN.

Environmental Systems Research Institute. 1984. ARC/INFO Users Manual. Redlands, CA.

Ganesa Group International, Inc. 1984. Statmap: A Statistical and Demographic Mapping System. McLean, VA.

Gilbert, E.W. 1958. Pioneer Maps of Health and Disease. Geogr. J. 124: 172-183.

Gore, M. and J. Stubbe. 1979. Elements of System Analysis for Business Data Processing, 2nd Ed. Wm. C. Brown Co., Dubuque, IA. 450 pp.

Henco Software, Inc. 1984. Info Prime Reference Manual. Rev. 9.2. Waltham, MA.

Marble, D., H. Calkins, K. Dueker, J. Gilliland, and J. Saloma. 1972. Introduction to the economics of geographical information systems. pp. 206-1239. In: Data Handling, Vol. 2, Symposium Edition. R.F. Tomlinson (ed.). Geographical Union Commission on Geographical Data Sensing and Processing, Ottawa, Ont.

Monmonier, M.S. 1982. Computer Assisted Cartography Principles: Principles and Prospects. Prentice-Hall, Inc., Englewood Cliffs, NJ. 214 pp.

Robinson, A., R. Sale, and J. Morrison. 1978. Elements of Cartography. John Wiley & Sons, Inc., New York. 448 pp.

Smith, P.W. 1979. The Fishes of Illinois. University of Illinois Press, Urbana, IL. 314 pp.

Stiner, D., T. Stanhope, B. Cook, K. Dueker, F. Hornton, K. Rogers, and B. Wellar. 1972. Data base development. pp. 37-116. In: Data Handling, Vol. 1, Symposium Edition. R.F. Tomlinson (ed.). Geographical Union Commission on Geographical Data Sensing and Processing, Ottawa, Ont.

Tufte, E.R. 1983. The Visual Display of Quantitative Information. Graphics Press, Cheshire, CT. 197 pp.

Ullman, J.D. 1980. Principles of Database Systems. Computer Science Press, Inc., Potomac, MD. 379 pp.

ACQUISITION AND MANAGEMENT OF REMOTELY SENSED DATA FOR ECOLOGICAL RESEARCH SITES AND ESTUARINE SANCTUARIES

V. Klemas

ABSTRACT

Global biosystem studies, as planned by NASA and other
agencies, will require extensive inventories of biosphere
resources and processes as well as repetitive, detailed
observations of specific ecological research sites or rapidly changing areas. It is now possible to accomplish both
missions by combining data from several different satellite
systems with appropriate ground observations. For instance,
the NOAA/AVHRR can be used to monitor global vegetation
cover and land use changes. The LANDSAT Thematic Mapper
offers detailed information on vegetation and biomass production at small ecological research sites. In some instances, the inclusion of remotely sensed data in field experiments can decrease long-term costs and provide data that
is more representative of average conditions at a test site.
As microcomputers become less expensive, more users are able
to analyze satellite and other data at their own institutions. This paper reviews the state-of-the-art and presents
a strategy for the acquisition and management of remotely
sensed data for ecological research sites and estuarine
sanctuaries.

INTRODUCTION

During the past decade, international programs (e.g.,
United Nations Environmental Program) and conferences (e.g.,
1972 Stockholm Conference) have produced a realization in
scientists, governments, and environmental action groups

that the Earth is a dynamic whole, and is functioning by the numerous ecosystems and biomes connected through the atmosphere, oceans, and sediments. The National Aeronautics and Space Administration (NASA), the National Science Foundation (NSF), and other scientific institutions are now focusing their attention on the fundamentals of the systems that support life on Earth: mineral cycling, ocean and terrestrial productivity, perturbation effects, and biospheric responses (Rambler et al., 1983).

Global biosystem research, as being planned by NASA, requires large area surveys from remote sensing platforms as well as detailed investigations at specific ecological research sites. The ecological research sites are needed for calibrating the remotely sensed data and for developing conversion factors, as well as for basic investigations. For instance, global biosystems research requires observations of global vegetation cover, biomass, and carbon flux. While vegetation can be mapped directly from satellites, biomass surveys require some ground data to calibrate or verify the satellite results. Since many key biological functions in natural ecosystems are dependent upon carbon fixation and flux through the system, conversion factors must be developed that allow scientists to estimate carbon flux from remotely sensed data.

Ecological research sites and estuarine sanctuaries are ideal for testing and calibrating models that use remotely sensed data, and for producing baseline data against which natural and man-made changes anywhere on Earth can be compared. In return, the techniques developed and data gathered will significantly contribute to the preservation and management of ecological research sites and estuarine sanctuaries.

REMOTE SENSING OF WETLANDS,
ESTUARIES, AND COASTAL WATERS

Remote sensing techniques, if properly employed, offer cost advantages and repetitive synoptic observations of the Earth's environment and resources. A summary of satellite systems and sensors providing remotely sensed data on a global scale is given by Klemas (1981). Descriptions of satellites useful for ecosystem studies are given in Tables 1-3. We realize that remote sensors cannot provide all the information required by global biosystem models; in particular, gas exchange rates, subsurface plant biomass, or

actual ocean productivity. However, the LANDSAT Multispectral Scanner (MSS) can be used to map vegetative land cover
and land use change on most land surfaces around the globe,
except for some tropical regions covered by persistent
clouds where radar must be used (NASA Earth Resources Survey
Symposium, 1975). Vegetation maps (with some additional information) can be converted to biomass maps that, in turn,
can be used to infer carbon and nitrogen production and
other information needed for ecosystem models. Land use
change maps showing deforestation will be useful in determining changes in production of biomass and related organic
substances.

A global biosystems research program must include all
major ecosystem types. Wetland ecosystems constitute a
minor portion of the Earth in an areal sense, yet they are
probably one of the most important ecosystems when considering the global cycling of some elements that have major
reserves in the atmosphere. Among the outstanding characteristics of wetlands are the anaerobic sediments associated

Table 1. LANDSAT 4 and 5 orbital and sensor characteristics

ORBITAL CHARACTERISTICS

Altitude	705 km
Inclination	93.2°
Sun Angle	9:30am at descending node
Coverage	16 days

SENSOR CHARACTERISTICS

	TM	MSS
Channels	0.45-0.52 μm	0.5-0.6 μm
	0.52-0.60	0.6-0.7
	0.63-0.69	0.7-0.8
	0.76-0.90	0.8-1.1
	1.55-1.75	
	10.40-12.50	
	2.08-2.35	
IFOV	30 m (Channels 1-5 & 7)	80 m
	120 m (Channel 6)	
Ground Swath	185 km	185 km
Data Rate	85 Megabits/Second	15 Mbs
Quantization Levels	256	64

Table 2.   TIROS-N orbital and sensor characteristics.

LAUNCH: October 1978

ORBIT: Sun-Synchronous   Near Polar
           830 km   14 Orbits/Day

HIRS/2 (HIGH RESOLUTION IR RADIOMETER SOUNDER)

           20 channels   Temp ± 1°C Surface to 50 km
                                Water Vapor to 15 km

AVHRR (ADVANCED VERY HIGH RESOLUTION RADIOMETER)

           0.55 - 0.90μm   Clouds
           0.725 - 1.0   Land/Water, Snow, Ice
           10.5 -11.5   Sea Temperature ($\Delta T = 0.5°C$)
           3.55 - 3.93   Cloud Thermal (NEAT = 0.20°C)

RESOLUTION = 1100 m x 1100 m

Table 3.   NIMBUS orbital and sensor characteristics.

LAUNCH: NIMBUS 1-6   1964-1975
               NIBMUS 7   1978

ORBITS: Sun-Synchronous   Near Polar
               NIMBUS 1-6   1100 km
               NIMBUS 7   955 km

COVERAGE: Twice Daily

SMMR (SCANNING MULTIFREQUENCY MICROWAVE RADIOMETER)

CZCS (COASTAL ZONE COLOR SCANNER)

Channel 1   0.433 - 0.453 μm   Chlorophyll Absorption
        2   0.51 - 0.53   Chlorophyll Absorption
        3   0.54 - 0.56   Gelbstoffe
        4   0.66 - 0.68   Chlorophyll Absorption
        5   0.70 - 0.80   Surface Vegetation
        6   10.5 - 12.5   Surface Temperature

RESOLUTION = 825 m x 825 m

---

with these areas. Within the anaerobic zone, fermentation,
denitrification, sulfate reduction, and methanogenesis provide the mechanisms for C, N, and S to be returned to the
atmospheric reservoir (Odum, 1961). When this is considered

in conjunction with the fact that wetland ecosystems are among the most productive (per unit area) on Earth, it is apparent that wetland ecosystems must be evaluated as extremely important areas in global ecology models.

Table 4 summarizes the present capability of aircraft and satellite remote sensors to provide data on coastal land and water properties. As shown in Table 4, the species composition and condition of coastal vegetation can be mapped from aircraft or satellites using color and color-infrared film or multispectral scanners. Identical techniques apply to land use change mapping. However, only multispectral scanners can be used successfully to map vegetative biomass and stress levels in vegetation, since the very fine spectral and radiometric discrimination that is required cannot be attained with film cameras (Klemas et al., 1975; Bartlett & Klemas, 1980, 1981). Although the multispectral scanner data must be enhanced and analyzed using digital computer techniques, this is less cumbersome than applying these techniques to data stored on photographic film.

Coastal erosion and coastal geomorphology have been studied successfully using aircraft film cameras and LANDSAT (Stafford & Langfelder, 1971; Dolan, 1973; Dolan et al., 1977). Aircraft photography provides the high resolution required for accurate measurement of beach erosion or accretion. LANDSAT, however, can provide a geologic overview of an entire coastline, including underwater features, if multispectral scanner band 4 is used in clear water.

Depth profiles can be obtained by laser profilers using wavelengths that penetrate clear waters (Hoge et al., 1980). Visible images from aircraft or satellites can help provide relative depth profiles, which then may be calibrated with airborne laser profilers. Laser profilers that do not penetrate into the water column (e.g., near infrared) can be used to obtain sea surface profiles from which wave spectra can be derived (Schule et al., 1971).

With appropriate surface truth data, suspended sediment concentrations have been mapped from aircraft and satellites (Maul et al., 1974; Johnson, 1975; Moore 1978; Munday & Alfoldi, 1979). To calibrate the imagery in terms of suspended sediment load, one must obtain not only concentration measurements on the ground, but also data on the grain-size distribution. Atmospheric corrections are required for both sediment and chlorophyll concentration mapping (Wilson et al., 1978). On the other hand, if one is looking only for suspended sediment patterns (qualitative

Table 4.   Performance of aircraft and satellite remote sensors used in coastal studies. Ratings are on a scale of 0-3: 3=reliable and operational; 2=needs additional field testing; 1=of limited value (future potential); 0 not applicable.

Sensor	Platform	Vegetation Type and Land Use	Vegetative Biomass	Coastline Erosion	Coastal Geomorphology	Depth Profiles	Suspended Sediment Patterns	Suspended Sediment Concentration	Chlorophyll Concentration	Oil Slicks	Surface Temperature	Water Salinity	Current Circulation Patterns	Ocean Wave Spectra	Sea Surface Roughness	Surface Wind Velocity
Film	AC	3	1	3	3	2	2	1	1	2	0	0	2	2	2	1
Cameras	SC	2	1	2	2	1	2	1	1	1	0	0	2	2	2	1
Multispectral	AC	3	2	3	3	2	3	2	2+	3	0	0	2	2	2	1
Scanners (Visible)	SC	2	1	2	2	2	3	2	2	2	0	0	2	2	2	1
Thermal Infrared	AC	1	1	1	0	0	1	0	0	3	3	1	2	0	1	1
Scanners	SC	0	0	0	0	0	1	0	0	1	3	0	2	0	1	1
Laser	AC	0	0	1	1	3	1	0	0	1	0	0	0	3	3	1
Profilers	SC	0	0	1	1	2	0	0	0	0	0	0	0	2	2	0
Laser	AC	1	0	1	0	1	1	2	3	3	1	1	1	0	0	0
Fluorosensors	SC	0	0	0	0	0	0	1	1	1	0	0	0	0	0	0
Microwave	AC	1	0	1	0	0	1	1	1	3	3	2	2	1	2	3
Radiometers	SC	0	0	0	0	0	0	0	0	1	2	1	1	0	1	2
Imaging Radar	AC	2	1	3	3	1	1	0	0	3	1	1	2	3	3	2
(SAR or SLAR)	SC	1	0	2	2	1	0	0	0	2	0	0	1	2	2	1
Radar	AC	1	0	1	1	0	0	0	0	1	1	1	2	2	2	1
Altimeters	SC	0	0	0	0	0	0	0	0	0	0	0	2	1	1	0

Platform

AC = Aircraft (Medium or low altitude)

SC = Spacecraft (Satellite)

information), film photography with appropriate filters and LANDSAT MSS bands 4 and 5 can be quite useful. The same discussion applies to mapping pollutant concentrations in coastal waters (Whitlock et al., 1981). To map pollutant concentrations, good ground measurements and complex data analysis techniques are required for the multispectral scanner data (Klemas & Philpot, 1981).

Highly productive coastal upwelling areas have an effect on the global C and N cycles, which exceed their fractional areas and contribution to global primary productivity. As shown in Tables 2 and 3, NOAA and NIMBUS satellites provide thermal infrared (Very High Resolution Radiometer) and ocean color (Coastal Zone Color Scanner) data that have a suitable scale and spatial and temporal resolution for quantitative studies of the extent and intensity of coastal upwelling, fronts, eddies, and other ocean regions of high biological activity (Legeckis, 1975; Wilson & Smith, 1981). Multispectral scanners, such as the Coastal Zone Color Scanner, are providing valuable data on chlorophyll-a and ocean color (Wilson et al., 1978; Gordon & Clark, 1980). Chlorophyll-a concentrations have been mapped reliably within accuracies of ±50%. Chlorophyll concentrations can also be determined using low-altitude airborne lasers operating in the fluorosensing mode (Jarrett et al., 1979).

Thermal infrared scanners have been very effective for mapping ocean surface temperatures (about ±1°C accuracy), and for studying coastal surface currents (Wyrtki, 1977; Legeckis, 1978). Thermal infrared scanners on NOAA satellites, together with multispectral scanners such as the Coastal Zone Color Scanner, have been used to study coastal upwelling and coastal fronts. Estuarine fronts and their effects on oil dispersion have also been investigated with LANDSAT (Klemas, 1980).

Several aircraft and ground-based radar techniques are being developed for measuring currents (Klemas, 1980). Currents can also be measured using dyes and drogues tracked from shore or from aircraft (Klemas et al., 1974; Klemas et al., 1977). There are also photogrammetric methods for surveying tidal currents (Keller, 1963).

One of the most difficult properties to sense remotely is water salinity. Microwave radiometers employed from low-altitude aircraft have been able to map salinity to ±1 ppt, which is sufficient for estuarine studies, but not for deep ocean work (Swift, 1980).

As shown in Table 4, sea state and wave spectra are best obtained using laser profilers from aircraft, radar mappers (SAR), and radar altimeters (Ross et al., 1970; Schule et al., 1971; Panicker, 1974; Born et al., 1979). Imagers such as synthetic aperture radar or film cameras are particularly effective for wave studies if the data are analyzed using optical Fourier analysis techniques (Stilwell, 1969). Since surface winds induce capillary waves which influence microwave emission and reflectance, microwave sensors (particularly radar scatterometers such as the one on Seasat) have been tested for surface wind determinations (Born et al., 1979).

The capabilities of various remote sensors for measuring coastal properties shown in Table 4 are only estimates. The ratings assigned are quite subjective, depending on environmental conditions, availability of ground truth, and mode of sensor operation.

TYPICAL IMAGE ANALYSIS AND FIELD EQUIPMENT

Most remotely sensed data is recorded on magnetic tape in digital form. Aerial photo-interpretation equipment has been available at many laboratories since World War II. However, until the late 1970's, dedicated digital computer systems for image analysis were available only at major remote sensing institutes. Low-cost microcomputers have recently made such systems available to small institutions and state agencies. While there are many different types of systems in use, I will describe the inexpensive systems we use at the University of Delaware for analyzing data from four major satellites sensors: LANDSAT Multispectral Scanner, LANDSAT Thematic Mapper, Simulated SPOT imagery, NOAA Satellite Advanced Very High Resolution Radiometer (AVHRR) and Nimbus 7 Coastal Zone Color Scanner (CZCS).

As shown in Figure 1, at the University of Delaware's Center for Remote Sensing, an LA-120 terminal is used with ORSER software to analyze satellite digital tapes on the University's B7700 and DEC KL-10 computers in a batch mode. An ERDAS 400 System, developed by Earth Resources Data Analysis Systems of Atlanta, GA, is used to analyze satellite data interactively. The system is based on a Cromemco processor with a Tele-Video CRT console, and features a high resolution Mitsubishi color display, a joystick-controlled crosshair cursor for screen coordinate sensing and location, and an Anadex graphic printer for hardcopy products and

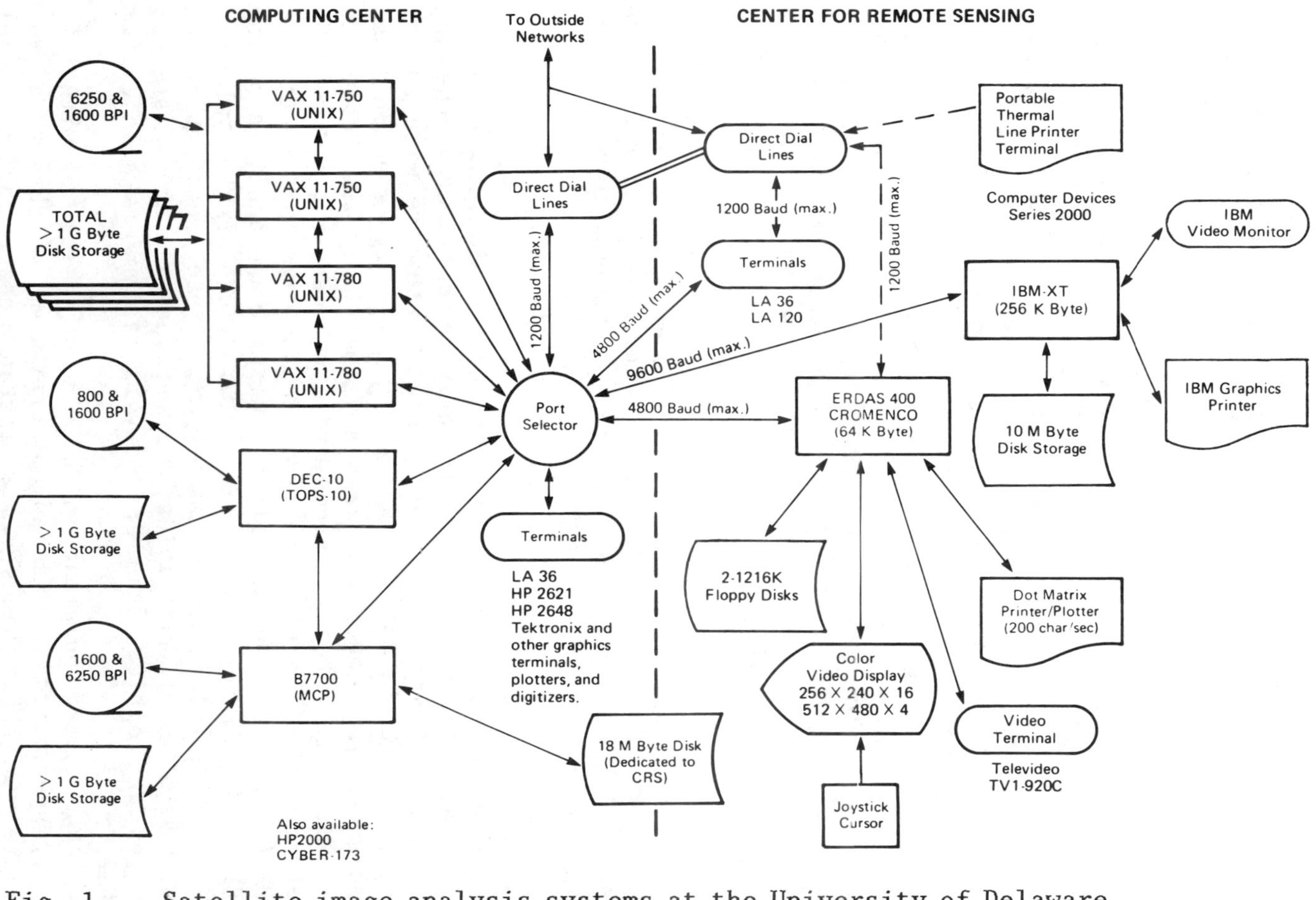

Fig. 1.   Satellite image analysis systems at the University of Delaware.

utility printing. The system is interfaced with the main-frames of the University's Computing Center through a dedi-cated data line.

The ERDAS' menu-driven "LANDSAT" software package enables the user to interactively display, enhance, and ana-lyze multiband digital imagery. Although designed to oper-ate with 4-band LANDSAT data, the ERDAS has been used effec-tively with TM, SPOT, and AVHRR imagery. Complete control over the display screen and color look-up tables enables the user to interactively explore and enhance the imagery. Band ratioing, filtering, edge enhancement, and geometric correc-tions are some of the options available. The ERDAS supports two major means of classification. Supervised classifica-tion entails defining signatures on the displayed image with the joystick. Unsupervised classification employs a power-ful spectral clustering algorithm. Images displayed on the 256 x 240 pixel display can be annotated and stored inde-pendently. The output of the classification routines are classified image files. These can be displayed on the color monitor in the 512 x 480 mode in 16 colors selected inter-actively from over 2 million. The GIS (Geographic Infor-mation System) package on the ERDAS facilitates proximity analysis, overlaying for multitemporal change detection, re-classifying and the creation of multilayer geographic databases to which logical and arithmetical functions can be assigned. The ERDAS is completely programmable in several high- or low-level programming languages and has a large applications software library, making it a very powerful and adaptable research tool.

The Center also owns IBM-AT and IBM-XT microcomputers. These stand-alone systems currently handle text processing and database management tasks. They interface with any of the University's VAX 11-780 mainframes. High-speed dial up and hardware modems are also maintained by the Center to transmit data between systems, research centers, and outside networks. Software programs are being developed to tie the IBM-AT into the ERDAS, and to perform image analysis on the IBM-AT at a higher speed and on a 1024 x 1024 pixel display.

A side selection of optical photo-interpretation equip-ment such as light tables, zoom transfer scopes, stereo-scopes and optical Fourier analysis systems is also avail-able.

The field equipment includes mapping cameras with fil-ters, spectrometers and radiometers, some matched to appro-priate satellite bands. For instance, our Exotech Model

100-A and Bendix Radiant Power Measuring instruments are matched to the LANDSAT MSS bands. Our GSFC Mark-II radiometers (biometers) have interchangeable bands that are presently matched to the LANDSAT Thematic Mapper bands. For more continuous spectral measurements in the field, we use the United Detector Technology Model 1100 A/B spectroradiometer, LI-COR Model LI-1800 spectroradiometer, and Biospherical Instruments Model MER-1000 spectroradiometer systems. The LI-1800 system includes a portable LCD terminal, plotter/printers, integrating sphere, remote cosine receptor, fiber-optic probe, and other accessories.

The MER-1000 measures upwelling and downwelling irradiance in 12 narrow bands (approximately 10 nm) between 400 and 700 nm. Measurements are made using an array of stable silicon photodiodes, each with its own interference filter and amplifier. There are no moving parts on the optical system, thus adding to the ruggedness and simplicity of the instrument. An internal 6802 microprocessor is used, controlling a 12-bit analog to digital converter with programmable gain, giving the system a dynamic range of 1,000,000 to 1. Scan time is typically 10 milliseconds, and the individual sensors have nominal time constants of 100 milliseconds. The MER-1000 is battery-powered and totally self-contained, allowing for easy and rapid data acquisition from ship, small boat, or diver. All data is stored internally on hardcopy using a thermal printer. If desired, the data may also be transmitted to the surface using an attached cable.

APPLICATION OF REMOTE SENSING TO ECOLOGICAL RESEARCH SITES AND ESTUARINE SANCTUARIES

In order to illustrate how the inexpensive remote sensing techniques and equipment described in previous sections could be applied to an ecological research site, I have chosen two estuarine sanctuary sites being established by the State of Maryland's Department of Natural Resources (DNR) in Chesapeake Bay (Maryland DNR, 1984). The Sanctuary Program, through DNR's Tidewater Administration, is making two sites available for field laboratory work, which will provide researchers with long-term opportunities to observe and explain changes to relatively undisturbed, representative sub-embayment systems in different ecological zones of the Bay.

   As shown in Figure 2, the Monie Bay site consists of
nearly 2000 acres of the Deale Island peninsula in Somerset
County on the Eastern Shore of Maryland. It is largely comprised of marshes that have been acquired over the years by
the State as a public wildlife recreation area for waterfowl
hunting and muskrat trapping. A variety of sport fish and
wildlife species, both resident and migratory, use the area
for food, shelter, nesting and/or breeding. The site is
composed of tidal creeks, open estuarine waters, marshes,
and pine forest areas. Most of the firm land is either high
or low marsh. The general terrain is flat, only a few feet

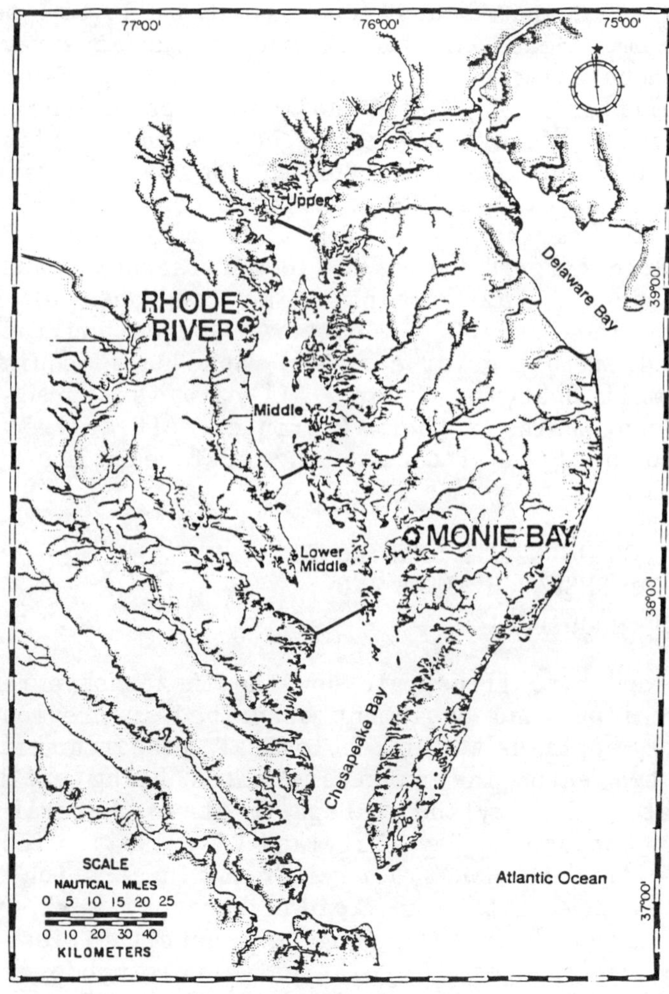

Fig. 2.   Regional setting and ecological zones for estu-
          arine sanctuary site selection.

above sea level, and has broad, expansive marshes. The salt marsh vegetation is characteristic of East Coast mid-salinity regimes. Low marsh zones are dominated by <u>Spartina alterniflora</u>, while high marsh areas have a mixture of <u>S. patens</u>, <u>S. cynosuroides</u>, <u>Distichlis spicata</u>, <u>Juncus roemerianus</u>, <u>Iva frutescens</u>, and <u>Scirpus</u>. The high and low marshes are brackish. <u>Panicum</u> is typical of the least brackish waters. Distribution of high marsh species is patchy, and no single species is dominant.

The Rhode River site encompasses 1185 acres of the lower Rhode River watershed, and includes most of the Smithsonian Institution's Environmental Research Center (SERC) at Edgewater. Tidal marshes at this site are located primarily along the Muddy Creek tributary. Marsh areas along the upper reaches of Muddy Creek form a narrow border between the upland forest and open water. They become much broader as the creek enters Rhode River. Low marshes are dominated by <u>Typha</u> or by <u>S. cynosuroides</u>. High marshes are more complex with patches of <u>I. frutescens</u>, <u>D. spicata</u>, <u>S. patens</u>, <u>Scirpus</u>, and <u>S. alterniflora</u>. Some high marsh areas are dominated by <u>Hibiscus moscheutos</u>. The salt marshes cover approximately 110 acres within the Rhode River test site.

Both test sites are relatively small and could be easily covered by aircraft or LANDSAT sensors. Since both sites have patches of different vegetation types and water types intermingled, this calls for a fine resolution system (such as aircraft cameras or the LANDSAT Thematic Mapper) that can provide better than 30 m resolution on the ground. Although film cameras might suffice for mapping vegetation types, land use and land/water boundaries, a multispectral scanner such as the Thematic Mapper (TM) is required to observe biomass and water properties. Furthermore, the 16-day TM cycle is sufficient for recording seasonal and most other changes, even if cloud cover negates two-thirds of the passes.

The Thematic Mapper spectral bands and resolution have been shown to be effective for mapping coastal land use, wetlands vegetation types, and above ground biomass at sites similar to the two sanctuary test sites (Hardisky <u>et al.</u>, 1983a). The seasonal variation of wetland biomass has also been determined using TM spectral bands (Hardisky <u>et al.</u>, 1983b). A comparison of remotely measured and field-harvested <u>S. alterniflora</u> biomass data over an entire growing season is shown in Figure 3. The Infrared Index is

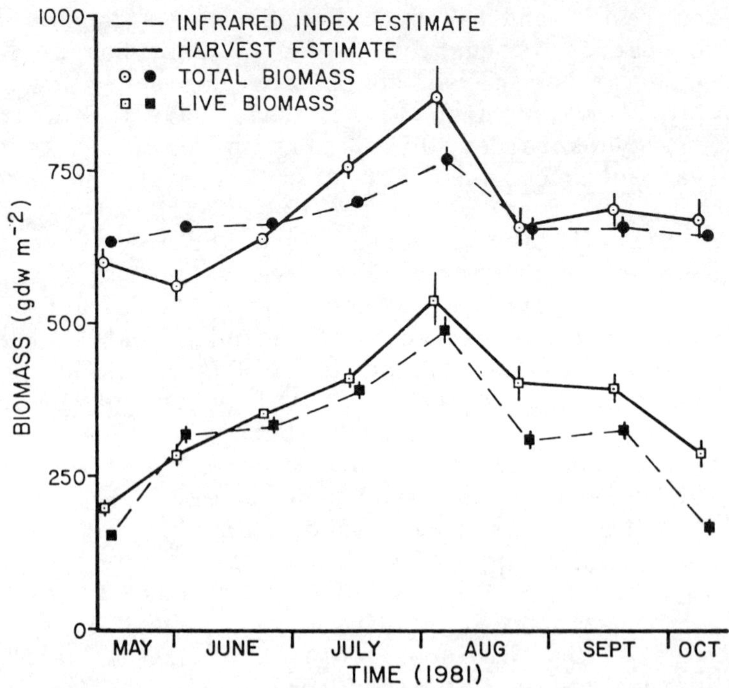

Fig. 3.   Comparison of remotely measured and field-harvested biomass of S. alterniflora over an entire growing season.

defined as the difference between the radiance measured in TM band 4 (0.76-0.90μ) and band 5 (1.55-1.75μ) divided by the sum of the radiances for normalization. Note that, in tidal marshes, both live biomass and dead biomass must be included in the analysis.

Submerged aquatic vegetation (SAV) has traditionally been mapped from low-altitude aircraft. Recently, the TM has been tested successfully for detecting SAV in Chesapeake Bay (Ackelson & Klemas, 1983). We have also found that the TM can be used to map suspended sediment concentrations and identify water masses dominated by organic and inorganic substances. However, in order to monitor various stages of the same tidal cycle, one would have to use aircraft with color film cameras. If a camera with color infrared film were added to the aircraft payload, patches of wetlands vegetation too small for the 30 m TM resolution could also be mapped. Thus, while the LANDSAT Thematic Mapper would be the primary sensor, aircraft with simple color and color IR film cameras would "fill in the gaps" by providing better temporal and spatial resolution as baseline data is being gathered. Once the baseline has been established, the Thematic Mapper, with its repetitive coverage, can be used to monitor changes at the test sites and compare them with ecological changes occurring over entire estuaries or other coherent ecological regions. If the French SPOT satellite is launched in 1985, its 10 m resolution and more frequent coverage may further reduce the need for aircraft missions (Ackleson et al., 1984).

Different mixing rates of fresh and salt waters create different habitats for biological communities. Available remote sensors cannot reliably measure salinity, even though they can determine flood stage and current circulation patterns. Microwave radiometers have been used to map salinity in estuaries to within 1 ppt (Swift, 1980). This accuracy may suffice for gross delineation of water mass types or estuarine fronts; however, the equipment is expensive and cumbersome to operate. Therefore, salinity will have to be measured from boats, at least for the next five years.

COST CONSIDERATIONS

There is considerable evidence that aerial photography is more cost-effective than ground surveys for mapping large areas; however, it is difficult to obtain detailed cost information between aircraft and satellite surveys.

The curves in Figure 4, obtained from Bartlett and Klemas (1979), compare the cost of mapping coastal vegetation in situ from aircraft and from satellites (LANDSAT). From these curves, two conclusions can be drawn: (a) field and aircraft surveys will be too expensive for data acquisition on a global scale; and (b) even with satellites, only areas undergoing rapid change or "representative" sites can be inventoried. As a result, old maps and data will have to be used in areas where no rapid changes are occurring, with new aircraft or satellite data filling in gaps or focusing on areas of rapid change.

Most of the global data on vegetative land cover, land use change, etc., already exists and needs to be "pulled together" from a wide range of projects and agencies (e.g., Food and Agriculture Organization, UNEP, AgRISTARS). Many agencies (FAP, UNEP, UN Man and the Biosphere Program) and many countries have sponsored land use and land use change inventories. These data will be collected and validated as a first step towards defining the most rapidly changing regions, or the regions where our current information is quantitatively or qualitatively inadequate. A data bank will have to be developed at some institution to collect, store, and digest all existing data so that it will be accessible in a usable form to global ecosystem modellers. Note that additional satellite or aircraft surveys need to provide only inventories, not cartographic maps. New maps meeting cartographic standards are far too expensive to produce and are not required for this global study.

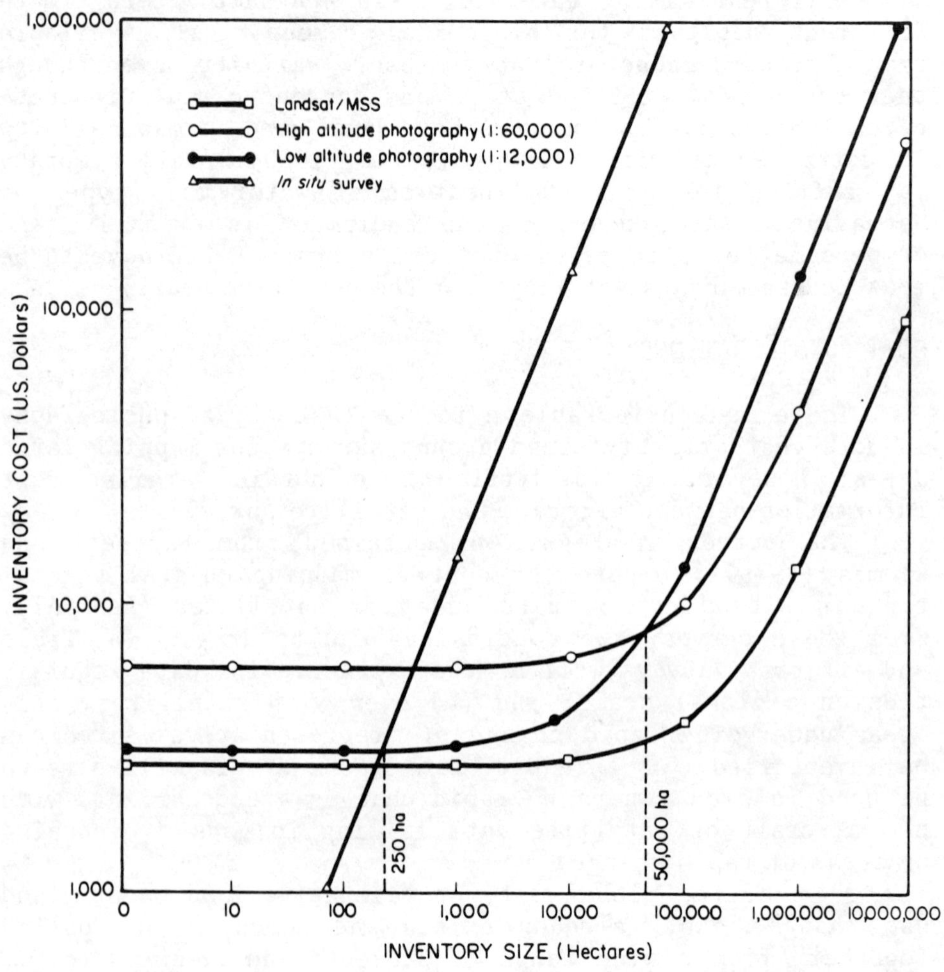

Fig. 4.   Inventory cost vs. total area of wetlands inventoried for three remote sensing platforms and in situ survey.

The cost of LANDSAT data has significantly increased since Figure 2 was prepared. Figure 4 suggests that for simple vegetation inventories of small test sites, aerial photography should be seriously considered. However, for repetitive observations of vegetative changes, biomass and water properties of medium-size test sites, the Thematic Mapper will give better results at a lower cost in the long run.

DATA COLLECTION AND MANAGEMENT PROTOCOL

There are several approaches for using remotely sensed data in studies of ecological research sites and in the management of estuarine sanctuaries. If at this stage we are willing to overlook certain details and refinements, a typical protocol could be as follows:

1. The location (coordinates) and boundaries of the test site should be defined clearly on an appropriate map.

2. Local, state, and federal agencies should be searched for appropriate aerial photographs and maps. For instance, the US Agriculture Department Soil Stabilization Service has mapped most of the area around the Chesapeake Bay every eight years since 1932. Historical maps and aerial photographs are important for studies of man-made and natural changes and for planning future tests.

3. A search of LANDSAT MSS and TM data for the test site coordinates should be requested from the USGS EROS Data Center in Sioux Falls, SD, or from the USGS National Cartographic Information Center in Reston, VA.

4. Researchers should review the computer printout of the LANDSAT imagery search provided by EROS to select suitable LANDSAT passes of test sites for appropriate season and year. The selection should be optimized by choosing minimum cloud cover.

5. Only transparencies and prints should be ordered. From transparencies, the best dates can be selected for which digital tapes should be ordered. By projecting transparencies onto a screen, one can carefully check contrast, haze, and cloud cover to ensure that the features of interest are indeed visible. Both contrast between vegetation species and water types should be checked.

6.   Only those LANDSAT MSS or TM digital tapes that are likely to show the required features (with sufficient spatial resolution and spectral discrimination) should be ordered.
7.   Assistance should be obtained from a local remote sensing group to perform the analysis of the digital tapes. Image analysis techniques for various applications are discussed in Moik (1980) and Klemas (1983).
8.   To obtain more frequent coverage of rapidly varying phenomena, such as the variation of suspended sediment patterns and fronts over a tidal cycle, employ a small aircraft from a local airport. In many cases, one can perform current circulation and drogue studies from a small aircraft using hand-held cameras (Klemas et al., 1974; Klemas et al., 1977).

CONCLUSIONS

Global biosystem studies will require satellite observations of the Earth at global scales. Such studies also will require ecological research sites in various climate zones for calibrating and verifying satellite data; for developing conversion factors to relate directly measured properties (e.g., biomass) to indirectly deduced properties (e.g., gas emission); to gather new data at specific test sites (e.g., methane gas emission from marshes); and for developing new measurement techniques.

New remote sensing systems such as the LANDSAT Thematic Mapper can provide synoptic observations of ecological research sites and estuarine sanctuaries, establishing baseline conditions and monitoring changes via repetitive coverage. The Thematic Mapper has the resolution and spectral bands for detecting land use, vegetation types, biomass, flood stage, current circulation patterns, water turbidity, and organic/inorganic substances in water. In some instances, remotely sensed data is more representative of average conditions than data collected along specified field transects.

New low-cost microcomputer systems now make it possible for most investigators to analyze satellite data at their own institutions. User-friendly software programs are being developed for this purpose.

Therefore, it is important that scientists and managers working with ecological research sites or estuarine sanctuaries seriously consider the inclusion of remotely sensed data in their ecological data banks. The long-term cost of including remotely sensed data may actually be lower than relying totally on data collected in the field.

LITERATURE CITED

Ackleson, S.G. and V. Klemas. 1983. Assessing Landsat TM and MSS data for detecting submerged plant communities. pp. 325-336. In: Proceedings of Landsat-4 Science Characterization Early Results Symposium, Vol. IV: Applications. J.L. Barker (ed.). NASA Goddard Space Flight Center, Greenbelt, MD, February 22-24, 1983. NASA Conference Publication 2355, Greenbelt, MD.

Ackleson, S.G., V. Klemas, H.L. McKim, and C.J. Merry. 1984. A comparison of SPOT simulator data with Landsat MSS imagery for delineating water masses in Delaware Bay, Broadkill River, and adjacent wetlands. pp.178-189. In: 1984 SPOT Symposium. Spot Image Corp., Washington, DC (ed.). Scottsdale, AZ, May 20-23, 1984. American Society of Photogrammetrics, Falls Church, VA.

Bartlett, D.S. and V. Klemas. 1979. Assessment of tidal wetland habitat and productivity. pp. 693-703. In: Proceedings of 13th International Symposium on Remote Sensing of Environment. J. Cook (ed.). ERIM, Ann Arbor, MI.

Bartlett, D.S. and V. Klemas. 1980. Quantitative assessment of tidal wetlands using remote sensing. Environ. Management 4(4): 337-345.

Bartlett, D.S. and V. Klemas. 1981. In situ spectral reflectance studies of tidal wetland grasses. Photogrammetric Engineering and Remote Sensing 47(12): 1695-1703.

Born, G.H., J.A. Dunne, and D.B. Lane. 1979. Seasat mission overview. Science 204: 1405-1406.

Dolan, R. 1973. Coastal processes. Photogrammetric Engineering 39(2): 255-260.

Dolan, R., D. Hayden, J. Heywood, and L. Vincent. 1977. Shoreline forms and shoreline dynamics. Science 197: 49-51.

Gordon, H.R. and D.K. Clark. 1980. Atmospheric effects in the remote sensing of phytoplankton pigments. Boundary Layer Meteorology 18: 299-313.

Hardisky, M.A., V. Klemas, and F.C. Daiber. 1983a. Remote sensing salt marsh biomass and stress detection. Adv. Space Res. 2(8): 219-229.

Hardisky. M.A., R.M. Smart, and V. Klemas. 1983b. Seasonal spectral characteristics and aboveground biomass of the tidal marsh plant, Spartina alterniflora. Photogrammetric Engineering and Remote Sensing 49(1): 85-92.

Hoge, F.E., R.N. Swift, and E.B. Frederick. 1980. Water depth measurement using an airborn pulsed neon laser system. Applied Optics 19(6): 871-883.

Jarrett, O., Jr., C.A. Brown, Jr., J.W. Campbell, W.M. Houghton, and L.R. Poole. 1979. Measurement of chlorophyll fluorescence with an airborne fluorosensor. pp. 703-712. In: Proceedings of 13th International Symposium on Remote Sensing of Environment. J. Cook (ed.). ERIM, Ann Arbor, MI.

Johnson, R.W. 1975. Quantitative suspended sediment mapping using aircraft remotely sensed multispectral data. pp. 2087-2098. In: Earth Resources Survey Symposium. Houston, TX. NASA, Washington, DC.

Keller, M. 1963. Tidal current survey by photogrammetric methods. Photogrammetric Engineering 29(5): 824-832.

Klemas, V. 1980. Remote sensing of coastal fronts and their effects on oil dispersion. International Journal of Remote Sensing 1(1): 11-28.

Klemas, V. 1981. Proceedings of National Research Council Ocean Policy Committee. La Jolla, CA.

Klemas, V. 1983. The use of computer graphics in remote sensing of coastal and marine resources. Presented at Harvard Computer Graphics Conference. Cambridge, MA, July 31 to August 4, 1983.

Klemas, V., D. Bartlett, and R. Rogers. 1975. Coastal zone classification from satellite imagery. Photogrammetric Engineering and Remote Sensing 41(3): 499-513.

Klemas, V., G. Davis, J. Lackie, W. Whalen, and G. Tornatore. 1977. Satellite, aircraft and drogue studies of coastal currents and pollutants. IEEE Transactions on Geoscience Electronics GE15(2): 97-108.

Klemas, V., D. Mauer, W. Leatham, P. Kinner, and W. Treasure. 1974. Dye and drogue studies of spoil disposal and oil dispersion. J. Water Poll. Control Fed. 46(8): 2026-2034.

Klemas, V. and W. Philpot. 1981. Drift and dispersion studies of ocean-dumped waste using Landsat imagery and current drogues. Photogrammetric Engineering and Remote Sensing 47(4): 533-542.

Legeckis, R. 1975. Application of synchronous meteorological satellite data to the study of time dependent sea surface temperature changes along the boundary of the Gulf Stream. Geophys. Res. Letters 2: 10.

Legeckis, R. 1978. A survey of worldwide sea surface temperature fronts detected by environmental satellites. J. Geophys. Res. 83(C9): 4501-4522.

Maryland Department of Natural Resources Tidewater Administration. 1984. Chesapeake Bay National Estuarine Sanctuary in Maryland: Multiple site sanctuary management plan. Final draft. 99 pp.

Maul, G.A., R.L. Charnell, and R.H. Qualset. 1974. Computer enhancement of ERTS-1 images for ocean radiances. Remote Sensing of Environment 3: 237-254.

Moik, J.G. 1980. Digital Processing of Remotely Sensed Images. NASA Scientific and Technical Information Branch, NASA SP-431. Washington, DC.

Moore, G.K. 1978. Satellite surveillance of physical water-quality characteristics. pp. 445-462. In: Proceedings of 12th International Symposium on Remote Sensing of Environment. J. Cook (ed.). ERIM, Ann Arbor, MI.

Munday, J.C. and T.T. Alfoldi. 1979. Landsat test of diffuse reflectance models for aquatic suspended solids measurement. Remote Sensing of Environment 8: 169-183.

Odum, E.P. 1961. The role of tidal marshes in estuarine production. New York State Conservationist 16: 12-15.

Panicker, N.N. 1974. Review of techniques for directional wave spectra. pp. 669-688. In: Proceedings of International Symposium on Ocean Wave Measurement. American Society of Civil Engineers (ASCE). ASCE, Washington, DC.

Proceedings of NASA Earth Resources Survey Symposium. 1975. NASA, Houston, TX.

Rambler, M.B. (ed.). 1983. Global Biology Research Program. NASA Technical Memorandum 85841. Washington, DC. 106 pp.

Ross, D.B., V.J. Cardone, and J.W. Conaway, Jr. 1970. Laser and microwave observations of sea-surface conditions for fetch-limited 17- to 25-m/s winds. IEEE Transactions on Geoscience Electronics GE8(4): 326-336.

Schule, J.J., L.S. Simpson, and P.S. DeLeonibus. 1971. A study of fetch-limited wave spectra with an airborne laser. J. Geophys. Res. 76: 4160-4171.

Stafford, D.B. and J. Langfelder. 1971. Air photo survey of coastal erosion. Photogrammetric Engineering 37(6): 565-575.

Stilwell, D. 1969. Directional energy spectra of the sea from photographs. J. Geophys. Res. 74(8): 1974-1986.

Swift, C.T. 1980. Passive microwave remote sensing of the ocean: A review. Boundary Layer Meteorology 18: 25-54.

Whitlock, C.H., L.R. Poole, J.W. Usry, W.M. Houghton, W.G. Witte, W.D. Morris, and E.A. Gurganus. 1981. Comparison of reflectance with backscatter and absorption parameters for turbid waters. Applied Optics 20(3): 517-522.

Wilson, W.H., R.W. Austin, and R.C. Smith. 1978. Optical remote sensing of chlorophyll in ocean waters. pp. 1103-1113. In: Proceedings of 12th International Symposium on Remote Sensing of Environment. J. Cook (ed.). ERIM, Ann Arbor, MI.

Wilson, W.H. and R.C. Smith. 1981. Ship and satellite bio-optical research in the California Bight. pp. 281-294. In: Oceanography from Space. J.F.R. Gower (ed.). Plenum Press, NY.

Wyrtki, K. 1977. Advection in the Peru Current as observed by satellite. J. Geophys. Res. 82(27): 3939-3943.

ROUTINES FOR COLLECTING AND SUMMARIZING HYDROMETEOROLOGICAL DATA AT COWEETA HYDROLOGIC LABORATORY

L.W. Swift, Jr. and G.B. Cunningham

ABSTRACT

An extensive series of hydrometeorological data, collected and available at Coweeta Hydrologic Laboratory, has been one factor supporting expansion of ecological research at this Appalachian Mountain site in western North Carolina. Continuous records of streamflow, precipitation, and other meteorological variables began in 1934. Since then, data collection, processing, and archiving have changed from tedious manual techniques to computerized methods that have significantly increased accuracy and availability of information. Methods developed at Coweeta have been adopted and applied by national and international research groups. Data are frequently shared with other investigators, in part because Coweeta is a site in the Long-Term Ecological Research program. This paper reviews the concepts and procedures developed to collect, edit, and summarize hydrometeorological data at Coweeta and is intended as a reference for users of these data.

INTRODUCTION

For two decades, watersheds have been recognized as useful vehicles for research on complex ecological systems (Bormann & Likens, 1967; Hewlett _et al_., 1969; Johnson & Swank, 1973). A watershed is the natural base for studying and modeling terrestrial systems, because inputs and outputs can be defined and quantified, and integrated system responses determined. Water moving within and through a basin links living and nonliving components. With water as the transfer agent, internal processes may be described, modeled, and related to the net response of the entire ecosystem. This natural unit is easily visualized in a mountain setting where boundaries for precipitation, evaporation, and streamflow are clearly defined by topography.

In the United States, several watershed research areas have become focal points for ecosystem research. Coweeta Hydrologic Laboratory in western North Carolina is such a place. Here, use of experimental and control (undisturbed) watersheds for integrated ecological research has burgeoned over the last 20 years. Scientific activity is supported by long-term records of climate, precipitation, streamflow, land use, and plant and animal populations. The questions asked most often by scientists conducting research at such sites are: How does the period of my measurements compare with other years? and How do differences in climate and streamflow affect the interpretation of my data? If long-term data are to be useful for addressing these questions, they must be obtained by acceptable and carefully documented procedures with strict attention to effective quality controls. This paper describes methods for collecting, processing, and reporting meteorological and hydrological data at Coweeta. Emphasis is given to quality-control practices and user-oriented outputs developed over the past 50 years. Separate sections are devoted to precipitation, climatic, and streamflow data.

Coweeta Hydrologic Laboratory is a 2185-ha forested area in the southern Appalachian Mountains of western North Carolina (Fig. 1). The majority of the hydrological and meteorological measurements have been made in an east-facing, bowl-shaped 1626-ha basin containing two third-order streams which join to form Coweeta Creek. Soils are deep, averaging 7 m, and elevation ranges from 678 m to 1592 m above sea level. Precipitation is relatively abundant throughout the year, about 150 mm per month. At latitude 35°N, winters are quite moderate for a mountain site while summers are cool, even in the valley bottoms. Multistoried deciduous forests thrive in a climate classifed according to Thornthwaite's (1948) system as perhumid and mesothermal with a water surplus in all seasons.

The Laboratory was established in 1934 by the USDA Forest Service to study effects of land-use practices upon the quantity and quality of water in southeastern mountain

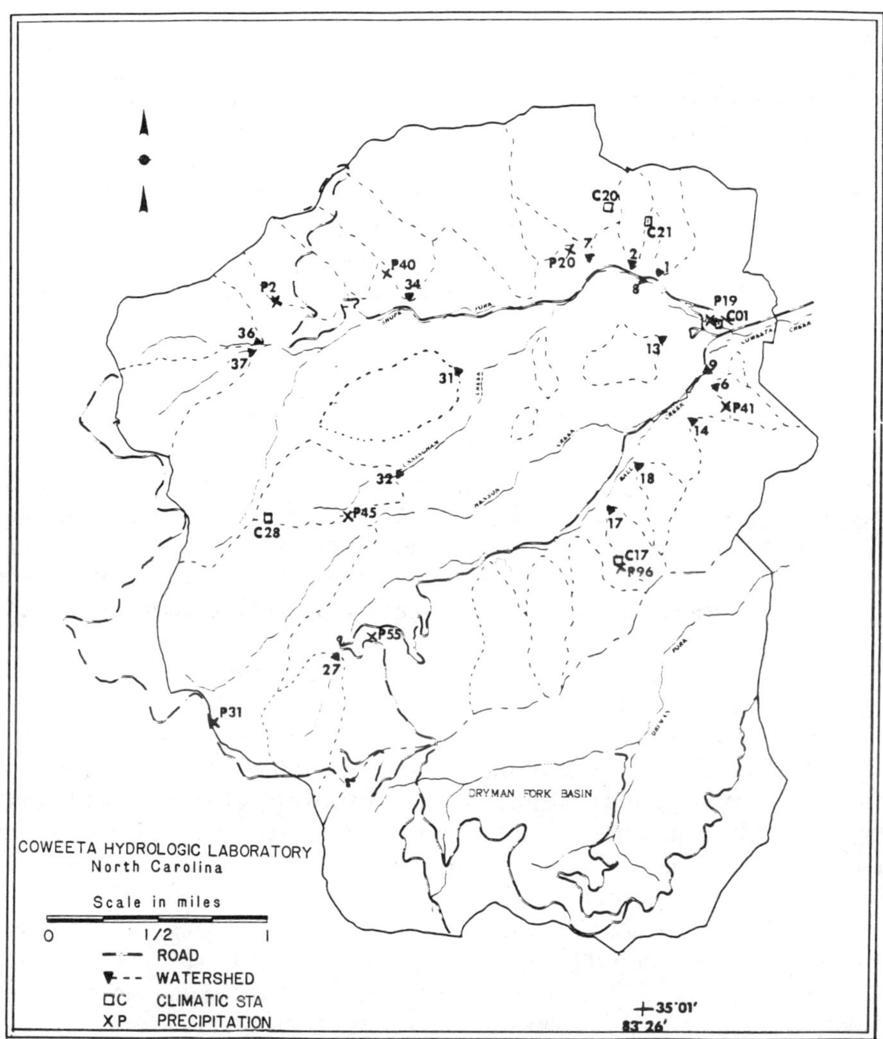

Fig. 1.   Map of 2185-ha Coweeta Hydrologic Laboratory, a
Forest Service research site in western North
Carolina, showing principal hydrometeorological
measurement points.

streams. By the start of World War II, networks of precipitation gages, climatic stations, and weirs had been
installed. Despite severe shortages of supplies and man-
power, measurements were maintained through the 1940's, so
an unbroken 50-year record exists for key sites. Maintenance of instruments and collection, processing, and
storage of long-term hydrometeorological data have always
been important functions at Coweeta.

QUALITY CONTROL

Throughout the various steps of data processing, quality control is an ever present concern. At Coweeta, checks are made for validity of routine data in several ways. As charts and tapes are collected, trained technicians check for time errors, indications that pen or recorder mechanisms are not moving freely, calibration shifts, and signs of vandalism. A permanent log is kept on instrument faults, servicing, modifications, and quality control observations (checks). Records from adjacent sampling sites are compared for consistency as they are collected and again when they are processed or summarized. To aid in summarizing and comparing, a midnight-to-midnight time base is used for all meteorological and hydrological data. A shifted 8am to 8am day is used only in reporting meteorological observations to the National Weather Service and Tennessee Valley Authority. As observations are entered or translated into computer files and during all stages of processing, programs check for inconsistencies in dates and times, illogical changes in values between successive observations, or values outside expected ranges.

Quality control checks do not end when the printout is filed. The most demanding check is made by the user, who may identify inconsistencies by plotting graphs, performing statistical routines, and using data in calculations and simulations. Users are encouraged to test and question the data. We have attained high-quality records at Coweeta by having the same individual supervising collection and processing since computerization began. In this way, consistent decisions have been made and standards maintained. Backup personnel are trained to sustain this consistency in the future.

Original, extracted, and complied records are protected by storage in an isolated vault. Copies of frequently used results are maintained in the office workspace and requests for shared data are met by duplicating rather than by loaning original records or computer files.

PRECIPITATION NETWORK

Precipitation measurements were initiated at Coweeta in 1934. The long-term precipitation network of nine gage sites gives a sampling density of one gage per 180 ha (Fig. 1). Each site consists of an 8-inch standard gage, a weighing recording gage (Brakensiek et al., 1979), and a collector for precipitation chemistry. For our brief snow season, the inner funnel is removed from the recording gage and an extra standard gage is exposed without inner cylinder or funnel. The contents of the snow gage are measured only when there is a possibility that snow has piled on the standard-gage funnel, causing part of the sample to be lost. Most gage sites are openings in the forest, cut with a radius equal to tree height. Once a week, recording charts are changed and standard gages read by dipstick to the nearest 0.25 mm. The standard-gage observation is accepted as our best estimate of the precipitation amount; the recording gage is used only to determine the time and rate of precipitation. Comparative measurements in Coweeta's climate show that standard-gage observations are not altered appreciably over 7 days by evaporation.

A base network of 50 standard gages was operated for 20 years, and a total of 135 sites have been gaged for various periods over the last 50 years. Until the late 1950's, standard gages were read by dipstick after nearly every storm. Manual observations are written on a pocket-sized "trail form," which is retained as a permanent original record. The individual observations for all gages are entered and summed in a monthly report, which is the primary reference document (Fig. 2). To ensure accuracy of this critical step, each entry on the monthly report is verified by an employee other than the one who created the report. The most frequently used precipitation file is the individual-gage summary, which lists only monthly precipitation totals and seasonal and annual sums for the total period of record. It is updated approximately biennially for all gages and monthly for the gage at Climatic Station 01. Other records are derived from this file, such as weighted monthly precipitation for experimental watersheds and the climatological display of monthly, seasonal, and annual totals by increasing magnitude (Fig. 3).

Initially, average precipitation for each watershed was calculated using Thiessen polygon weighting factors (Wisler & Brater, 1959). Watershed precipitation is now estimated using factors calculated from map zones based upon the mean isohyetal pattern. Isohyetal lines were drawn using information from all 135 gage sites. Rainfall distribution patterns such as reduced precipitation next to ridges or local zones of high precipitation on wind-sheltered (lee) sides of gaps in ridgelines could be incorporated. Thus, the new

MONTHLY RECORD
OF
STANDARD RAIN GAUGES

Coweeta Hydrologic Laboratory

Month September Year 1982

Computed by M. Payne

Checked by N. Buchanan

Date		Rain Gauge Number													Remarks
Read	Rain of	19	2	20	31	40	41	45	55	96					
1	1	.05						1.99							
2	1	3.78													
7	1		3.69	3.74	2.69	356	3.89		2.88	3.23					
8	1							1.46							
10	10	.04													
13	11	1.51													
14	13	.02													
14	10,11,13		1.72	1.61		1.62									
14	10-13				1.98				1.69	1.54					
14	10,11						1.51								
15	10,11,13							1.78							
21	20		.02	.02	.08	.02	.01		.03	.04					
22	20							.02							
27	25,26	1.04													
28	25,26		1.05	1.07	1.39	1.04	1.09		1.28	1.31					
29	25,26							1.20							
Monthly Totals		6.44	6.48	6.44	6.14	624	6.49	6.45	5.88	6.12					

Form 7 (Rev. 7/44)

Sheet 1 of 1

Fig. 2.   Monthly precipitation standard-gage record for September 1982.

MONTHLY PRECIPITATION BY INCREASING AMOUNTS   (INCHES)
GAGE 19 COWEETA

FALL QUARTER

SEQUENCE		AUG YR	AUG AMT	SEP YR	SEP AMT	OCT YR	OCT AMT	4THQ YR	4THQ AMT
1	50	81	1.63	40	0.82	63	0.03	54	4.89
2	49	51	1.76	54	1.08	38	0.20	38	5.26
3	48	38	2.19	39	1.09	78	0.27	63	6.69
4	47	82	2.29	78	1.09	54	0.87	39	8.35
5	46	63	2.46	35	1.16	48	1.03	53	8.63
6	45	56	2.55	34	1.98	53	1.11	73	9.07
7	44	71	2.70	58	2.47	39	1.15	78	9.34
8	43	57	2.75	61	2.60	52	1.27	58	9.36
9	42	53	2.78	37	2.60	44	1.40	41	9.54
10	41	41	2.81	55	2.66	58	1.42	52	9.95
11	40	54	2.94	38	2.87	73	1.65	81	10.36
12	39	83	3.08	41	3.20	74	1.89	51	11.30
13	38	59	3.39	74	3.27	40	1.94	71	11.32
14	37	75	3.52	52	3.39	61	2.12	55	11.69
15	36	45	3.70	73	3.47	35	2.48	74	12.15
16	35	42	3.73	72	3.65	43	2.60	43	12.32
17	34	80	3.88	48	3.69	80	2.80	82	12.37
:	:	:	:	:	:	:	:	:	:
37	14	77	6.63	60	6.19	66	6.01	57	19.04
38	13	47	6.71	82	6.44	60	6.02	65	19.16
39	12	44	6.83	59	6.49	83	6.56	66	19.22
40	11	74	6.99	80	6.53	72	6.95	79	19.70
41	10	78	7.98	44	7.20	57	7.35	36	20.22
42	9	48	8.09	66	7.36	34	7.51	67	20.47
43	8	49	8.69	45	7.91	47	8.02	69	20.47
44	7	60	8.99	77	8.16	75	8.40	77	20.59
45	6	35	10.02	65	8.55	59	8.50	70	21.19
46	5	50	10.08	57	8.94	37	8.59	60	21.20
47	4	67	11.35	79	9.63	76	8.87	50	21.65
48	3	61	12.03	36	10.11	49	9.77	49	24.25
49	2	69	12.06	64	10.46	70	10.55	75	25.02
50	1	40	13.38	75	13.10	64	11.46	64	28.48
MEANS			5.56		5.04		4.38		14.97

Fig. 3.   Portions of the table of precipitation by increasing amounts at Gage 19 for three fall months.

weighting factors are more realistic for mountain sites than those obtained by the inflexible Thiessen polygon system.

The computerized data set for storm duration and precipitation intensity is based upon the recording gage record, adjusted to agree with the standard-gage observations. Storms are defined as events separated by at least 4 h without measurable precipitation (defined as <0.25 mm for manual observation, whereas <0.5 mm is the practical sensitivity for a good chart recorder trace). The 4-h time period best delimits those storms for which streamflow hydrographs do not overlap. Detailed storm data are extracted from charts (as described below) for seven of the nine gages in the long-term network.

One technique is to divide a storm into 60-min increments, ignoring changing rainfall rates within the hour. At

Coweeta, the chart record for each storm is divided into constant intensity periods. The duration of an intensity period will range from the minimum time sensitivity of the chart-reading operation to several hours. Time and accumulated amount are read from the chart at each selected breakpoint. Before 1958, points were picked by hand from charts with a time precision of 5 to 10 min. The electro-mechanical oscillograph chart reader used through 1983 had a precision of 90 sec; its replacement is set to read to the nearest minute. This level of precision is valuable for determining the relative lengths of precipitation intensity periods, but it does not imply that real time is known to this accuracy. Most clock-driven chart recorders in use at Coweeta can be set to the nearest 15 min and are accurate to the nearest 30 min at the end of a week. Manual point-picking can determine precipitation amount to the nearest 0.5 mm, whereas chart readers are able to digitize repeatedly within 0.25 mm. Extracted data are obtained with a precision that exceeds the accuracy of the precipitation gage solely for the purpose of better defining the individual intensity rates within a storm.

Approximately once a year, recording gages are cleaned and calibrated to within 0.5 mm. Charts are checked weekly to ensure that chart time and real time agree and that the precipitation amount on the chart agrees with that measured in the standard gage. Times should agree within 30 min and precipitation within 4 percent. We have retained the pen-and-chart recording gage because the sensitivity of the available punch-tape gage was only 0.1 inch. An important part of our total annual precipitation falls in small storms that would be missed by the less sensitive gage. Tipping bucket gages with 1-mm sensitivity are being added to the network as part of a conversion to data loggers at climatic stations. At present, the tipping bucket record provides only mean 60-min intensities; we will continue to rely on the chart recorders for detailed precipitation records.

Since computerization, precipitation intensities have been printed in a two-part format. The first part is an index to all storms, listing the time and date the storm began, its duration and total amount, and the 5-, 15-, 30-, and 60-min maximum intensities (for all storms over 6.35 mm or 0.25 inch). A serial number is assigned to each storm, beginning with 1 at the start of each water-year (November through October). The detailed intensity record (Fig. 4) lists all intensity periods within each storm. An intensity

COWEETA RRG NO.   20   PRECIPITATION INTENSITY RECORD   WATERYEAR 1982   PAGE 1

STORM NO.	DATE YR MO DAY	TIME OF DAY (HRS-MIN)	CUMULATIVE STORM DURATION (HRS)	TIME INCREMENT (MIN)	INTENSITY (IN/HR)	RAINFALL INCREMENT (IN)	RATIO OF MONTHLY TOTAL RAIN	CUMULATIVE STORM RAINFALL (IN)	STORM OR RAINFALL (IN)	DAILY TOTAL RATIO OF MONTHLY TOTAL
155	82 9 20	1107							0.02	0.0179
NO MAXIMUM INTENSITIES CUMULATIVE DEPTH LESS THAN 0.25 INCHES										
		1124	0.28	17	0.04	0.01	0.0089	0.01		
		1151	0.73	27	0.02	0.01	0.0089	0.02	0.02	0.0179
156	82 9 25	857							0.05	0.0446
NO MAXIMUM INTENSITIES CUMULATIVE DEPTH LESS THAN 0.25 INCHES										
		919	0.37	22	0.03	0.01	0.0089	0.01		
		940	0.72	21	0.09	0.03	0.0268	0.04		
		1007	1.17	27	0.02	0.01	0.0089	0.05		
157	82 9 25	2224							0.80	0.7143
STORM MAXIMUM INTENSITIES 60 MIN 0.41 IN/HR 30 MIN 0.61 IN/HR 15 MIN 0.77 IN/HR 5 MIN 0.77 IN/HR										
		2241	0.28	17	0.04	0.01	0.0089	0.01		
		2306	0.70	25	0.24	0.10	0.0893	0.11		
		2327	1.05	21	0.77	0.27	0.2411	0.38		
		2400	1.60	33	0.16	0.09	0.0804	0.47	0.52	0.4643
	82 9 26	46	2.37	46	0.14	0.11	0.0982	0.58		
		114	2.83	28	0.04	0.02	0.0179	0.60		
		151	3.45	37	0.24	0.15	0.1339	0.75		
		240	4.27	49	0.04	0.03	0.0268	0.78		
		432	6.13	112	0.01	0.01	0.0089	0.79		
		453	6.48	21	0.03	0.01	0.0089	0.80		
158	82 9 26	1322							0.22	0.1964
NO MAXIMUM INTENSITIES CUMULATIVE DEPTH LESS THAN 0.25 INCHES										
		1349	0.45	27	0.02	0.01	0.0089	0.01		
		1437	1.25	48	0.09	0.07	0.0625	0.08		
		1559	2.62	82	0.01	0.01	0.0089	0.09		
		1700	3.63	61	0.05	0.05	0.0446	0.14		
		1853	5.52	113	0.01	0.02	0.0179	0.16		
		1925	6.05	32	0.11	0.06	0.0536	0.22	0.55	0.4911

Fig. 4.   Precipitation intensity record for a portion of September 1982 for Recording Gage 20.

period record includes beginning date and time, duration and precipitation increments, and calculated intensity rate. Cumulative storm duration and amount are given for each period and totals extended for each date and storm. When a storm overlaps the end of the month, a subtotal for each month is listed. Less detailed listings of hourly or daily precipitation amounts are derived from the intensity period file.

Weighted watershed precipitation is normally computed as a monthly total. To assist in dividing monthly totals into shorter period amounts, the intensity output also lists intensity-period, day, and storm amounts as ratios of the monthly amount.

CLIMATIC STATIONS

Other meteorological observations began in 1935 at the main climatic station (CS 01) in the valley bottom near the Laboratory headquarters and, in 1936, at six sites distributed throughout the basin. Initially, manual observations were made once a day at the remote sites and twice daily at CS 01. Parameters included air temperature, relative humidity, wind travel, evaporation, soil temperature, and stream temperature. Measurements were abandoned by 1940 at many sites, and recording hygrothermographs were substituted for daily observations at others. In 1965, solar radiation observations began and, since 1969, four climatic stations have been reestablished on the slopes above the valley floor (Table 1). In 1984, battery-operated electronic data loggers were installed at all climatic stations within the Coweeta basin.

Table 1. Location of active climatic stations at Coweeta.

Station Name	Elevation (meters)	Slope Type	Type of Site
CS01	685	bottom	large opening, grassed
CS17	887	north-facing	forest opening
CS20	866	south-facing	regrowing forest after logging
CS21	817	south-facing	forest opening
CS28	1189	east-facing	forest opening

The primary records for older climatic stations are the observer's log sheets, which are retained in permanent files. Each meteorological parameter was summarized manually and listed as monthly maximums, minimums, and means or totals. With computerization, past and ongoing data from selected stations were edited, keypunched and henceforth machine summarized. The format of these summaries will change somewhat to accommodate hourly data from the new loggers, but the format of the principal user-document will remain on a daily scale.

Each climatic station in the forest is an opening similar to the precipitation sites. CS 01 is in a 2-ha grassed field. Air temperature and relative humidity are measured 1.2 m above the ground. At the forest stations, these parameters are also measured at the same height under vegetation. Temperatures of litter and soil (at 10 cm) are measured only under the forest. Initially, wind was measured at 1.7 m in the opening, but now the anemometers are approximately 2 m above the forest or 6 m above the ground at CS 01.

Vapor pressure is calculated from air temperature and relative humidity. Two sets of minimum and maximum temperatures are taken over a 24-h period--one set for day and one for night (Fig. 5). A mean is calculated for each 12- and 24-h period. For simplicity, daytime is defined as 0600-1800. Where wind was measured as a 7-day total, the total is partitioned by daily windspeed observations from CS 01. New data loggers will provide peak and mean windspeed and mean wind vector for each hour.

These electronic data loggers monitor up to 9 sensors every 60 sec and can be programmed with calibration equations for each sensor. Totals, means, maximums, and minimums may be calculated as data are acquired. At the end of each hour, the logger stores the accumulated summaries and resets for the next hour. Data for up to 9.5 days are stored in the logger, and the output copied weekly onto cassette tape for transport to a data processing center. At Coweeta, the tape is read into a microcomputer and stored in a disk file from which daily and monthly reports and tape files are later created.

The report format lists daily values with a monthly mean, minimum, and maximum for each variable. With the addition of more measurements, a multipage report will be required or special reports printed for each variable.

COWEETA HYDROLOGIC LABORATORY
Climatic Station 20
Air Temperature --Deg.C
In Forest Opening

Sept.

1982

DAY	MIN	DAY MAX	MEAN	NIGHT MIN	MAX	MEAN	24-HR MEAN
1	17.2	20.0	18.6	17.2	18.9	18.1	18.3
2	17.8	25.6	21.7	17.2	20.0	18.6	20.1
3	15.6	28.3	21.9	14.4	21.1	17.8	19.9
4	12.2	25.0	18.6	12.2	18.9	15.6	17.1
5	12.2	26.1	19.2	12.8	17.8	15.3	17.2
6	11.1	27.2	19.2	11.1	20.0	15.6	17.4
7	12.8	28.3	20.6	12.2	19.4	15.8	18.2
8	12.2	26.7	19.4	12.2	20.6	16.4	17.9
9	13.3	24.4	18.9	13.3	18.9	16.1	17.5
10	14.4	22.2	16.3	15.0	18.9	16.9	17.6
11	15.6	18.3	16.9	15.6	17.2	16.4	16.7
12	16.1	21.1	18.6	16.1	18.9	17.5	18.1
13	17.8	23.9	20.8	16.7	20.0	18.3	19.6
14	17.2	30.0	23.6	16.7	23.3	20.0	21.8
15	15.6	30.0	22.8	15.6	23.9	19.7	21.2
16	15.6	30.0	22.6	15.6	23.3	19.4	21.1
17	12.8	29.4	21.1	12.8	22.2	17.5	19.3
18	13.9	29.4	21.7	13.9	22.2	18.1	19.9
19	15.0	24.4	19.7	15.0	19.4	17.2	18.5
20	12.8	23.3	18.1	12.8	17.8	15.3	16.7
21	11.7	23.3	17.5	6.3	14.4	11.4	14.4
22	7.2	18.9	13.1	7.2	11.1	9.2	11.1
23	3.9	21.7	12.8	3.9	14.4	9.2	11.0
24	6.1	21.1	13.6	6.1	15.6	10.8	12.2
25	10.0	18.9	14.4	9.4	13.9	11.7	13.1
26	10.6	17.8	14.2	8.3	11.7	10.0	12.1
27	7.2	20.6	13.9	7.2	13.3	10.3	12.1
28	7.8	25.0	16.4	7.8	17.2	12.5	14.4
29	9.4	25.6	17.5	10.0	17.8	13.9	16.7
30	11.1	24.4	17.8	11.7	19.4	15.6	16.7
Mean	12.5	24.4	18.5	12.3	18.4	15.3	16.9
Low Value	3.9	17.8	12.8	3.9	11.1	9.2	11.0
High Value	17.8	30.0	23.6	17.2	23.9	20.0	21.8

Fig. 5.   A portion of Climatic Station 20 air temperature report for September 1982.

Solar radiation will continue to be measured only at the main station, with solar energy on experimental watersheds calculated by the method developed by Swift (1976).

Computer programs are also used to control the quality of meteorological data. Maximums must equal or exceed minimums, vapor pressure should not change beyond a specified range within a day, and variations in temperature and evaporation should track solar radiation.

Periodically, specialized summaries are prepared such as mean monthly solar radiation, evaporation, and temperature listed by increasing magnitude.

STREAMFLOW

Streamflow records began on 9 weirs at Coweeta in 1934, and by 1941, flow was being measured on 28 streams draining watersheds ranging in size from 2.8 to 760 ha. Thirty-two streams have been gaged, and 16 weirs are currently in service (Fig. 1). Until 1964, water depth was recorded on strip charts; thereafter, 16-channel punched paper-tape recorders have been used. A strip chart recorder is retained on one weir--its record serves as a pictorial index to storm hydrographs. The strip chart is changed weekly, while tapes are checked weekly and changed at the first of the month. The head or water depth recorded on chart or tape is verified each week by a hook-gage reading (Wisler & Brater, 1959, Fig. 128).

Approximately once a year, each active weir is inspected and surveyed. When the weir pond is drained for cleaning, surface flow is bypassed around the structure. The streambed immediately below the weir wall is checked for flow, a sign of leakage. The weir blade is cleaned and refinished to maintain the sharp edge. A level is used to verify that the relative elevations of both ends of the weir blade and the hook-gage bracket have not changed. A permanent record of these checks is kept. Calibrations of most of the smaller weirs were verified from 1953 to 1955 by placing a large tank below each weir and directing the flow into it. Knowing the volume and measuring the time required to fill the tank, the flow rate (volume per time) could be calculated for a range of stream depths. The measured flow for each sharp-crested weir fit the theoretical equation for the weir type. Installed weir types include 90- and 120-degree V-notch, 5- and 6-foot rectangular, and 6-, 8-, and 12-foot Cipolletti blades.

Streamflow was originally calculated by numerical integration after laborious point-picking from charts. From 1955 though 1960, a mechanical discharge integrator was used. Hibbert and Cunningham (1966) discuss in detail the improvements attained thereafter by digitizing charts with an oscilloscope chart reader, converting to punch-tape flow recorders, and numerical integration of digital records by computer. Although the computer programs discussed in that paper have been modified and those specific computers are no longer used, the plan and logic of the streamflow processing programs are unchanged. Point-picking from charts is now done by a digitizing planimeter, which has a software translater for curvilinear charts. This X-Y digitizer can operate a card punch or input directly into a microcomputer. Data from punch tapes are likewise translated onto punch cards or into microcomputer files. The storage medium depends upon the editing and computer processing facilities available to the data user.

The X-Y output of the chart digitizer consists of time-head data pairs at irregular time intervals. The operator selects breakpoints on the chart trace that bracket periods of either constant or smoothly changing streamflow rates. The computer or punch card file is initialized with the stream gage identification, month, and year. Processing programs presume each subset of data begins with midnight of the first day of a month. A flow value must be digitized each midnight to force the day number to advance. Processing programs verify that the number of days in the month is correct as a check on the digitizing.

The operator of the digitizer selects and digitizes pairs of time-head points. In contrast, the punched tape only contains head values; to keep track of time, all observations must be read in sequence. For a non-storm period, punchout at 5-min intervals yields many redundant observations that are later deleted by a data condensing program. This program calculates and inserts time values for those observations that are retained. The starting time and date are entered manually before the tape is translated, and the machine-calculated ending time and date are compared with notations on the end of the tape to check for recorder, translater, and operator errors. Thus, both the tape translater and digitizer files go through manual and computer editing and condensing steps that add weir identification, date and time; reduce the number of observations retained; and check for inconsistencies. Depending upon the hardware used, either a punch card, disk, or magnetic tape file is generated.

The edited and condensed time-head file is next processed by a flow computation, integration, and frequency program (Hibbert & Cunningham, 1966). Streamflow head values are converted to flow rate using either a weir equation or rating table. Instantaneous time-flow values are printed, and card, disk, or magnetic tape files created. At the same time, the instantaneous flow values are integrated (trapezoidal rule) and a printed table and computer file of daily mean flows are created with monthly, seasonal, and water-year totals. The instantaneous flow values are also used to generate a flow frequency file by months for 90 flow classes. Class limits approximate a logarithmic scale (Fig. 6). This file has double the number of flow classes than the original seasonal report described by Hibbert and Cunningham (1966). A check sum of total minutes is calculated for each month. The flow frequency file is summarized for 3-, 6-, and 12-month periods. Specialty programs are under development to summarize flow frequency data across years. These data are expected to provide useful parameters for characterizing watersheds and aquatic ecosystems or documenting treatment effects.

The streamflow processing programs described to this point are fairly routine procedures. The flow separation program described by Hibbert and Cunningham (1966) established hydrograph analysis concepts that have been employed widely by other hydrologists. Huff and Begovich (1976) compared two techniques and found the Coweeta flow-separation method superior. Woodruff and Hewlett (1971) used the method to develop a response map for the Eastern United States.

The advantage of the Coweeta flow-separation technique is its consistent and unique division of each hydrograph into quick and delayed flow components (Fig. 7). Quickflow (v) is an estimate of the additional water in the stream during and directly due to the storm event. To accomplish separation, time-flow coordinates are sequentially tested and the start of a storm is marked when successive coordinates show an increase in flow rate exceeding 0.000547 $m^3/s/km^2/h$ (0.05 csm/h). This rate defines the flow separation line. Within the storm hydrograph, the area above this line represents the quickflow volume and the area below the line is the delayed flow volume (d_1). The second intersection of the line with the hydrograph defines the end of the storm event. Other parameters determined by the flow-separation program are the duration of the storm (t_{d1}), the duration and flow volume between events (T_{d2} and d_2), the initial and final flow rates for the event (q_i and q_f), the peak flow rate (q_p), the time to peaking (t_p), and the proportions of quickflow volume before and after peak flow. An option prints the monthly and annual sums of flow in these various categories. Normally the series of streamflow computation programs is run on a 12-month block of time-flow observations.

FLOW FREQUENCY IN MINUTES BY CSM CLASSES

WATERSHED 7   WATERYEAR 82   COWEETA HYDROLOGIC LAB.   OTTO, NC

CLASS	LIMIT	NOVEMBER	DECEMBER	JANUARY	FEBRUARY	MARCH	APRIL	MAY	JUNE	JULY	AUGUST	SEPTEMBER	OCTOBER
1	.049	0	0	0	0	0	0	0	0	0	0	0	0
2	.099	0	0	0	0	0	0	0	0	0	0	0	0
:	:	:	:	:	:	:	:	:	:	:	:	:	:
15	.749	0	0	0	0	0	0	0	0	0	0	0	0
16	.799	2940	0	0	0	0	0	0	0	0	0	0	0
17	.849	17250	0	0	0	0	0	0	0	0	0	0	0
18	.899	12425	7500	0	0	0	0	0	0	0	0	0	0
19	.949	3025	1565	0	0	0	0	0	0	0	0	0	0
20	.999	2075	6930	0	0	0	0	0	0	0	0	0	0
21	1.049	1195	4025	0	0	0	0	0	0	0	0	0	0
22	1.099	705	2695	0	0	0	0	0	0	0	0	0	0
23	1.149	570	2175	0	0	0	0	0	0	0	0	0	0
24	1.199	410	715	0	0	0	0	0	0	0	0	0	12495
25	1.249	195	1920	0	0	0	0	0	0	0	1740	2940	15130
26	1.299	265	2110	0	0	0	0	0	0	0	5410	8640	7460
27	1.349	195	1180	0	0	0	0	0	0	0	4800	7965	2900
28	1.399	220	1160	0	0	0	0	0	0	0	5235	6705	1545
29	1.449	165	1320	0	0	0	0	0	0	0	3405	3270	890
30	1.499	70	1205	0	0	0	0	0	0	995	5205	2750	715
31	1.749	550	4270	2825	0	0	0	0	215	29605	14700	6155	1150
32	1.999	315	1545	9990	0	0	0	0	15745	11325	2760	1265	805
33	2.249	445	1085	8870	0	0	0	1375	22200	1000	715	915	625
34	2.499	95	845	5295	0	0	0	12490	3025	255	510	690	320
35	2.749	0	440	3500	0	0	3465	16840	700	425	15	315	165
36	2.999	5	540	3165	0	0	21545	9890	430	275	20	290	80
37	3.249	0	355	2510	0	9535	8975	3520	150	225	30	270	40
38	3.499	0	205	1800	0	10595	3545	155	240	60	55	125	50
39	3.749	5	120	985	2415	10505	1690	125	60	80	30	55	30
40	3.999	50	135	640	4340	5940	1135	40	115	85	0	105	20
41	4.249	5	65	680	3485	3945	1140	85	90	115	10	35	20
42	4.499	5	60	235	6495	3010	350	35	60	60	0	35	10

FLOW FREQUENCY IN MINUTES BY CSM CLASSES

WATERSHED 7   WATERYEAR 82   COWEETA HYDROLOGIC LAB.   OTTO, NC

CLASS	LIMIT	NOVEMBER	DECEMBER	JANUARY	FEBRUARY	MARCH	APRIL	MAY	JUNE	JULY	AUGUST	SEPTEMBER	OCTOBER
43	4.749	10	65	395	3175	430	365	75	25	5	0	40	15
44	4.999	5	50	230	3540	285	240	10	30	5	0	85	45
45	5.249	5	55	430	3720	175	160	0	15	30	0	55	10
46	5.499	0	40	375	2785	100	110	0	55	15	0	75	10
47	5.749	0	25	265	1280	40	95	0	0	20	0	25	0
48	5.999	0	25	115	1590	60	85	0	15	20	0	20	0
49	6.249	0	20	110	785	10	230	0	20	0	0	15	25
50	6.499	0	50	120	1080	10	20	0	0	5	0	10	0
51	6.749	0	10	105	855	0	50	0	5	0	0	5	0
52	6.999	0	10	135	595	0	0	0	5	15	0	15	45
53	7.499	0	15	245	930	0	0	0	0	20	0	0	15
54	7.999	0	35	290	440	0	0	0	0	0	0	45	25
55	8.499	0	15	205	495	0	0	0	0	0	0	20	0
56	8.999	0	30	140	305	0	0	0	0	0	0	80	0
57	9.499	0	30	100	195	0	0	0	0	0	0	40	0
58	9.999	0	0	100	135	0	0	0	0	0	0	40	0
59	12.499	0	0	400	605	0	0	0	0	0	0	105	0
60	14.999	0	0	290	360	0	0	0	0	0	0	0	0
61	17.499	0	0	95	300	0	0	0	0	0	0	0	0
62	19.999	0	0	0	305	0	0	0	0	0	0	0	0
63	22.499	0	0	0	110	0	0	0	0	0	0	0	0
64	24.999	0	0	0	0	0	0	0	0	0	0	0	0
:	:	:	:	:	:	:	:	:	:	:	:	:	:
89	749.999	0	0	0	0	0	0	0	0	0	0	0	0
90	999.999	0	0	0	0	0	0	0	0	0	0	0	0
NO. OF MINS.		43200	44640	44640	40320	44640	43200	44640	43200	44640	44640	43200	44640

Fig. 6.   A portion of the flow frequency table for Weir 7 for water-year 1982.

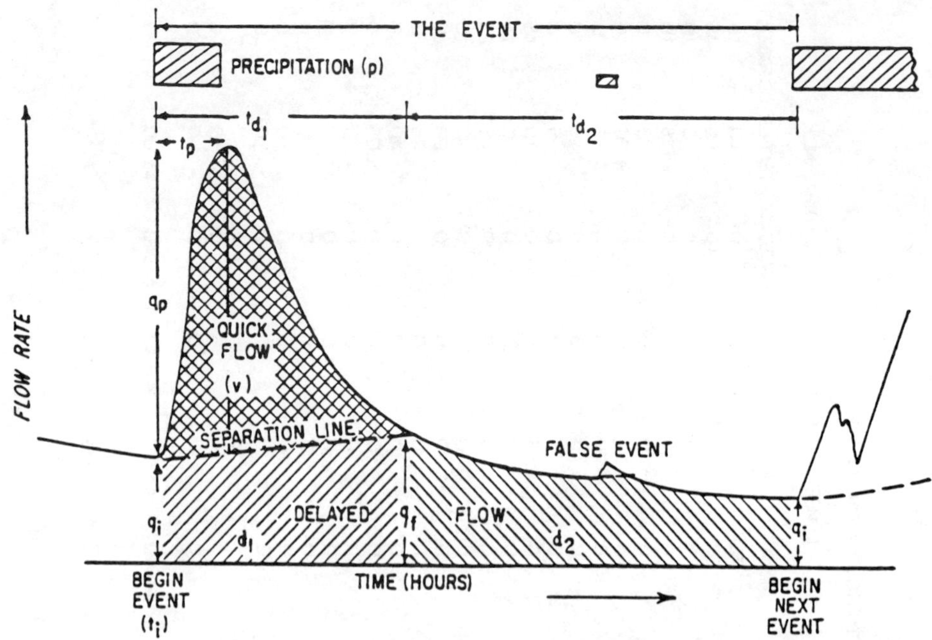

Fig. 7.   Flow separation of streamflow hydrograph. v = quickflow volume, d_1 = delayed flow volume during event, d_2 = delayed flow between events, q_i = initial flow rate, q_f = final flow rate, q_p = peak flow rate, t_{d_1} = event duration, and t_p = time to peak.

SUMMARY

The data collection and processing group at Coweeta has two goals: to acquire accurate data and to prepare and present it in ways to best serve the investigator. A crystal ball estimate is often needed to provide appropriate data for unknown future requirements.

At Coweeta, approximately 150 station-years of data are collected and processed by three employees each year. Over the 50 years of Coweeta's existence, 6000 station-years of data have been made available in computer-accessible files. An appreciable number of additional records are available only in handwritten form.

The data collection history at Coweeta spans developments from the days of manual observations and handwritten data sheets to automatic loggers, and from manual point-picking and hand calculations with log tables and slide rules to digitizers and computers. Many of the exacting methods presently used at Coweeta reflect the heritage of the days when all work and calculations had to be double- and triple-checked. The ease and speed with which such work can be done today could lull us into complacency and a false sense of accuracy. The opportunities for error, human and machine, still exist. The lesson we have learned and wish to share is that each and every observation, and each manipulation of the data, deserves an honest rechecking.

ACKNOWLEDGMENTS

We are indebted to all the employees and cooperating investigators at Coweeta Hydrologic Laboratory who instituted the data collection program, established standards, kept the program functioning, and contributed to the data handling and archiving methods described in this report. We also thank Drs. Judy L. Meyer, Jackson R. Webster, Martin E. Gurtz, William K. Michener, and other reviewers for their useful comments.

LITERATURE CITED

Bormann, F.H. and G.E. Likens. 1967. Nutrient cycling. Science 155(3761): 424-429.

Brakensiek, D.L., H.B. Osborn, and W.J. Rawls. 1979. Field Manual for Research in Agricultural Hydrology. Agricultural Handbook, 224, US Department of Agriculture, Beltsville, MD. 550 pp.

Hewlett, J.D., H.W. Lull, and K.C. Reinhart. 1969. In defense of experimental watersheds. Water Resour. Res. 5(1): 306-316.

Hibbert, A.R. and G.B. Cunningham. 1966. Streamflow data processing opportunities and application. pp. 725-736. In: Forest Hydrology, Proceedings of International Symposium, Pennsylvania State University, Aug. 29-Sept. 10, 1965. W.E. Sopper and H.W. Lull (eds.). Pergamon Press, Oxford, UK.

Huff, D.D. and C.L. Begovich. 1976. An Evaluation of Two Hydrograph Separation Methods of Potential Use in Regional Water Quality Assessment. ORNL/TM-5258. Oak Ridge National Laboratory, Oak Ridge, TN. 112 pp.

Johnson, P.L. and W.T. Swank. 1973. Studies of cation budgets in the southern Appalachians on four experimental watersheds with contrasting vegetation. Ecology 54(1): 70-80.

Swift, L.W., Jr.   1976.   Algorithm for solar radiation on mountain slopes. Water Resour. Res. 12(1): 108-112.

Thornthwaite, C.W.   1948.   An approach toward a rational classification of climate. Geog. Rev. 38: 55-94.

Wisler, C.O. and E.F. Brater.   1959.   Hydrology.   2nd Ed. John Wiley, NY.   408 pp.

Woodruff, J.F. and J.D. Hewlett.   1971.   Predicting and mapping the average hydrologic response for the eastern United States. Water Resour. Res. 6(5): 1312-1326.

OPTIMAL UTILIZATION OF REAL-TIME DATA
IN AN ONGOING RESEARCH FRAMEWORK

Franklin B. Schwing and Jackson O. Blanton

ABSTRACT

Recent developments in microprocessors have made the collection of real-time data a fairly inexpensive reality. A system for obtaining real-time hydrographic and meteorological data, developed by personnel at Skidaway Institute of Oceanography, presently operates on a fixed platform off the Georgia coast. Data are telemetered to a microcomputer system at the Institute, where the data are used to update ship operations and sampling schedules. The data can also be accessed on a real-time basis by other agencies such as the National Weather Service, which uses the data to update marine forecasts. Shipboard real-time data acquisition systems are also used in conjunction with field operations.

The collection of real-time data and the methods employed for storing, analyzing, and presenting data on a real-time and time-series basis are described. The techniques for processing real-time data into finite data sets also are described. Using a software package, data sets are reduced to a standard format that can be readily combined with data series from other sources. Efforts are made to rapidly modify and disseminate data in a format easily interpreted by a variety of users. Simultaneously, data are stored for subsequent analysis. Data are effectively used to optimize ongoing experimental research, as well as provide a real-time look at conditions in the coastal environment for scientists and the general public.

INTRODUCTION

With the advent of microprocessors, computers and data processing systems that once filled rooms are now quite portable and fairly inexpensive. Such advances have made possible the development of compact oceanographic instrumentation complete with data acquisition and telemetry systems. These instruments are now being used by oceanographers to collect real-time data from the marine environment and transmit these data over vast distances. The instantaneous collection of environmental data has many applications, including the ability to remotely sample environmental conditions and quickly adjust field sampling in response to changing physical conditions.

Several points must be considered when designing and using a real-time system. The system itself, of course, must be small enough to fit into appropriate field packaging and be easily transported. The power supply is generally self-contained, therefore the system must also be very power efficient to maintain a small physical size. Instruments are often lost or damaged due to hostile environmental conditions, therefore units should be inexpensive and components easily replaced. Telecommunications are preferably fast and inexpensive, and receiving units need to be portable and self-contained to operate in the field (e.g., shipboard). Signals should be capable of being transmitted over very long distances. UHF or satellite transmissions accomplish this most effectively. Receiving equipment should be readily available to all potential users. A simple terminal, connected with a modem, is one possible solution.

To get the maximum use of real-time data, the data stream needs to be quickly and efficiently interrogated, and modified so that all users can easily interpret the results. Data must be stored in a minimal amount of space, especially if several parameters are being sampled, or data are being collected for very long time periods or at a high frequency. The storage and output of data should be standardized to that of existing data or made compatible to other data acquisition systems. Finally, in the case of long, continuous time-series, considerations have to be made about dividing the data into discrete time-series for final analysis and presentation. The storage system must also accommodate data that are subsequently appended to an existing series.

Such a real-time data acquisition system is operated by personnel at Skidaway Institute of Oceanography (SKIO) on the Savannah Navigational Light Tower (SNLT), near Savannah, GA. The SNLT system provides an instantaneous and continuous monitoring of several meteorological and oceanographic parameters, and telemeters this information to SKIO, where it is accessed on demand and stored by a microcomputer system.

In addition to the SNLT system, SKIO personnel access several other systems that provide real-time data. The use of these systems will be described, accompanied by a discussion of how data from these systems are used on a nearly real-time basis in various research projects. The multidisciplinary nature of research at SKIO requires the widespread distribution of physical data to scientists with varied backgrounds, as well as to the general public. Emphasis is placed on the rapid and efficient modification of data to a format that is compatible with existing data and easily interpreted by various users.

The term "real-time," as used in this paper, is not instantaneous in a strict engineering sense. Rather, it refers to the collection and processing of data and the ability to act in response to those data before conditions change. This time is very short for turbulent features. For information from SNLT, the response time is on the order of an hour or more. Large-scale, synoptic features have a real-time response base of several hours, or even days. Thus, the use of information on a real-time basis is highly dependent on the scale and the process being studied.

SNLT DATA ACQUISITION SYSTEM

Savannah Navigational Light Tower (SNLT), a fixed platform located about 17 km offshore of Savannah, Georgia (Fig. 1), is one of several permanent platforms along the US Atlantic coast that replaced US Coast Guard Lightships in the 1960's. They serve as navigational beacons for ships and are usually unattended. SNLT has a large cabin located about 20 m above the water (Figs. 2 & 3), which houses power-generating equipment that operates the navigational light and fog horn. The dredged channel for the Savannah River entrance is situated west-northwest from SNLT, and the major axes of the tidal current ellipses at SNLT lie approximately along this bearing. Total water depth is 16 m at mean low water, and the tidal range at SNLT varies between 2 and 3 m, the largest range on the US Atlantic coast south of Cape Cod.

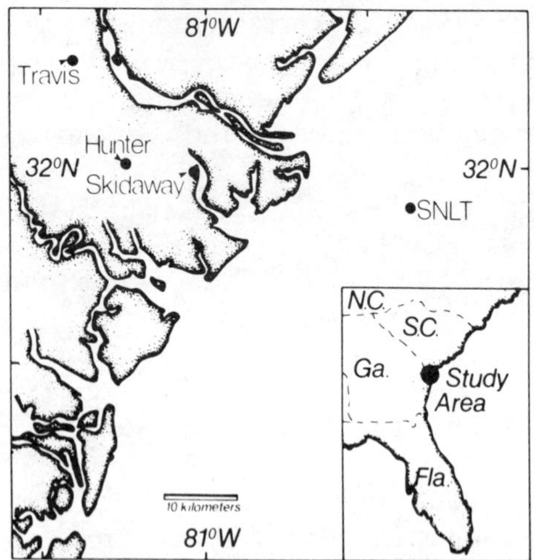

Fig. 1.   Map showing location of Savannah Navigational Light Tower (SNLT), Skidaway Institute of Oceanography, Hunter Army Air Field, and Travis Field (National Weather Service). Meteorological data from Hunter and Travis were used to calibrate SNLT instrumentation.

Fig. 2.   Savannah Navigational Light Tower (SNLT).

The area around SNLT is typical of the nearshore region from Cape Romain, SC to Fernandina Beach, FL. The coastline is indented with tidal inlets spaced 10-20 km apart, each of which feeds an extensive network of interconnected sounds and waterways. Major rivers such as the Pee Dee, Savannah, and Altamaha discharge freshwater through some of the inlets. Other inlets are little more than pocket estuaries where freshwater input is almost zero. The large tidal range and the extensive network of shallow sounds and salt marshes act together to form a 10-20 km wide band of turbid and low salinity water along the coast. The low salinity water forms a frontal zone along the inner shelf which greatly influences circulation in the region (Blanton, 1980, 1981; Blanton & Atkinson, 1983). SNLT is located within this regime.

Dr. D.W. Hayes, an oceanographer at the Savannah River Laboratory, E. I. Dupont de Nemours, Aiken, SC, originally conceived and initiated the concept of using SNLT as a monitoring platform for meteorological and oceanographic information of the coast. The tower is well built and likely to survive the worst of storms. It also has ample electrical power to operate a wide variety of sensors. At Dr. Hayes' suggestion, the US Atomic Energy Commission (now Department of Energy) signed an interagency agreement with the US Coast Guard in 1975 to allow contractors access to the tower and to permit the deployment of oceanographic and meteorological sensors. The Savannah River Laboratory, under Dr. Hayes' direction, operated the sensors on SNLT from 1976 to 1978. Two data reports issued by Skidaway Institute of Oceanography summarize the data obtained during that period (Blanton et al., 1978, 1980).

In 1979, the Department of Energy transferred the responsibility to acquire data at SNLT from Savannah River Laboratory to Skidaway Institute of Oceanography. At that time, the data acquisition system was changed from a magnetic tape data storage system located in the cabin of SNLT to a telemetering system that broadcasts real-time data to a computer located at the Institute. From 1979 to the present, SNLT has been operated in this configuration to support several large oceanographic experiments sponsored by the Department of Energy. Schwing et al. (1983, 1984) describe in detail the design and operation of the data acquisition system at SNLT and give examples of how SNLT-based data have been used in scientific research.

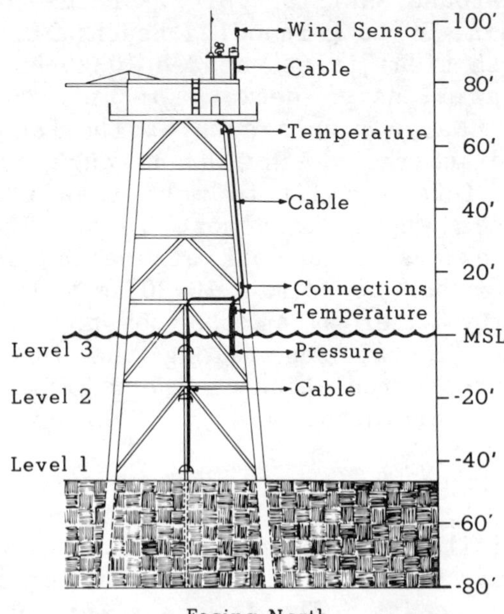

Fig. 3. Schematic of SNLT showing location of oceanographic and meteorological instrumentation. Current, conductivity, and temperature are measured
at each underwater level, located 1, 8 and 13 m
from the bottom.

The cabin located at the top of the tower houses the
sensor power supplies, the microprocessor, and the radio
transmitter (Fig. 4). The main power supply comes from an
electrical generator which is operated by the Coast Guard
and is backed up for short periods by a bank of batteries.
The present data acquisition system ties together a
suite of meteorological and oceanographic sensors. Although
the types and number of sensors deployed varies greatly from
time to time, a typical deployment is shown in Figures 3 and
4. Wind velocity, air and cabin temperature, and barometric
pressure are measured at the top of the tower. The wind
sensor is located 30 m above the water surface. Below
water, sensors are deployed to measure ocean currents, temperature, and salinity at three levels: 1, 8, and 13 m off
the bottom; and water level (tides). We have had varying
success with this deployment. Although above-water sensors
usually operate over long periods of time, underwater sensors frequently suffer from the effects of storm damage and
fishermen, requiring frequent recalibration and replacement.

Each sensor (Fig. 3) is connected by electrical cable to an associated deck unit located in the upper cabin of SNLT. The deck units (Fig. 4) contain electronics which adapt the sensor's voltage output signal for input to the microprocessor. The microprocessor, custom manufactured by Gary Howell, Consulting Engineer, Gainesville, FL, includes a small computer which produces a coded data stream. This data stream is transmitted by a VHF transmitter to SKIO at a

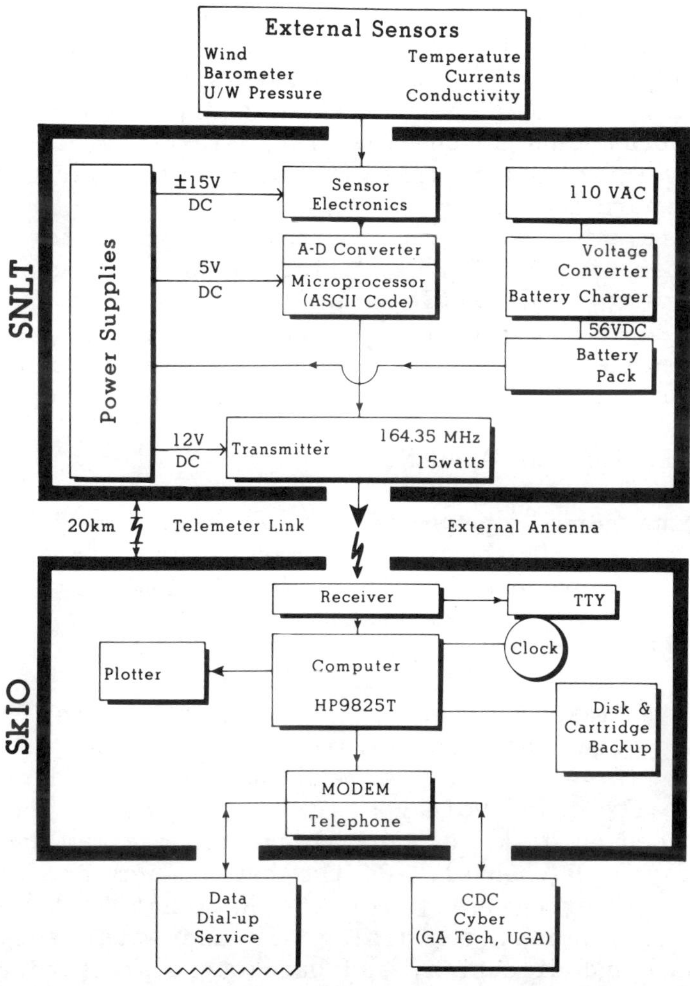

Fig. 4.   Schematic of data acquisition system showing instrumentation at SNLT, telemetry link, and computer hardware at Skidaway Institute.

frequency of 164.35 MHz. The transmitter, microprocessor, and deck units are all powered from a 110V AC wall outlet converted to 56V DC. A pack of batteries is continuously recharged, and provides a backup power source to the system should the generator powering the AC electrical system fail. The VHF transmission from SNLT is received at SKIO by either a mono-frequency or scanning receiver which interfaces with a Hewlett-Packard 9825 computer.

Prior to any field installations, all instrumentation is calibrated by factory and SKIO personnel. After each instrument is individually calibrated, the entire system is assembled and tested in the laboratory under constant environmental conditions. A regression analysis is performed to compare actual and sensor-measured parameters, and calibration curves are established to adjust the various sensor readouts where necessary.

Instrumentation is periodically checked for accuracy during operation at SNLT. Collected data are checked statistically for reasonable mean and variance values. Data are plotted to visually inspect files for suspicious values. The instantaneous data output is also compared to actual observations made by personnel while working on or near SNLT.

DATA COLLECTION AND MANAGEMENT

During normal operation, the data stream is sampled on the hour for 150 sec. Flow charts for data collection, storage, and processing are included (Figs. 5 & 6). The output from the pressure transducer is 2 values/sec to provide sufficient data for wave analysis. All other sensors output 1 value/sec. At least 96 of the 150 sec must be sampled to make a statistically valid reading possible. Otherwise, the loop is reinitiated and the data stream sampled again (Fig. 5). A mean and standard deviation of the digitally coded voltages are calculated for each sensor and recorded on disk and tape along with maximum and minimum values for each channel. On the hour, a summary is printed out in engineering units for the National Weather Service and other agencies, containing air and water temperature, wind speed and direction, and barometric pressure means for that hour (Figs. 5 & 7). Significant wave height is calculated and also printed, along with a spectrum of wave variance. Additional printouts of recorded data or non-hourly sampling of the data stream are available on demand, using the appropriate software. All the software used in collecting, storing, editing, and preliminary analysis of data has

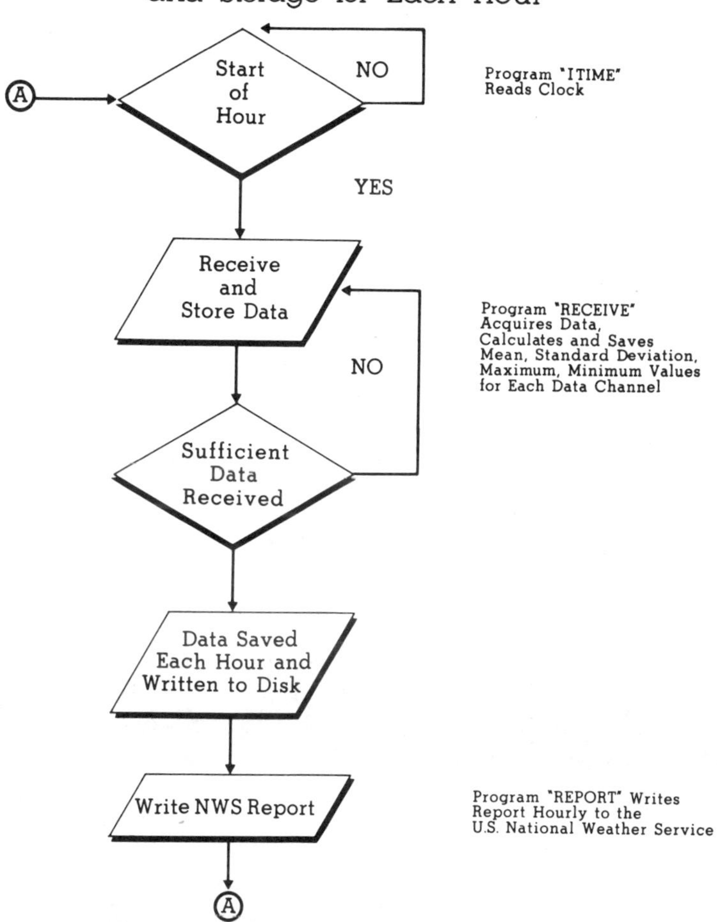

Fig. 5.   Flow chart of procedure for collecting and storing
           SNLT data.

been specifically written for Hewlett-Packard 9825 computers
and various peripheral devices available at SKIO. The HP
9825 is an older system and the newer supermicrocomputers
can manage these tasks even more rapidly and efficiently.
     Once data are properly collected from the data stream,
they are stored in a standard NODC (National Oceanographic
Data Center) format for further analysis. The majority of
oceanographic data collected by researchers at SKIO and
other sites are stored in this format, thus facilitating
the reading and processing of different data sets. Data

collected subsequently may be easily appended to existing NODC files. Major editing of the data is done periodically from NODC-formatted data. Data can be represented in several ways, including weekly (Fig. 8) and monthly plots, and can be output in either voltage or engineering units. These plots can be quickly distributed to other oceanographers requiring a qualitative view of the time series.

SNLT Data Processing

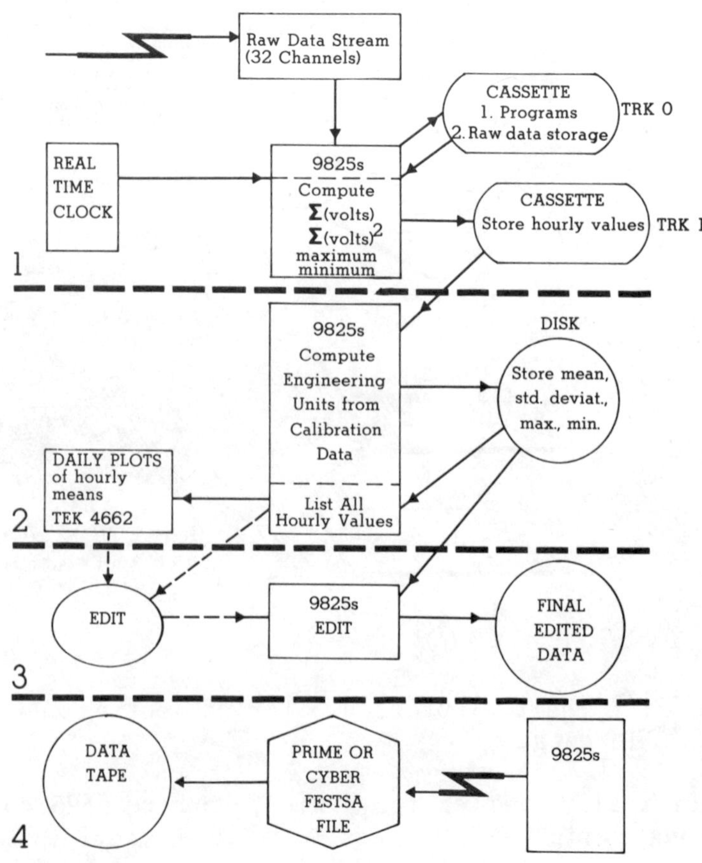

Fig. 6.   Flow chart of data collection, reduction, and storage by Hewlett-Packard system at Skidaway Institute. Data stream interrogated on the hour, coded data stored, and statistics calculated during Step 1. Data converted to engineering units, stored, and plotted during Step 2. Data edited during Step 3. Data transferred to main computer and stored during Step 4.

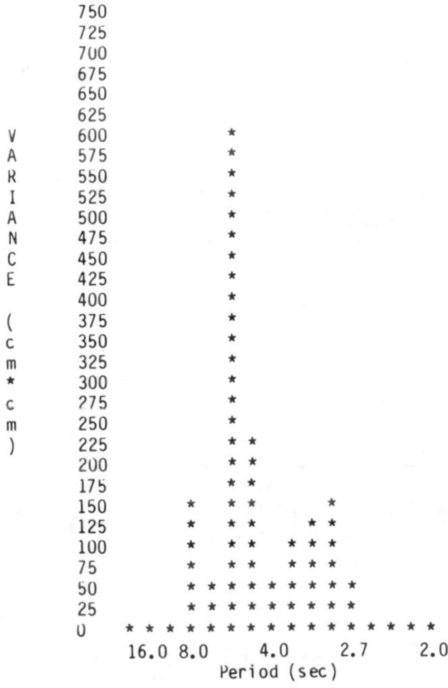

Fig. 7.   SNLT data summary printed for National Weather Service and other agencies. Summary includes air and water temperature, wind speed and direction, barometric pressure, and significant wave height and wave spectrum.

From these plots, missing data can be interpolated and inserted, and unreasonable values adjusted. The final edited data are saved on disk at SKIO (Fig. 6). Time series of a finite time-span are converted to engineering units, reformatted into a packed BCD file, and transferred to and

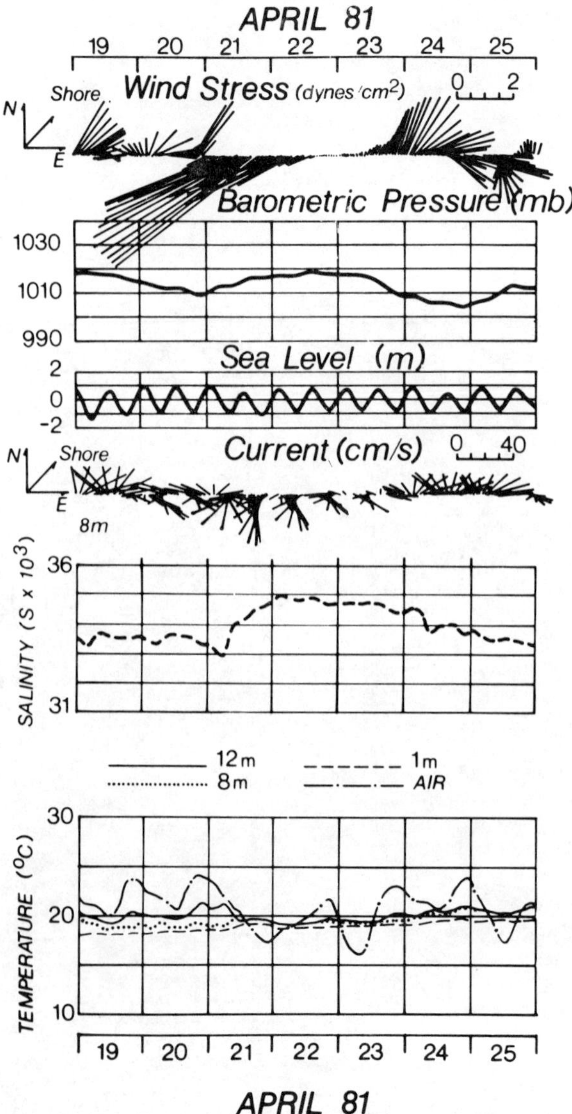

Fig. 8.   Example of SNLT weekly data summary, from the week of 19 April 1981. Data were smoothed with a 3-hour low-pass filter.

saved on tape either on the Cyber computer (the University of Georgia or the Georgia Institute of Technology) or the Prime computer (SKIO). The BCD files are used for subsequent analysis. The advent of the latest microcomputers makes possible the use of one system for these analyses and makes the transfer of data to larger computers unnecessary. A flow chart showing the processing of SNLT data at SKIO is outlined in Figures 6 and 9.

A software package of FORTRAN programs is available to spectrally analyze time series on the Cyber or the Prime computer (Fig. 9). This package (FESTSA, Fast and EaSy Time Series Analysis) performs statistical operations on time series and produces a documented set of files (Table 1). FESTSA-formatted files use much less storage space, and are stored internally labeled and ready for analysis using subroutines from the package. Other institutions use FESTSA as well, facilitating the exchange of data. Up to 99 time-series can be assembled into a single FESTSA unit with common or unique start times and data intervals. This feature enables us to store several simultaneous series together for subsequent analysis. Data can be added to the end of previously created FESTSA files after their conversion.

Table 1.   Example of directory generated by FESTSA software package. File titles refer to various data series corresponding to a specific time (or space) frame. Length gives number of values in the series. Start times refer to Julian hour relative to 00Z, 1 Janaury 1900. Data interval defines spacing between values in a predetermined time or space unit, generally hours and meters.

DIRECTORY FOR FESTSA UNIT NAMED SAVWIND.2

FILE	TITLE	LENGTH	START	INTERVAL
1	SNLT U WIND COMP. IN M/S	3273	7.09413E+05	3.00
2	SNLT V WIND COMP. IN M/S	3273	7.09413E+05	3.00
3	SNLT SL IN M	3273	7.09413E+05	3.00
4	BUOY 41005 U WIND COMP. IN M/S	3273	7.09413E+05	3.00
5	BUOY 41005 V WIND COMP. IN M/S	3273	7.09413E+05	3.00
6	20-1 U CURRENT COMP. IN CM/S	3273	7.09413E+05	3.00
7	20-1 V CURRENT COMP. IN CM/S	3273	7.09413E+05	3.00

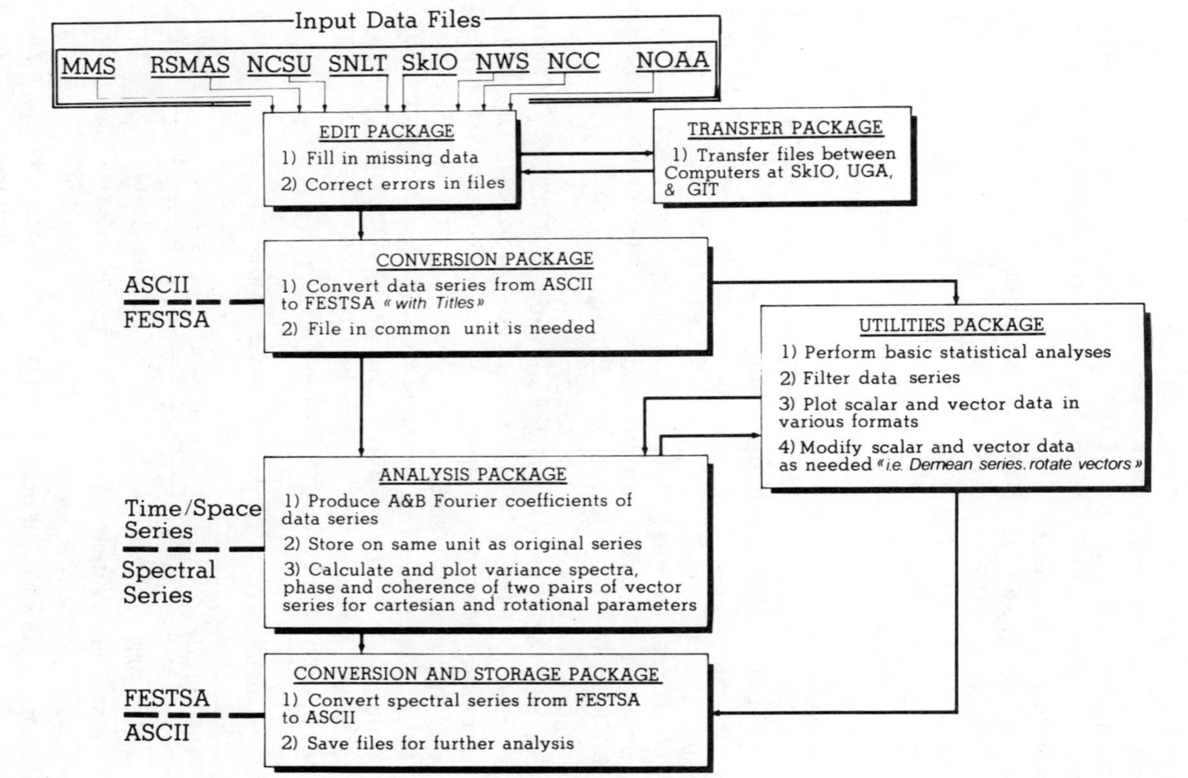

Fig. 9. Flow chart of data analysis showing various software packages used to convert data between ASCII and FESTSA format, and perform spectral analysis on finite time/space series. Top of figure shows various government agencies and academic research facilities that provide data in addition to SNLT.

After discrete time-series of SNLT data have been transferred to the Cyber or Prime computers and converted to FESTSA format, the data undergo a standard reduction and interpretation (Fig. 9). At this point, other data sets from instrumentation other than real-time are added to the appropriate FESTSA units. These data may come from current meters; salinity, temperature, and depth recorders; or various other oceanographic and meteorological instruments deployed in conjunction with SNLT for a specific experiment. In most cases, the length of these series is dependent on when the instruments were deployed and removed.

A detailed analysis is performed on FESTSA files using standard time-series analysis routines. To remove "noise" from the signal, each series is low-passed with a Lanczos filter with a quarter power cutoff of 3 hours. The filter allows the integrity of a tidal signal to remain in the series while removing signals with a period of 3 hours or less. A second low-passed series is generated with a Lanczos-squared filter with a 40-hour cutoff for further analysis of subtidal, weather-related events. Many of the series tend to contain a very low frequency trend, due to annual variations or sensor drift, extending through the entire series, thus causing potential problems with aliasing or excessive "red noise" on the lower end of the spectrum. Because of this, data series are usually detrended and tapered with a 10% cosine filter before analysis (Bendat & Piersol, 1971).

A standard statistical summary is produced for all time series, giving the series mean, standard deviation, and range. Time-series plots of all data are generated. The data are made available in printout and plot form, as well as on 9-track tape to other oceanographers.

Spectral analysis is performed on the 3-hour and 40-hour low-passed time series using the techniques described by Mooers (1973). Tidal and diurnal events are examined in the 3-hour series, while meteorological and other important low-frequency events are defined from the 40-hour data sets. The A and B Fourier coefficients of each series are calculated with a Fourier transform and stored for further computations and analyses. For scalar series ($\underline{i}.\underline{e}.$, salinity, temperature, pressure), these coefficients are used to derive spectral densities for each series, and covariance, phase, and coherence spectra between any two scalars.

Rotary spectral analysis is an efficient method of analyzing a vector time-series ($\underline{i}.\underline{e}.$, wind, current) and comparing it to a scalar or another vector series. The x and y components of a vector are cross-correlated in the frequency domain to produce cospectra, phase, and coherence spectra between the components. In addition, the principle axis of rotation, and ellipse shape and stability are determined. Clockwise and counterclockwise rotary spectra with the associated rotary statistics are also derived. A Fast Fourier Transform is performed on two sets of vector pairs to produce cartesian and rotary correlation statistics. Correlations between a vector and scalar, such as wind stress and sea level, can be made as well. The analysis identifies by frequency the vector angle which produces the best correlation with a scalar. Phase and gain spectra between the vector and scalar are also generated. All spectral output files are saved for subsequent plotting and comparison.

Using Real-Time Data From SNLT

One of the most effective uses of SNLT real-time data is for the analysis and short-term forecasting of weather and sea conditions. Ship operations at SKIO are very dependent on marine weather, particularly smaller vessels. A shore-based station cannot accurately describe winds and seas in the coastal ocean, where wind speeds may be twice the magnitude of those over land (Schwing & Blanton, 1984). Information from SNLT provides a much more accurate and updated description of the conditions mariners can expect offshore. In addition to SKIO, agencies such as the National Weather Service, US Coast Guard, and private mariners can also access the data using a computer telephone linkup with a modem and a printing terminal.

SNLT data are valuable to weather forecasting agencies. SNLT is unique in that it provides the only continuous real-time monitoring of coastal marine weather in the Southeastern US. The National Weather Service, National Hurricane Center, and local media meteorologists all access the SNLT hourly summaries on a nearly real-time basis as part of their forecasting analysis. This is particularly important during the hurricane season, when SNLT can safely and effectively monitor atmospheric changes hours before they would be detected by the nearest land station. Changes in the paths of recent tropical storms and hurricanes such as Dennis in 1981 and Diana in 1984 have been quickly recognized and predicted with the use of SNLT data, thus better enabling coastal residents to take necessary precautions.

A more basic use of SNLT data on a real-time basis occurs during ongoing scientific experiments on the continental shelf. With the help of other real-time data collection devices such as those described below, SNLT data provide a way of monitoring and predicting changes in the physical structure of shelf waters. Scientific vessels can access SNLT information either by dialing up the system directly or having the summary relayed via radio. With this information, scientists can adjust sampling strategies and cruise tracks. Data from SNLT also aid in the prediction of conditions elsewhere on the shelf. Ship costs run several thousand dollars a day, therefore a better understanding of the physical environment can greatly reduce the amount of sampling time needed to describe oceanographic features.

REAL-TIME SHIPBOARD PROFILING SYSTEMS

Although SNLT is an effective real-time system, it has a limited spatial application since it provides information from a single location only. Other types of systems are used to provide expanded spatial coverage. The use of real-time shipboard profiling systems allows scientific crews to monitor hydrography while underway and sample intensively in

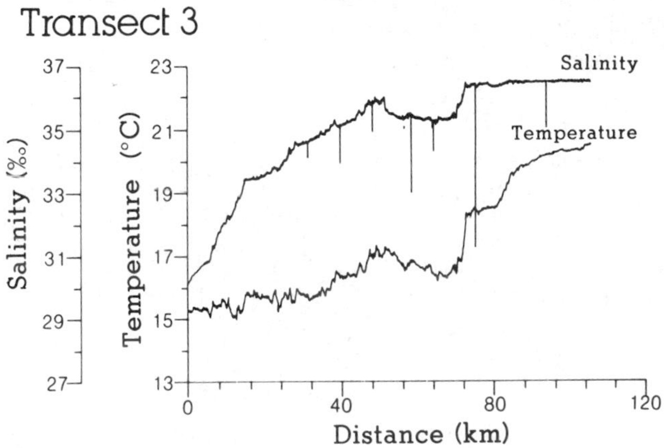

Fig. 10.   Transect of surface salinity and temperature taken with shipboard continuous surface-profiler. Distance offshore from the coastline is shown. The depression in salinity and temperature between 40 and 80 km indicates the presence of the coastal frontal zone, running parallel to the coast, where coastal and continental shelf water meet.

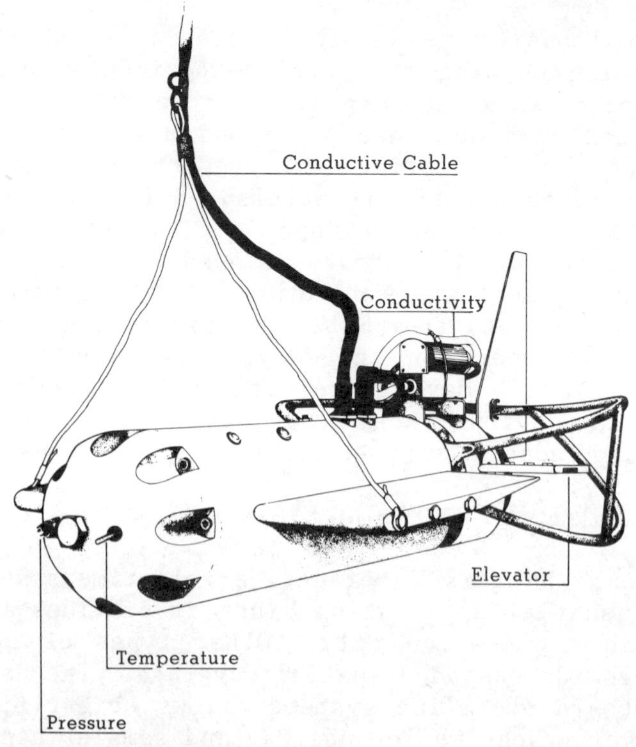

Fig. 11.   Schematic diagram of towed submersible vehicle. Temperature, pressure, and conductivity sensors are indicated. Elevators, controlled from the deck unit on ship, "fly" the vehicle to specified depths or within a certain range of depths. Conductive cable transmits information between vehicle and deck unit.

areas of physical, chemical, and biological interest.

Two different systems operated by SKIO profile hydrography from a moving ship, yielding continuous real-time series in a spatial domain. A continuous surface profiler measures salinity, temperature, chlorophyll-a concentration, and turbidity in surface water; plots and prints out the data on shipboard; and logs it onto a Hewlett-Packard microcomputer identical to that used to preserve SNLT data. A typical transect of temperature and salinity is shown in Figure 10.

A second unit, a submersible vehicle (Fig. 11), which provides vertical profiles of conductivity, temperature, and depth, is towed behind the ship by an armored conductive cable. This towed "fish" features a pair of elevators which, by adjusting their angle, can actually "fly" the vehicle. This is done from the onboard microprocessor, which can be used to fly the "fish" manually or can be programmed to profile within a desired depth range. These

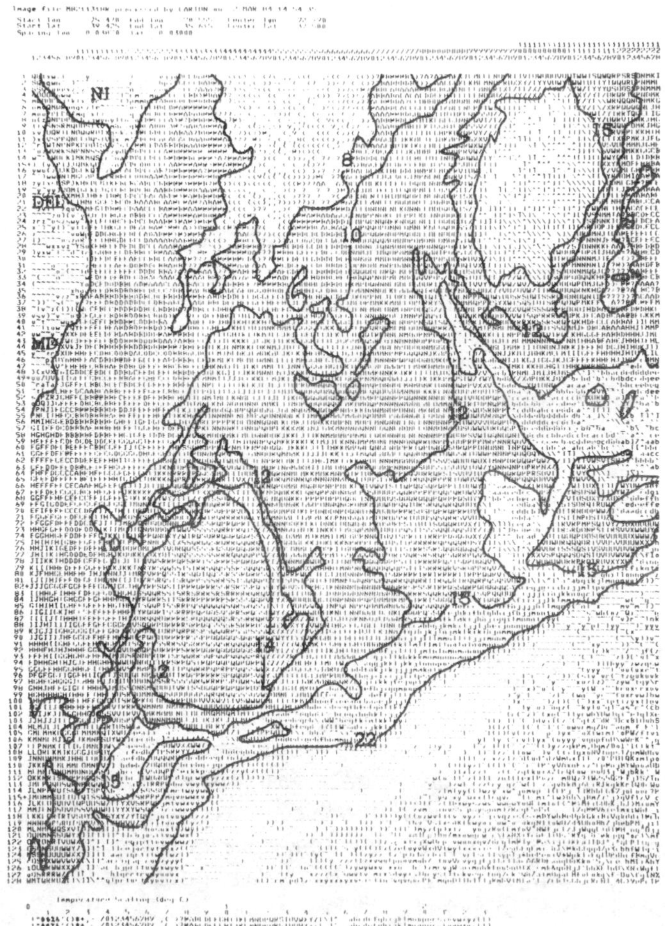

Fig. 12.   Computer-generated map of surface sea temperature (SST) for a portion of the continental shelf off the Eastern United States, compiled from NOAA-7 satellite data. The data were transmitted from ground stations to ship via the ATS-3 communications satellite. Each character represents a certain surface temperature, at 0.25°C intervals. Isotherms are hand-fitted to characterize water masses, such as coastal water (<8°C) and the Gulf Stream (>22°C). Image taken on 23 April 1982.

units provide real-time information on water characteristics and are used, in combination with SNLT data, to appropriately modify other sampling. The output from these units is used to produce 3-dimensional plots of water masses, which are used as indicators of specific chemical and biological processes.

Recent developments in real-time data acquisition and management center on lagrangian current measurements and mapping the ocean's surface by satellite. Lagrangian, or drifting, current drogues are now being developed that contain Loran-C navigational systems capable of transmitting their real-time position over a range of 300 miles. Other drogue systems presently use satellite navigational techniques, although their transmission interval, dependent on satellite location, is much less frequent than that for the Loran-C units. Satellite imagery is now processed to provide sea surface temperature on a nearly real-time schedule (Fig. 12). However, there still exists a problem in accessing and relaying data from its source to shipboard. Software and hardware packages are presently being assembled to accomplish this.

The management of these spatial series is a different challenge than the time-series obtained from SNLT. However, the same software and analysis techniques can be applied (Fig. 9). In the case of the aforementioned profilers, the data can be assembled into a "space" series and processed as a time-series. Thus, much of the software used in the analysis of SNLT data (described above) can be used to describe the spectral properties of hydrographic structure. Standard statistics such as the mean, variance, and range are calculated. Data are smoothed with a low-pass filter to remove high-frequency turbulence and irregularities. Files are converted to FESTSA format with a specified length and spatial sampling interval. The data series undergo spectral analysis and simultaneous series are cross-correlated. Integrating time and space series is difficult and must be done subjectively, as they exist in different dimensions. However, such integration is very important as it provides a combined temporal and spatial description of ocean features.

Lagrangian and satellite data collection systems have a unique set of data management problems that need addressing, particularly in terms of how position data over time are stored, presented, and statistically analyzed. Particularly challenging is the need to present geometrically corrected data in a standard format so that it may be compared to time-series and other associated oceanographic data.

Future efforts will concentrate on this area of real-time
data acquisition.

SUMMARY

Real-time information provides researchers with the
ability to adjust experimental strategies in response to
changes in environmental conditions. In the design and
development of the systems described above, several considerations were made to optimize the use of data both on a
real-time basis and for further analysis in association with
other data. Software allows for error checking as well as
visual inspection of the data. Analyses are also designed
to accommodate time or space series. Attempts are made to
get data as rapidly as possible from source to real-time
users, while concurrently preserving the original data on
storage media. To maintain efficient use of data, a reduced
output is generated that can be quickly and easily interpreted for many applications by a variety of users. Additionally, data processing and analysis have been streamlined
and standardized so that printed and plotted versions of the
data are available during all phases of processing, and data
from various sources can be easily compared. A data management package, FESTSA, is used both to catalog and preserve
data in time-series form and perform a majority of the analyses. For very long data files, decisions must be made on
how to truncate the data into finite lengths for time-series
analysis. This is often controlled by the length of a particular field experiment.
Our experiments typically include oceanographic data
obtained by two or more ships, arrays of moored instruments,
satellite imagery in the form of surface maps of temperature
and other properties, and weather information from SNLT and
the National Weather Service. To optimize the use of expensive ship-time, we are presently working on a system which
will combine microcomputers that collect files of these data
at ship and shore locations and transmit them through
modems with satellite links to other sites. For example,
SNLT data reports (Fig. 7) can be transmitted to ship, and
shipboard maps can be transmitted to shore stations such as
SKIO, as well as other ships involved in the experiment.
The object of these efforts is to minimize the time taken
for data exchange between various investigators and facilitate the making of "real-time" decisions that would optimize
ship tracks and sampling patterns.

ACKNOWLEDGMENTS

The authors thank Lee Knight, Carroll Baker, Berry Beckham, Lester Lamhut and other personnel at Skidaway Institute for their assistance in the development of the instrumentation and software described in this paper. We also thank Anna Boyette and Susan Salyer for drafting and typing the manuscript. We express our gratitude to the US Coast Guard for permission to work on Savannah Navigational Light Tower. This study was sponsored by Department of Energy, Contract No. DE-AS09-80EV10331.

LITERATURE CITED

Bendat, J.S. and A.G. Piersol. 1971. Random Data: Analysis and Measurement Procedures. John Wiley and Sons, Inc., New York. pp. 323-325.

Blanton, J.O. 1980. The transport of freshwater off a multi-inlet coastline. pp. 49-64. In: Estuarine and Wetland Processes. P. Hamilton and K.B. MacDonald (eds.). Plenum Publishing Corp., New York.

Blanton, J.O. 1981. Ocean currents along a nearshore frontal zone on the continental shelf of the southeastern United States. J. Phys. Oceanogr. 11: 1627-1637.

Blanton, J.O. and L.P. Atkinson. 1983. Transport and fate of river discharge on the continental shelf of the southeastern United States. J. Geophys. Res. 88: 4730-4738.

Blanton, J.O., L.L. Bailey, W.S. Chandler, D.W. Hayes, and A.S. Dicks. 1978. Data Report No. 1: Oceanographic and meteorological data 15 km off the coast of Georgia. Georgia Marine Science Center Technical Report Series, No. 78-6. Savannah, GA. 49 pp.

Blanton, J.O., L.L. Bailey, D.W. Hayes, and A.S. Dicks. 1980. Data Report No. 2: Oceanographic and meteorological data 15 km off the coast of Georgia. Georgia Marine Science Center Technical Report Series, No. 80-5. Savannah, GA. 56 pp.

Mooers, C.N.K. 1973. A technique for the cross spectrum analysis of pairs of complex-valued time series, with emphasis on properties of polarized components and rotational invariants. Deep Sea Res. 20: 1129-1141.

Schwing, F.B. and J.O. Blanton. 1984. The use of land and sea based wind data in a simple circulation model. J. Phys. Oceanogr. 14: 193-197.

Schwing, F.B., J.O. Blanton, L. Lamhut, L.H. Knight, and C.V. Baker. 1983. A real-time telemetry system for hydrographic and meteorological data collection in the South Atlantic Bight. pp. 44-50. In: Proceedings, 1983 Working Symposium on Oceanographic Data. C.D. Tollios, M.K. McElroy, and J. Syck (eds.). IEEE Computer Society Press, Silver Spring, MD.

Schwing, F.B., J.O. Blanton, L. Lamhut, L.H. Knight, and C.V. Baker. 1984. Ocean circulation and meteorology off the Georgia coast - Part 1. Real-time telemetry system for data collection. Georgia Marine Science Center Technical Report Series, No. 84-1. Savannah, GA.

PREMOD AND MODAID: SOFTWARE TOOLS FOR THE CONSTRUCTION AND ANALYSIS OF SIMULATION MODELS

Thomas B. Kirchner

ABSTRACT

PREMOD and MODAID are software tools that simplify the process of writing simulation models, provide interactive support for generating graphical and tabular output, and facilitate validation of models. PREMOD is an interactive precompiler that prompts users through the construction of models and reduces the programming aspects of model construction to a minimal entry of essential information. It is designed to be usable by people with little or no programming experience and is also flexible enough to allow construction of quite complex models. MODAID is a library of software utilities that are often required by simulation models. PREMOD writes code that makes use of the MODAID library. This paper describes versions of PREMOD and MODAID that are written in Microsoft BASIC and can be run on an IBM PC microcomputer.

INTRODUCTION

Simulation modeling is a powerful technique that can be used to study ecological systems. A simulation model is a program that numerically solves the rate equations of a mathematical model by stepping through simulated time from a starting time to an ending time, with the rate equations solved and state variables updated after each time step. Simulation models are valuable because they help researchers evaluate hypotheses about the functioning of systems, stimulate the generation of new ideas, facilitate testing of experimental approaches, identify areas where research would be most beneficial, and reduce the amount of ad hoc experimentation. Simulation modeling is often considered an "art" that should be attempted only by one highly trained in mathematics and computer programming. While these skills are important, simulation languages and other programming tools have made it much easier for anyone to build and analyze relatively sophisticated simulation models, in much the same way as standardized statistical packages have simplified statistical analyses.

PREMOD is a precompiler used to help build simulation models. (A precompiler is a program that uses information supplied to write computer code.) PREMOD allows users who have a minimum amount of programming experience to build simulation models. It also can be used to construct large, complex simulation models. The simulation models are coded in FORTRAN on mainframe or minicomputers, and in Microsoft BASIC on the IBM PC microcomputers. The models are designed to be used with MODAID, a set of software tools for simulation modeling which includes a library of subprograms that (1) facilitate the generation of printed or plotted output, (2) facilitate making changes in the values of parameters used in the model, (3) facilitate the updating of state variables during a simulation, (4) help perform sensitivity, uncertainty, and validation analyses on models, (5) provide easy access to a variety of mathematical functions, and (6) provide pseudorandom numbers for a variety of distribution types.

The first versions of PREMOD and MODAID were written in FORTRAN 66 (Kirchner & Vevea, 1983) and are currently being revised and rewritten in FORTRAN 77 by Kirchner and Vevea. The BASIC versions of the software incorporate the revisions in design of PREMOD and MODAID but at present are subsets of the FORTRAN 77 versions. The BASIC version of PREMOD and the models built by PREMOD are designed to run on an IBM PC or XT. The software requires 256K of RAM, two double-sided, double-density diskette drives or one DSDD diskette drive and fixed disk, and DOS 2.0 or a later release of the operating system. A graphics display adapter and monitor is useful but not essential.

PREMOD

The general information required to build a simulation model using PREMOD consists of the names of the state variables, the names of parameters used in the rate equations of the model, and the rate equations themselves. A state variable represents a compartment or component of the system being modeled. For example, a state variable may represent the amount of carbon in the leaves of a plant. The rate equations define mathematically how the states of a system change through time. Thus, the processes of photosynthesis, translocation, and metabolism could all be represented by rate equations. Parameters are variables that may represent rate constants, threshold values, or any other variables needed to describe the system mathematically. One must define initial values for all state variables and parameters in every model.

Entry of Information

PREMOD issues a series of prompts, or instructions, to guide the user in entering the information required to build a simulation model (Figs. 1 & 2). When possible, PREMOD supplies a default response for the prompt, which is shown in brackets ([]) at the end of the prompt. The default response is used when an empty line is entered (i.e., if one presses the RETURN key before typing an entry). If the prompts are found to be insufficient, one can obtain further information by entering the word HELP in response to a prompt. Specific examples for constructing simple simulation models are provided in the HELP information at each appropriate step. In addition, one is allowed to select from a menu of several other topics related to the operation of PREMOD including BASIC conventions, a description of the general structure of the simulation model produced by PREMOD, and information about the various types of mathematical models that can be built using PREMOD. Thus, the HELP feature of PREMOD provides one with an on-line user's manual.

PREMOD is an interactive program and therefore normally receives its input from the keyboard; however, PREMOD can also be instructed to read information from a file. The user initiates this process by entering COPY FROM followed by the name of the file that is to be read. PREMOD will return to taking its information from the keyboard once the file has been read. A similar command, EDIT FROM, can be used if one wishes to modify the entries that are being read from a file. In this case, the facilities of PREMOD's line editor are made available to modify the entries of the file.

OPTIONS

The following options may be specified by the user:
1)   Solve a system of differential equations
2)   Solve a system of difference equations
3)   Solve a Markov model
4)   Solve an age dependent population growth model
5)   Add statements to any of the following 5 subroutines:
     USTART, a subroutine called once per run prior to
     executing the first simulation
     UINITL, a subroutine called prior to the execution of every simulation
     UCYCL1, a subroutine called once each time step before solving the rate
     equations for the time step
     UCYCL2, a subroutine called once each time step after solving the rate
     equations for the time step
     FINIS, a subroutine called once after the exection of each simulation
6)   Change the default names of the predefined variables TIME,TEND,DT,DTPL.

Enter the numbers of the options that you want to choose, or the word
<HELP> if you want more information [1]: **1**

The following solution techniques can be used to solve the system of
differential equations :
     1)   Runge-Kutta method of order 2
     2)   Runge-Kutta method of order 4
     3)   Adams-Bashforth method
     4)   Euler's method of order 1
Choose the one that you want to use by entering the menu number, or
enter the word <HELP> for more information [2] : **2**

Fig. 1.   PREMOD provides prompts to guide one in entering
the information required to build a simulation
model. This example (continued in Fig. 2) shows
the prompts and responses (bold face type) used
to construct a model for the logistic growth of a
population. One starts the model building process
by choosing the type or types of models to be
built. If the model is to be described by differential equations, as in this case, then one must
also choose the algorithm to be used to solve the
derivatives.

The simulation model produced by the BASIC version of
PREMOD is coded in BASIC. Interpreted BASIC on an IBM PC
requires that each line be numbered. However, when entering
code in PREMOD, only those lines that require line numbers
because of branching (e.g., GOTO's and GOSUB's) need to be
numbered, and they do not need to be ordered sequentially.

```
                        ENTER DERIVATIVES
Enter the name of the state variable (6 characters or less),
       <HELP> if you want more information, or
       <END> if there are no more names of state variables to be entered.
POP
 Enter the names of new parameters (constants), or
       <HELP> if you want more information.  Enter
       <END> after the last name to terminate the list.
R,K,END

The derivative of POP must be defined.  Enter statements to
define the derivative, or <HELP> if you want more information.
The derivative must be defined by <D:POP=> in at least one
statement.  Enter <END> after entering your last statement.
D:POP=POP*R*(K-POP)/K
END


                        ENTER DERIVATIVES
Enter the name of the state variable (6 characters or less),
       <HELP> if you want more information, or
       <END> if there are no more names of state variables to be entered.
END


You will be able to plot the following variables:
 1)* TIME                2)* POP
The following variables are parameters:
 51)*TEND               52)*DT               53)*DTPL               54) R
 55) K
Do you want to make any additions or deletions to this list [N] ? N


You must now enter a value for each of the following variables:
TIME is the initial value assigned to the simulated time
TIME= 0
POP= 2
TEND is the time at which a simulation will end.
TEND= 50
DT is the time step or integration interval to be used.
DT= 1
DTPL is the time step on which simulated values are stored for plotting or
printing
DTPL= 1
R= .2
K= 200
```

Fig. 2.   PREMOD requires the user to enter the names of the
state variables and parameters to be used in the
simulation model. In this example for logistic
growth, the number of individuals in the population is POP, the intrinsic rate of growth is R,
and the carrying capacity is K. The operator D:
is used to identify the derivative of POP. PREMOD
completes writing the BASIC code for the simulation model after the last entry is made.

In addition, subroutines can be called using a CALL <u>name</u> syntax, rather than a GOSUB <u>number</u> syntax. The first <u>line</u> of the subroutine that is called must be a SUBROUTINE <u>name</u> statement. Although this syntax is similar to that used in FORTRAN, there presently is no mechanism to pass arguments to the subroutine as formal parameters. The line numbering and CALL features are made available to simplify the entry of structured code and to allow the conversion of BASIC models to FORTRAN models. PREMOD will prompt one to enter subroutines if CALL or GOSUB is included in any code.

PREMOD performs several data entry error checks. For example, all variable names that are entered are checked to insure that they conform to the American National Standards Institute (ANSI) conventions for variable names in FORTRAN 77. FORTRAN naming conventions are used even in the BASIC version to facilitate the conversion of the models to FORTRAN. PREMOD also checks to insure that variable names are not duplicated and are not reserved names in FORTRAN or BASIC. However, the present version of PREMOD does not check equations or other code for syntax errors. Entries which are found to be in error are rejected by PREMOD; the error is described, and one is given the opportunity of making a new entry.

All entries that are accepted by PREMOD are saved on a "keystroke" file. PREMOD can be instructed to read this file rather than requesting information to be entered from the keyboard. Thus, one can rebuild a model starting from the keystroke file. This feature is useful where disk storage is limited, since the keystroke file for a model is significantly smaller than the model program created by PREMOD. In addition, a utility is planned for the FORTRAN version of PREMOD that will convert a BASIC keystroke file into a FORTRAN keystroke file. Thus, one will be able to do initial model development on a microcomputer, then upload and mechanically convert the model into FORTRAN. The FORTRAN version of the simulation model will have access to the more extensive library of MODAID utilities, will be more portable, and will run on faster mini- and mainframe computers.

One often needs to modify the structure or rate equations of a model during its development. There are three options for modifying a PREMOD generated model: (1) edit the model program, (2) edit the keystroke file, or (3) instruct PREMOD to allow editing of the entries from the keystroke file as they are read. Consider the example of adding a parameter to the model. Editing the model program to add a parameter would involve making changes to the code in several locations, adding an initial value for the variable to the data file, and modifying a set of tables required by MODAID. Editing the keystroke file is significantly easier because one only needs to insert the new variable name and its initial value into the appropriate places within the file. However, identifying the appropriate places takes some care. Using the editing feature of PREMOD is by far the easiest method of adding a variable, because one can locate the appropriate places to make entries by following PREMOD's prompts. PREMOD will check new entries for errors, and automatically prompt the user to enter initial values for newly added variables. The PREMOD editor allows one to change, add, or delete lines in an existing keystroke file. The corrected entries of the keystroke file are used to construct a new program.

Options in PREMOD

PREMOD can currently produce simulation models for four types of mathematical models: (1) a system of ordinary first-order differential equations, (2) a system of ordinary first-order difference equations, (3) a Markov transition matrix, and (4) age-dependent natality and mortality rates for a population. One can create a single simulation model that includes code to simulate all four types of mathematical models. In such a case, PREMOD allows one to define a time step for each type of mathematical model, although the time steps must be defined in the same units. PREMOD can also be used to build models of different levels of complexity including stochastic simulation models.

At the beginning of the model construction process, PREMOD provides a menu of options which allows one to chose the mathematical models to be included and determine the structure of the model (Fig. 1). Options can also be chosen that allow renaming of certain predefined variables, and that cause PREMOD to include up to five additional subroutines in the model. The optional subroutines allow code to be entered that will be executed at specific times during a simulation.

Models Using Differential Equations

The first option of the PREMOD menu directs it to construct a model that is defined by first-order ordinary

differential equations. Ordinary differential equations
often are used to simulate the change in the state variables
with respect to time. The primary assumption in such models
is that the states change in a continuous fashion. PREMOD
uses the operator D: to represent the derivative operator
d/dt. The variable used to represent time is assigned the
default name TIME by PREMOD. For example, the differential
equation for the logistic growth of a population can be
written

$$\frac{dPOP}{dt} = R \times POP \times \frac{K-POP}{K}$$

where POP is the number of individuals in the population, R
is the intrinsic rate of natural increase, and K is the
carrying capacity of the population. The state variable for
this model is POP, and R and K are parameters. The equation
would be entered into PREMOD as

D:POP=R*POP*(K-POP)/K

PREMOD normally places all differential equations into a
subroutine called DESOLV.

There presently are four solution algorithms that can
be used with differential equations in PREMOD-generated
models. These algorithms are the Euler method of order one,
a Runge-Kutta method of order two, a Runge-Kutta method of
order four, and an Adams-Bashforth method of order four.
The algorithm order corresponds to the number of derivative
estimates that are used in making a single projection of the
functions over the interval. The Runge-Kutta method of
order four (RK4) is used as the default algorithm because it
is relatively stable and provides good precision for reasonably large time steps. However, RK4 requires that the differential equations be solved four times for each time step.
If a short integration interval, dt, is required in order to
provide a high frequency of data for plotting, the RK4 routine may result in more computations than are necessary for a
reasonable level of precision. In such a case, one of the
other algorithms may be sufficient.

The Runge-Kutta method of order two is similar to RK4
but requires only two estimates of the derivatives per time
step. However, the estimate of the local error for each
time step is usually over twice as great as that for RK4.

The Euler method of order one is a very simple algorithm. It requires only one estimate of the derivatives per time step. The method projects the solution based on the system state at the interval beginning and assumes that the function can be represented by a straight line having a slope equal to the derivative estimate.

The Adams-Bashforth method of order four requires only one computation of the differential equations per time step because it retains the estimated derivatives from previous time steps. This method is potentially a fast, reasonably precise algorithm; however, the Adams-Bashforth method is unstable and can lead to erroneous results. Therefore, care should be exercised when using this method.

Models Using Difference Equations

Choosing the second option of the PREMOD menu results in the inclusion of ordinary first-order difference equations in the model. Difference equations are used to describe discrete changes in the values of state variables over a specified interval, rather than describing a continuous process. For example, the removal of ten individuals from a herd due to hunting could be simulated as a discrete change in herd size. The rate equations define the quantities that flow between compartments during a time interval. Such a change can be defined most easily using the MODAID routines FLOW and FLOWL. The calls to these routines are

CALL FLOW(FROM,TO,AMOUNT,WHEN)

and

CALL FLOWL(FROM,TO,AMOUNT,WHEN,IERR)

where FROM is replaced with the name of the variable from which the flow comes, TO is replaced with the name of the variable to which the flow goes, AMOUNT is the quantity to be transferred between compartments, WHEN is replaced with the time at which the flow is to occur, and IERR is an error flag. The difference between FLOW and FLOWL is that FLOW permits a flow to take place regardless of the value of the FROM variable, while FLOWL never permits a flow to exceed the quantity in the FROM variable. IERR is used to show whether or not the flow could be made. For the hunting example, one could schedule the harvest of animals using

CALL FLOW(HERD,HARVST,10,11)

where HERD is the state variable for the number of individuals in the herd, HARVST is a state variable used to accumulate the harvested animals, 10 is the number to be removed by hunting, and 11 is the time at which hunting takes place. PREMOD normally places the rate equations for difference equation models in the subroutine FLOWS. Whereas FLOW and FLOWL are used to schedule flows, the routine FLOWUP must be called to actually update the state variables by the quantities indicated. PREMOD provides the call to FLOWUP as the last statement in the FLOWS routine.

Markov Models

The third operation of the PREMOD menu allows one to define the Markov transition matrix model. A Markov transition matrix gives the probabilities of changing from one state to another in a time unit. Markov models have been used in ecology to represent processes such as forest succession (e.g., Horn, 1975). A vector defines the system states, that is, the frequencies or amounts in each state of the system. To define a Markov model using PREMOD, one only needs to enter the names of the Markov states; the model can have up to ten states. PREMOD names the matrix of the transition probabilities. A screen-oriented editor provides a labeled grid to facilitate the entry of the probabilities. PREMOD insures that the entries of the matrix are reasonable before the matrix is accepted.

In a simulation, a subroutine called MARKOV is used to multiply the transition matrix by the vector of the system states once per time step. The vector of the system states then receives the results of the multiplication and thus the state of the system at the beginning of the next time step.

Age-Specific Population Models

The fourth option of the PREMOD menu is to construct models based upon age-specific population dynamics. These models subdivide a population into a set of age classes. For each age class one must define both a per-capita natality rate and a per-capita mortality rate. The age-specific population sizes are computed by first calculating births for a time step as the product of age-specific population size and the natality rate. Second, the remaining age classes are updated by multiplying each age class by its survivorship rate and then putting the result into the next age class. The newborn are placed into the first age class after computing survivorship. The time step for such models is equal to the interval between age classes. For example, if a population is broken down into cohorts by years, then the model would require a yearly time step. This method of computation requires that only individuals capable of reproduction be considered (usually females), or that one adjust the natality rates to reflect the sex-ratio of the population.

PREMOD creates three arrays for each population to be simulated. The length of each array equals the number of age classes to be simulated. One assigns names to the arrays and defines the number of age classes in response to prompts from PREMOD. The first array contains the number of individuals in each age class of the population. For example, DEER(3) would be the number of deer in age class 3. The second array stores the age-specific per-capita rate of natality for the population (_i.e._, the number of individuals produced per time step per individual in each age class). The third array contains the age-specific per-capita rate of survival, or the proportion of individuals that survive from one time step to the next. PREMOD also computes the total size of the population which is assigned to a variable named by the user.

Although natality and survivorship rates often are considered to be constants, one can also define such rates based upon functional relationships. PREMOD asks whether or not one would like to define either or both variables by functions. If one responds 'Yes' to that prompt, PREMOD allows one to enter any code required to specify the functions. This code is executed each time step just prior to computing births and survivorship. One is also given the opportunity to enter the names of additional parameters that might be required by the functions.

Additional Subroutines

Simulation models often require computations in addition to solving the rate equations of the model. For example, the initial values of some parameters or states may need to be computed at the start of a simulation, or the values of some external driving variables may need to be read on a periodic basis during the simulation. If the fifth option is chosen, PREMOD will include up to five predefined subroutines in the simulation model to handle such tasks. These five subroutines, named USTART, UINITL, UCYCL1, UCYCL2, and FINIS, are executed, respectively, once prior to the first simulation, once prior to each simulation, once per time step before solving the rate equations, once per time step after solving the rate equations, and once after each simulation. Only those subroutines to which code is added will appear in the final model. PREMOD provides the calls to these subroutines automatically.

Predefined Variable Names

The last option of the PREMOD menu allows one to rename certain of the variables defined by PREMOD. One may want to rename variables so that they have more descriptive names. For example, if a population model for deer requires a yearly time step, one may want to rename the time variable, TIME, to YEAR. The other variables which can be renamed are DT, DTMARK, DTPOP, DTFLOW, DTPL, and TEND. DT, DTMARK, DTPOP, and DTFLOW are the time steps for differential equations models, Markov models, age-specific population models, and difference equation models, respectively, when the simulation model includes two or more types of mathematical models. DT is used as the time step parameter for any simulation model that includes only one mathematical model. DTPL is the time step on which data are saved for plotting, and TEND is the time at which the simulation ends.

MODEL STRUCTURE

All simulation models built by PREMOD can be subdivided into three sets of routines (Fig. 3). The subroutines START, READIT, and USTART make up the first set of routines. These subroutines initialize variables used in the model or by MODAID and include the declarations of the dimensions of arrays used in the program. USTART is only present if the user selects option 5 of PREMOD when the model is being constructed. READIT reads the data file created by PREMOD and initializes the state variables and parameters. START is used to initialize variables that are required by MODAID.

The second set of routines constitutes the bulk of the simulation model (Fig. 4). The number and nature of the routines in this set is highly dependent on the options that the user selected when building the model. Subroutine INIT is called prior to each simulation to re-initialize the values of the state variables. Subroutine DTSET is called

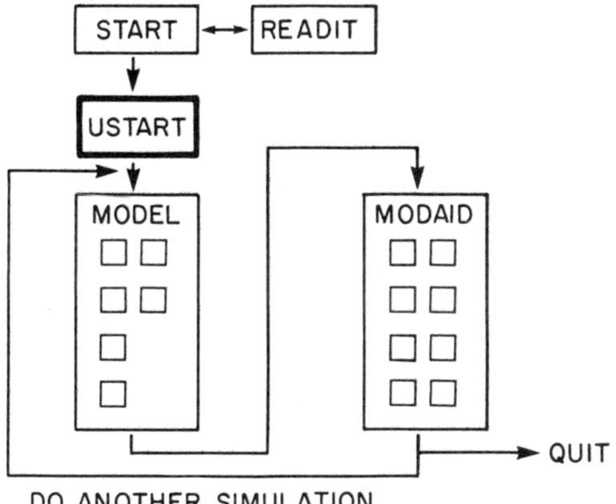

Fig. 3.   Models constructed by PREMOD can be subdivided
into three sets of routines. The subroutines
START, READIT, and USTART initialize variables
used in the model. The routines in the second
set, labeled MODEL, are involved with the solution
of the rate equations of the model. The third set
of routines, labeled MODAID, facilitate plotting
and printing of the results, changing the values
of variables used in the model, and saving results
as files of data.

to insure that the interval on which data are saved for
plotting (DTPL) is no smaller than the smallest time step
used in the simulation. PLTSET is called to initialize the
data file saved for plotting (the plot file). The plot file
is a random access file on which are written the values for
all of the state variables. PLTSAV writes the values of the
state variables to the plot file. PREMOD provides a call to
subroutine PLTSAV within the time loop of the simulation
model.

The routines to solve the rate equations are called
each time step. These routines are FLOWS if difference
equations are used in the model, one of the numerical solution algorithms for differential equations, MARKOV if a
Markov transition matrix is included in the model, and
LIFTAB if an age-specific population submodel is included.
Two optional routines, UCYCL1 and UCYCL2, may be called each
time step just before and just after solving the rate equations, respectively.

Two routine may be called at the end of the simulation. The optional subroutine FINIS is called if it is

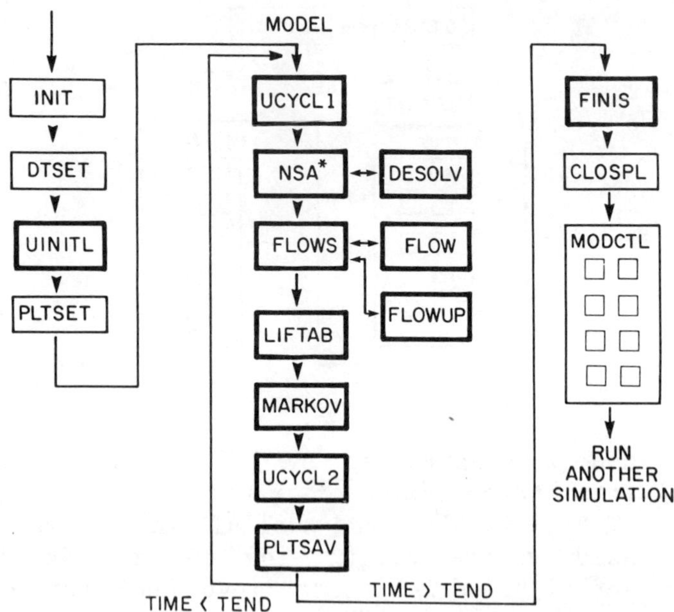

Fig. 4.   The routines that form the core of the simulation
model consist of a set that are executed prior to
each simulation, a set that are executed every
time step during the simulation, and a set that
are executed after each simulation. The routines
identified by thick walled boxes are optional, and
occur only if the user constructs a model that
requires them.

present in the model, and CLOSPL is called to close the plot
file before transferring control to MODAID. Control is
passed to the MAIN menu of MODAID at the end of each simulation.

MODAID

The BASIC version of MODAID implements only the input
and output (I/O) functions of the software. These utilities
are designed primarily to make it easy to plot the values of
the state variables, to print tables of those values, to
change values of parameters and initial values of state
variables, and to run additional simulations. In addition,
one can save the results from one or more simulations and

then directly compare the effects of changing the values of various variables by creating composite plots or tables of those results.

MODAID interacts with the model through three files. The first file, the plot file, is a random access file which contains the values of the state variables from different times during the simulation. The second file is a temporary file used to pass the initial values of parameters and state variables between the model and MODAID. The third file is a table of variable names and related information that is used by MODAID to access the variables from the first two files.

The user controls the functioning of the MODAID I/O routines through six menus: MAIN, PLOT, PLOT CHARACTERISTICS, VALIDATION, CHANGE, and LIST. One gains access to the PLOT, LIST, and CHANGE menus through the MAIN menu, which is displayed immediately after a simulation is completed. The PLOT CHARACTERISTICS and VALIDATION menus are reached from the PLOT menu. One selects the items of these menus by entering either the number of the menu item or a keyword associated with the menu item. Keywords are capitalized to set them off from the text of the menus. Keywords are also assigned to the function keys (F1-F10) and therefore appear in the function key descriptor at the bottom of the display. MODAID provides the option to use keywords with its menus so that it may be used conveniently in a batch mode (i.e., so that the user can create a file of keyword commands to control MODAID). A default entry is provided with each menu. The default entry can be changed by pressing the cursor (arrow) keys.

MAIN MENU

The MAIN menu of the MODAID I/O routines contains 14 selections which provide access to the remaining menus, allow one to return to the simulation model, to quit, and to perform some file management operations (Fig. 5). MODAID provides two ways of running another simulation. The first is executed by the command RUN and causes a simulation to start from the beginning. Normally, this command would be used only after some of the parameter values or initial values of state variables have been changed via the CHANGE menu. The second method is executed using the EXTEND command, which allows the user to restart a simulation from any

---

MAIN MENU

1) CHANGE values of parameters

2) EXTEND a simulation from a
   specified time

3) See the directory of FILES

4) Program KEYS to type alternative
   phrases

5) Go to the LIST menu (list state
   variables)

6) List the NAMES of state variables
   and parameters

7) Go to the PLOT menu (plot state
   variables)

8) RUN another simulation

9) SAVE the current plot file

10) STOP the program, return to DOS

11) STORE the current graphics display
    on a diskette file

12) GET a graphics display that was
    saved previously on a diskette file

13) KILL a specified diskette file

14) RENAME a diskette file

Enter the number of the menu item that you want to execute, or the
corresponding command [**PLOT**]:

Fig. 5.   The MAIN menu of MODAID allows one to run another
           simulation, interact with diskette files, and
           reach the other MODAID menus. The default option
           for the menu is enclosed in brackets at the end of
           the prompt. The default options in all of the
           menus can be changed by using the cursor keys.
           The default option is used as the response if one
           enters an empty line (i.e., if one presses RETURN
           with no entry).

time within the last run. EXTEND allows one to run part of
a simulation, look at the results, and then continue the
simulation if it seems reasonable. EXTEND can also be used
to introduce discrete changes in parameters or state variables during a simulation through the use of the CHANGE
menu.

The MODAID I/O utilities identify all model variables
by name. Two commands in the MAIN menu, NAMES and KEYS, are
provided to facilitiate the use of variable names. The
NAMES command prints the names of all variables saved for
plotting or printing, and all of the variables declared as
parameters. The KEYS command provides a mechanism of
assigning an alternate meaning to the character keys. The
alternate keys are initialized to contain the names of the
state variables and parameters of the model, the names are

assigned to the key corresponding to their first character. The KEYS command allows the user to enter additional phrases to be assigned to the character keys, including those keys already assigned by MODAID. This alternate key utility is provided to speed up the entry of variable names and to make it easy to enter frequently used phrases, such as labels for plots.

The remaining commands of the MAIN menu are used to interact with diskette files. The MAIN menu permits one to list the files on the diskette in a specified drive, to remove files from a diskette, or to change the names of diskette files. One may also save a copy of the current plot file or of the graphics image from the color graphics adapter. The plot file that is saved can be used to produce plots or tabular output at a later time. By saving plot files from two or more simulations, one can directly compare the results of using different parameter values or different initial conditions. Image files are particularly useful for demonstration because they can be loaded rapidly. Thus, one can quickly display graphs from simulations without needing to change variables or titles and without waiting for each plot to be drawn.

PLOT MENU

The PLOT menu controls the construction of plots showing the results of one or more simulations (Fig. 6). The PLOT menu allows one to specify one independent (X-axis) variable and up to five dependent (Y-axis) variables for each plot. The PLOT menu lists the names of the variables to be plotted along with a label for each variable. The labels for the variables are entered when one enters the names of the variables to be plotted. There are three ways to modify plot variables after an initial set has been defined. One can change the name of a single variable to be plotted or its label, one can enter a new set of X- and Y-variables, or one can add variables and their labels to the set to be plotted, up to a total of five Y-variables.

If one runs two or more simulations and saves the plot file for each simulation under a unique name, then one can choose the files from which data are to be read. The user may plot the results from the most recent simulation (which is the default case) or specify that the results from previous simulations be plotted. One may also create a composite plot for a single pair of variables (the X- and Y-variables) from up to five different simulations.

PLOT MENU

The label and variable for the X-axis are:
1) **Month**   **TIME**
2) **Foxes**   **FOX**
3) **Rabbits**   **RABBIT**
4) **Predator**   **PRED**
5) **Prey**   **PREY**

41) The plot FILE(s) to be read are: **PLOTF**
51) Create the PLOT
52) Change or enter VALIDATION
    variables
53) Return to the MAIN menu

54) Enter a NEW set of variables
55) ADD additional variables to be plotted
56) Change the CHARACTERISTICS of the
    plot(scaling, titles, etc.)

Enter the number of the line that you want to change (1 to 5, or 41),
the number of the menu item that you want to execute (51 to 56), or
the corresponding command [**PLOT**]:

Fig. 6.   The PLOT menu allows one to specify the names of
          the variables to be plotted, assign labels to
          those variables, and choose one or more plot files
          from which to get the data to be plotted.

The standard or default plot is drawn using linear-
linear axes scaled to include all of the data. If a
graphics adapter is used, the plot is drawn in high resolution mode and lines connect all data points. The lines for
each Y-variable are labeled at the top of the plot using the
appropriate label and symbol. The resolution of the plot to
be drawn, additional labels and headings, scaling, and other
characteristics of the plot can be modified by using the
PLOT CHARACTERISTICS menu. The VALIDATION menu allows one
to specify the validation data to be plotted along with the
model results.

PLOT CHARACTERISTICS MENU

The PLOT CHARACTERISTICS menu gives the user the opportunity to change the appearance of a plot (Fig. 7). There
are 23 items in this menu, although some of the 23 items
appear only when a graphics display is specified for a plot.
An additional set, related to color, is only displayed when
a color monitor is specified and a medium resolution plot is
to be drawn. The user is not required to set options in

```
                    PLOT CHARACTERISTICS MENU
The top title is:
 1)
 2)
The bottom title is:
 3)
 4)
The Y-title is:
 5)
 6)
 7)  Plotting DEVICE: Graphics display
 8)  INTERVAL of time to be plotted: Entire interval simulated
 9)  MODIFY scaling: NO              15)  Increment between SYMBOLS: 5
10)  Plot validation DATA: NO        16)  BACKGROUND color: 1
11)  Plotting INCREMENT: 1           17)  PALETTE: 1
12)  POINTS or LINES: LINES          18)  RESOLUTION: MEDIUM
13)  Graph coordinates CURSOR: OFF   19)  COLOR or MONOCHROME monitor: COLOR
14)  Character SIZE: 2               20)  Scaling of AXES: Linear-linear
51)  Return to the PLOT menu (create a plot, etc.)
52)  Return to the VALIDATION menu
53)  Return to the MAIN menu
Enter the line of the title that you want to change (1 to 6), a command, or the
menu number that you want to change (7 to 20) or execute (51 to 53)
[PLOT] :
```

Fig. 7. The PLOT CHARACTERISTICS menu is used to add titles to plots and to change the scaling, colors, and other factors affecting the appearance of the plot. In addition, the plotting of validation data and the presence of the graph coordinates cursor are enabled and disabled from this menu.

this menu in order to draw a plot; MODAID defaults to a set
of options on plot characteristics that are usually sufficient to produce an adequate plot.

The default labeling of a plot consists only of the
labels of the variables that are plotted along with the line
type or character used to designate the lines. The PLOT
CHARACTERISTICS menu allows one to add titles to the plot.
The user may add two lines of titles to the top of the plot,
to the bottom of the plot, and along the left side of the
plot (Fig. 8). The length of the titles depends upon
whether low, medium, or high resolution plots are to be
drawn.

A present, MODAID directly supports two plotting
devices: the IBM monochrome display adapter and the color
graphics display adapter. One can choose to use one of
these two devices or to specify the name of an alternate
device driver. A device driver is a program provided by the
user to create plots on devices other than the standard IBM
monochrome and graphics displays. For example, a device
driver could be used to draw plots on a pen plotter. MODAID

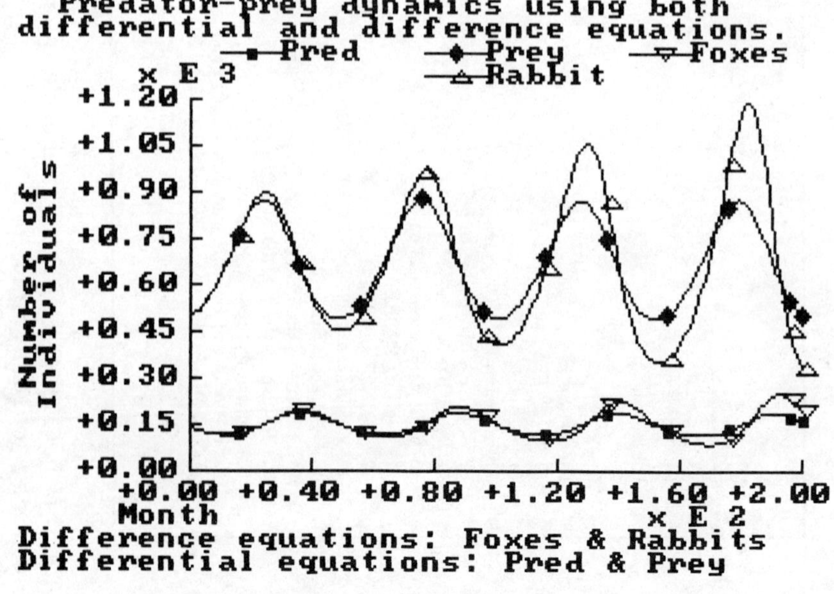

Fig. 8.   Plots produced using the IBM color graphics adapt-
          er can be drawn in either medium or high resolu-
          tion. This example of a medium resolution plot
          was produced on an IBM Graphics printer using the
          DOS GRAPHICS screen dump utility.

creates a file containing the data to be plotted, labels, titles, and other information each time a plot is to be drawn using an alternative device driver. MODAID then executes the device driver to create the plot.

The monochrome display only can be used to produce low resolution, character-oriented plots. The color display adapter will produce either medium resolution (200 v x 320 h) or high resolution (200 v x 640 h) plots. The use of a color display adapter does not require that a color monitor be present. Color plots are only enabled when a color monitor is indicated to be present, and color plots only can be drawn in medium resolution mode. If a medium resolution plot is to be drawn and a color monitor is available, one may set both the background color and the palette of colors. In high or medium resolution plots the user may specify that either points or lines be drawn. Only points are drawn when low resolution (IBM monochrome display) plots are drawn.

Normally, one would plot every pair of X- and Y-values saved in the plot file. However, when the data have been saved frequently over a long simulation, plotting the results can be slow. The PLOT CHARACTERISTICS menu allows the user to specify that only every nth data record be plotted, thus speeding up the plotting process. The plotting increment, n, defaults to 1.

Special characters may be used to identify lines on all high resolution and medium resolution plots and to represent point data. Characters are used to identify lines when a monochrome monitor is attached to the graphics display adapter or when more than three lines are drawn on color plots. Point data may represent either simulation results, which are to be plotted as points rather than lines, or validation data. Although the names of arrays containing the validation data must be specified from within the VALI-DATION menu, the plotting of validation data is performed from within the PLOT CHARACTERISTICS menu. One can change both the character size and the frequency of symbols drawn on the lines. The default frequency is every five points, which may be too frequent if a large number of points are to be drawn.

The PLOT CHARACTERISTICS menu has three options which modify the scaling of the axes of the plot. Either or both axes can be plotted using logarithmic scaling; non-positive values are ignored when doing logarithmic scaling of data. One also may specify a period of simulation time from which to get the data to be plotted. The default scaling of the plots is based upon the data within this time interval. One can override the default scaling by setting the <u>MODIFY</u> option. If the <u>MODIFY</u> option is chosen, then prior to creating each plot MODAID will display the actual maximum and minimum of the data along with the default maximum and minimum for each axis. One can then accept the defaults or enter a new maximum and minimum.

A graph-coordinates cursor can be enabled or disabled using the <u>CURSOR</u> command. The cursor is used to identify the values of points on the graph. The graph-coordinates cursor is moved on the graphics display using the four cursor control keys. The coordinates of the cursor location are printed whenever a character key is pressed.

VALIDATION MENU

The VALIDATION menu allows the user to name up to four sets of variables to be plotted as validation data (Fig. 9).

```
                            VALIDATION MENU
The validation variables are:
1)  X: OTIMES
2)  Y: ODATA1
       3)   Label: Obs. Rep1        4)   Length: 10             5)   Step:1
6)  X: OTIMES
7)  Y: ODATA2
       8)   Label: Obs. Rep2        9)   Length: 10            10)   Step:1




51)  Return to the PLOT menu (create a plot, etc.)
52)  Return to the plot CHARACTERISTICS menu (change scaling, titles, etc.)
53)  Return to the MAIN menu              54) Enter NEW sets of validation data
55)  ADD another set to the current sets of validation data
Enter the command or the number of the item that you want to change
(1 to 10) or the number of the menu item that you want to execute (51 to 55)
[NEW] :
```

Fig. 9.   The validation menu is used to specify up to four sets of data to be plotted along with the simulated results. The data are stored pairwise in two arrays, one array for the X-coordinates and the other array for the Y-coordinates. "Length" gives the number of points in the array to be plotted, and "Step" gives the increment when plotting the data.

Validation data are usually observations from the real world system to which the model predictions can be compared. These data must be stored in arrays that have been declared as parameters. Each set of data to be plotted consists of an array for the X-values, an array for the Y-values, a label for the Y-values, the number of points in the arrays, and a step size to use when plotting these data. Normally, every value in the arrays would be plotted; however, the user may choose to plot only every nth pair of values, where n is the step size. One may enter a label for each set of validation data to be plotted, with the default label being the name of the Y-variable. As in the PLOT menu, one may change individual items, add new sets of validation data to those shown, or specify entirely new sets of validation data.

CHANGE MENU

The purpose of the CHANGE menu is to facilitate changing the initial values of state variables and parameters prior to running another simulation (Fig. 10). A variable

CHANGE MENU

1)   CHANGE the value of a parameter

2)   FLAG a set of variable names

3)   LIST the values of one or more
     parameters

4)   Return to the MAIN menu

5)   Enter a NEW set of variables
     to be flagged

6)   RESTORE a file of parameters saved
     previously

7)   STOP the execution of the model

8)   RUN another simulation using the
     current parameter values

9)   Save the current set of parameter
     values on a disk file

10) List the names of parameters

Enter the number of the menu item that you want to execute, or
the corresponding command [**CHANGE**] :

Fig. 10.   The CHANGE menu allows one to change the values of all parameters or the initial values of all state variables used in the model. The current values for the variables can be saved on a diskette file and recovered later.

can be changed simply by entering its name and a new value. The new values will be used during subsequent simulations. The CHANGE menu also allows one to list the value of an individual variable or to list the values of all parameters used in the model.

Often one may want to frequently change a subset of the parameters used in a model. The CHANGE menu allows one to specify a list of variables to be flagged. Variables names that are in the flagged list are displayed with their values if the user enters the command LIST. The flagged variables also are easily changed. By entering CHANGE $, one is shown the name and current value of each variable in the list, and allowed either to enter a new value or to accept the old value. The list of flagged variables can be deleted and replaced with a new list.

It is often useful to save a set of parameter values so that they can be easily recovered at a later time. The CHANGE menu permits the current set of parameter values to be saved on a file. Thus, one can experiment with using different values of parameters and save those sets which produce interesting results. One can also direct MODAID to read these files to restore a particular set of parameters values.

LIST MENU

The LIST menu allows the user to create tables that list the values of the state variables through time (Fig. 11). LIST takes the data to be printed from the plot file; up to 30 variables may be listed in a single table. One has the option of specifying two lines of text to be used as a heading on the tables. Tables can be displayed on either of the two types of monitors or routed to a printer or a diskette file. The tables written as diskette files may include headings and labels or be formatted so that they can be read from other software, such as Lotus 1-2-3 .

One may print the results from the most recent simulation (the default case) or specify that the results from previous simulations be listed. One can specify the names of up to five plot files to be read. If more than one file is listed, the tables for each plot file are printed.

Normally, one would list every value saved in the plot file. However, when the data have been saved frequently over a long simulation, listing the results can be slow.

The LIST menu allows the user to specify that only every <u>nth</u> data record be printed, thus speeding up the listing process.   The LIST menu also allows one to delineate a period of simulation time from which to get the data to be printed.

```
                               LIST MENU
            Label              Variable
    1)   Month               TIME
    2)   Rabbits             RABBIT
    3)   Foxes               FOX
    4)   Predators           PRED
    5)   Prey                PREY
    6)   Soil                SOIL
    7)   Grass               GRASS
    8)   Shrubs              SHRUB
    9)   Trees               TREES
    The heading for the listing is:
    31)
    32)
    41)   The plot FILE(s) to be read are: PLOTF
    42)   Output DEVICE: IBM monochrome display
    43)   Listing INCREMENT:  1
    44)   INTERVAL of time to be listed: Entire interval simulated
    51)   PRINT the listing              52) VIEW remainder of variables to be listed
    53)   Return to the MAIN menu        54)  Enter a NEW set of variables
    55)   ADD variables to the current set
    Please enter a command or number of the line that you want to change (1 to 9 or
    31 to 44) or the number of the menu item that you want to execute (51 to 54)
    [PRINT]:
```

Fig. 11.   The list menu allows one to specify up to 30 variables to be printed in tabular form on either the IBM monochrome or color displays, a printer, or as a diskette file. The diskette files can be written using a format suitable for printing, or in a format that allows them to be read by other software, such as Lotus 1-2-3 .

SUMMARY

PREMOD is an interactive software tool that helps one easily construct and modify simulation models. The models are coded in FORTRAN on mainframe and minicomputers, and in BASIC on IBM-PC microcomputers. PREMOD will construct simulation models for four types of mathematical models that use (1) differential equations, (2) difference equations, (3) a Markov transition matrix, and (4) age-specific population dynamics. PREMOD minimizes the amount of code that the user must enter in order to build a model. Nevertheless, PREMOD can be used to construct relatively complicated models.

MODAID is a set of software utilities that are often used in simulation models. The BASIC version of MODAID is presently a subset of the FORTRAN version. The BASIC version of MODAID includes routines which facilitate the creation of plots and tables of state variable values, store the results of simulations, change parameter values, and run additional simulations. These MODAID input and output utilities make use of a system of menus to guide one in their use. The menus can be controlled by entering either a number or a keyword associated with each menu item. Thus, MODAID can be used conveniently in either an interactive or batch mode.

Although the BASIC versions of PREMOD and MODAID are subsets of the FORTRAN versions of the software, further extensions of the BASIC versions are planned. The next additions to MODAID will be utilities to help perform sensitivity and uncertainty analyses. The libraries of mathematical functions and pseudorandom number generators available in the FORTRAN version will be added thereafter. In addition, a utility is planned for a revised FORTRAN 77 version of MODAID that will facilitate the conversion of simulation models from BASIC to FORTRAN. Thus, one will be able to do some initial model development on a microcomputer and then upload and mechanically convert the model to FORTRAN so that it can be run on a mainframe or minicomputer.

ACKNOWLEDGMENTS

PREMOD and MODAID have evolved over several years as a cooperative effort by many people. James M. Vevea is primarily responsible for the FORTRAN versions of MODAID and has contributed greatly to the design of PREMOD. Vicki Kirchner contributed much to the design and implementation of the sensitivity and uncertainty analysis routines and provided invaluable support during the development of the BASIC versions of the software. Thanks also to Ward Whicker, Charles Van Baker, Bill Lauenroth, and the many students who helped test, evaluate, and criticize the software during its development. Support for the revision of the FORTRAN versions of the software is provided by contract DE-AS08-79NV10057 from the US Department of Energy. The BASIC versions of MODAID and PREMOD were developed under the support of grant SPE-8263297 from the National Science Foundation. The author will be pleased to supply information on how to obtain a copy of PREMOD and MODAID.

LITERATURE CITED

Kirchner, T.B. and J.M. Vevea. 1983. PREMOD and MODAID: Software tools for writing simulation models. pp. 173-183. In: Analysis of Ecological Systems: State-of-the-Art in Ecological Modeling. W.K. Lauenroth, G.V. Skogerboe, and M. Flug (eds.). Elsevier Scientific Publishing Co., New York.

Horn, H.S. 1975. Markovian processes of forest succession. pp. 196-211. In: Ecology and Evolution of Communities. M.L. Cody and J.M. Diamond (eds.). Belknap Press, Cambridge.

ADVANCES IN SCIENTIFIC SOFTWARE PACKAGES

Channing H. Russell

ABSTRACT

Early scientific software packages focused on compilers, individual applications, and specific aspects of computer support such as statistics. More recently, software packages provide a broad, integrated, easy to use, and extensible set of capabilities to support research data management. RS/1 (TM) is described as an example of modern scientific software.

INTRODUCTION

Most publicity about the software industry focuses on applications in office and home settings; however, there is also significant development taking place in packaged software specifically designed for research data management. The use of commercially available scientific data management software, either as a stand-alone tool or as the base for building applications, is increasingly contributing to the productivity of research organizations.

It is indicative of the growing importance of the scientific software market that a major corporation, Bolt, Beranek, and Newman Inc. (BBN), has recently formed a new business activity called BBN Software Products, focused entirely on developing software for science.

The development of scientific software at BBN grew out of a long history of support of a variety of scientific applications, including their corporate experience in supporting their own research and development projects as well as a number of government-sponsored activities. For example, in the 1960's, BBN developed a hospital information system for the Massachusetts General Hospital using an early minicomputer, the PDP-1, to support clinical research activities. Under the sponsorship of the National Institutes of Health, they developed a system called PROPHET[1], which provides a national time-sharing service to scientists studying the effects of chemical substances on biological activity. PROPHET operates as a kind of automated laboratory notebook, oriented toward data structures (such as tables and graphs) that are familiar to scientists (Castleman et al., 1974). A similar orientation toward data representations familiar to scientists is used in the CLINFO system, which provides specialized support for the study of time-oriented clinical data in the general clinical research center setting (Gottleib et al., 1979).

The concepts and technologies developed through this history of support of scientific applications have been incorporated into BBN's leading software product, RS/1. RS/1 represents a distinct contrast to earlier scientific software. In general, early software used by scientists took three primary forms: 1) languages such as FORTRAN, BASIC, and PASCAL were helpful in supporting highly computational applications, but were not designed to support research data management; 2) specific scientific programs were also developed, which were limited to supporting a single narrow applications area and were usually operable only using particular computer system equipment; and 3) specialized software packages began to appear in specific areas, such as SAS (SAS, 1982) and BMDP (Dixon, 1975) in statistics and MLAB (Knott, 1979) in curve fitting.

Some of the earlier software, such as SAS or APL, has since been extended to provide broader support, often with limitations that reflect their history. In addition, integrated business packages, such as Lotus 1-2-3 in the microcomputer world and some of the data management packages, have also been used for scientific work.

BBN took a more systematic approach than had been used in the development of earlier software packages for the scientific market; the company's goal from the start of the development process was to design an integrated general purpose package to provide broad-based interactive support for research data management. The software package, RS/1, that

---

[1]The PROPHET system is sponsored by the Biotechnology Resources Program, Division of Research Resources, National Institutes of Health under contract \#NO1-RR-8-2118.

was the result of this development process is an integrated information-handling environment supporting data management, analysis, graphics, statistics, modeling, and the generation of reports. The package is suitable for use by a broad spectrum of scientists engaged in a wide variety of research and development activities.

CHARACTERISTICS OF MODERN SCIENTIFIC SOFTWARE

In order to understand the characteristics of state-of-the-art scientific software, a description of the facilities of the RS/1 system is presented in this section as an example of the kind of software now available to support scientific data management needs. All facilities are based in a single integrated system. No extra steps are needed, for example, to use tabular data as the basis for statistics or to graph the results of modeling.

Data management in RS/1 is based on two-dimensional tables. Each cell of a table can contain data representing fixed or floating point numbers, dates, times, or free text. Cells in a particular column are not all constrained to the same type; it is possible, for example, to include a note about some missing data in a column of numerical results. A user can work with many hundreds of tables. Tables are based on disk files, accessed through a kind of paging scheme, so there is no limit on table size. Some users work with tables containing hundreds of columns and tens of thousands of rows.

Data analysis makes direct use of these tables. A new column can be created as a transformation of existing data. No additional steps are needed to create the column; a single command defines the transformation and transfers the data (Fig. 1). The results can be seen on the screen immediately, again without additional commands or setup steps.

RS/1 supports the major kinds of analytical graphics needed in a scientific setting, including scatterplots, fitted curves, bargraphs, histograms, piecharts, three-dimensional displays (Fig. 2), and contour plots. Graphs are stored as permanent data objects that can be edited and redisplayed. This makes it possible to transform a graph made for analysis into a graph suitable for publication. An advanced terminal-independent graphics support capability permits the output of pictures on more than a hundred different graphics devices.

Drill Test Data

0 Bit Code	1 Hardness	2 Feet Drilled	3 Final Height	4 Lower Bound	5 Upper Bound
1 N22	0.80	218	0.50	0.450	0.550
2 N32	0.84	309	0.48	0.432	0.528
3 N44	0.80	662	0.41	0.369	0.451
4 N81	0.80	512	0.46	0.414	0.506
5 N82	0.84	114	0.59	0.531	0.649
6 S14	0.82	319	0.49	0.441	0.539
7 S17	0.80	404	0.47	0.423	0.517
8 S19	0.83	844	0.33	0.297	0.363
9 S22	0.83	56	0.74	0.666	0.814

Collected 4/18/82
Measurements of height +/- 10%

COLUMN 5 OF DRILL = COLUMN 3 * 1.1 <RET>

Fig. 1.   Single command transform and transfer data in RS1 table.

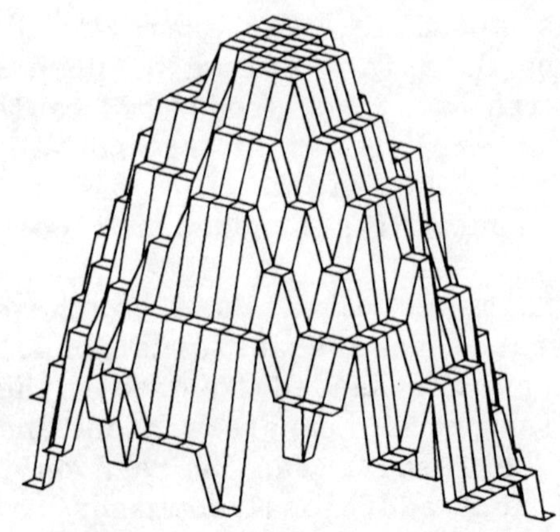

Fig. 2.   RS1 three-dimensional display.

Statistics available in the system include a large set of commonly used analysis techniques, as well as advanced nonlinear curve-fitting techniques. Statistical results can be displayed numerically or graphically.

Modeling offers a spreadsheet-like capability, which permits the integration of several different spreadsheets and the inclusion of general data manipulation commands in cells.

An optional extension to the system makes it simple to produce documents containing mixed text and graphics illustrations, which can be output on a laser printer.

Ease of use is a major theme in modern commercial software. The best packages offer several styles of interaction suitable for advanced or beginning users. In RS/1, both a command-based and a menu-based style of interaction are available. For example, using an English-like command, a bargraph can be constructed from a table in a single step as illustrated in Figure 3.

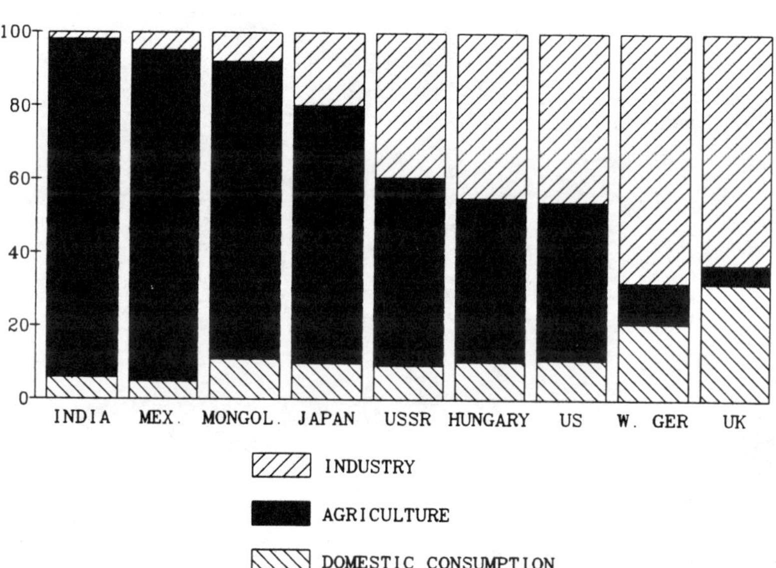

Fig. 3.   Bar graph created from RS1 table.

A nonlinear fit can be accomplished in much the same straightforward way (Fig. 4). Menu-oriented interactions are also supported to make complex editing easy to understand, and to provide a way for beginning users to start using the system. Thorough documentation and optional direct and videotape training are available.

A key attribute of software intended for use in science is extensibility. Research data management, almost by definition, involves special purpose information handling beyond standard data management techniques. These specialized needs are often confined to one particular phase of analysis; ideally, scientific software should support the

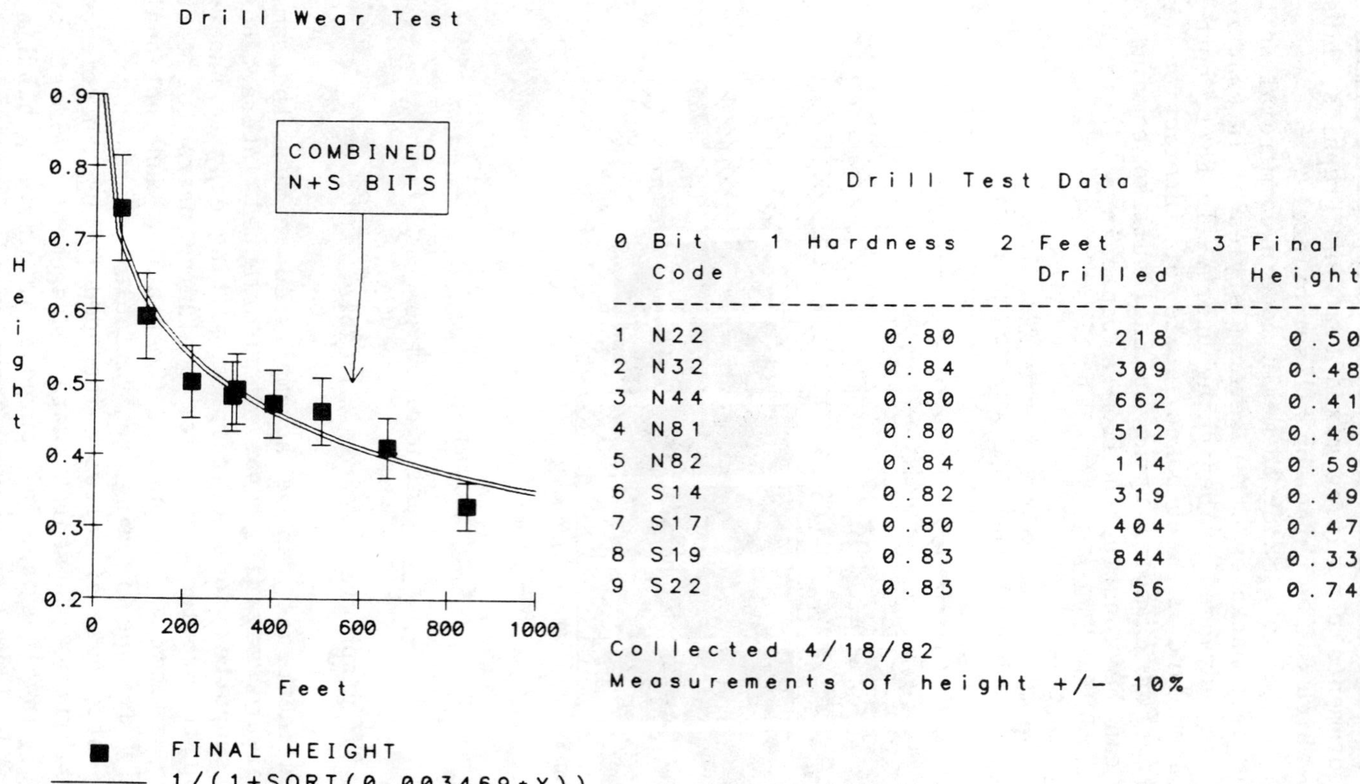

Drill Test Data

0 Bit Code	1 Hardness	2 Feet Drilled	3 Final Height
1 N22	0.80	218	0.50
2 N32	0.84	309	0.48
3 N44	0.80	662	0.41
4 N81	0.80	512	0.46
5 N82	0.84	114	0.59
6 S14	0.82	319	0.49
7 S17	0.80	404	0.47
8 S19	0.83	844	0.33
9 S22	0.83	56	0.74

Collected 4/18/82
Measurements of height +/- 10%

Fig. 4.   RS1 table and graph showing nonlinear fit.

smooth integration of special-purpose programming into an application which also makes maximal use of what is already available in the system. Too often in the past, entire new systems had to be constructed in a low-level language like FORTRAN because the existing higher-level data management systems did not support the needed kinds of customization.

There are several kinds of extensibility built into RS/1. A full-structured programming language called RPL is a part of the system. This language, stylistically similar to PL/1, allows direct access to data objects such as tables and graphs, and the intermixture of high-level data management commands with traditional programming constructs.

The RPL language is designed for easy-to-write, compact programs. Like APL, it supports a run-time environment in which variables can represent different data types at different times. There is no need for the kind of data declarations that make programming awkward in traditional languages like Pascal.

In the scientific world, maximum flexibility is needed in interfacing to a variety of different programs, in accessing various databases, and outputting information to different kinds of graphics devices. This goes far beyond the kinds of flexibility needed in business-oriented packages. RS/1 supports an ability to call programs written in other languages, and has interfaces to a growing set of commercial database systems. Especially powerful applications can be constructed using information from a large database as the basis for analysis.

Output can be directed to a variety of printers, plotters, and display terminals. It is even possible for sophisticated users to add support for new kinds of devices, through a facility called the terminal data table, without a need for system programming.

APPLICATIONS OF SCIENTIFIC SOFTWARE

Scientific software packages are beginning to have a large impact on productivity in research organizations. In a major pharmaceutical research and development organization at Merck, Sharp and Dohme, RS/1 is used at several different laboratories in the US, Canada, the UK, and continental Europe to share data. A large VAX installation supports users in different sites through a computer network. Collaborative projects between different laboratories, using data shared through the central system, have become common.

The semiconductor industry makes heavy use of manufacturing data to provide ongoing quality control information. RS/1 is used as the central software tool to tie together a number of different databases and software systems to make the data available for graphing and analysis.

More than a thousand scientists and engineers use RS/1 at DuPont's Experimental Station in Wilmington, DE. It was particularly attractive to DuPont that the software was available on microcomputers as well as superminicomputers; this enabled them to adopt RS/1 as a standard for remote research laboratories and engineering stations in production plants throughout the Eastern seaboard.

In the new field of genetic engineering, scientific data management software is used to manage the long alphabetic codes that represent genetic sequences, as well as more traditional numeric and text applications. At Genentech, scientists use the software for these tasks as well as for laboratory data analysis.

FUTURE DEVELOPMENTS

Advances in computer science continue to serve as the basis for new extensions to software products. In particular, artificial intelligence techniques have begun to mature to the point at which they can play a role in scientific software. In the future, scientific software will incorporate expert systems technology in order to provide a new level of assistance to scientists in applying statistical and graphical techniques to data analysis. Two major areas are likely to be the focus of expert systems in the scientific software area: assisting users without extensive statistical training in starting to use statistics, and helping design multifactor experiments.

LITERATURE CITED

Castleman, P.A., C.H. Russell, F.N. Webb, C.A. Hollister, J.R. Siegel, S.R. Zdonic, and D.M. Fram. 1974. The implementation of the PROPHET System. AFIPS Conference Proceedings 34: 457-468.

Dixon, W.J. (ed.). 1975. BMDP Biomedical Computer Programs, 3rd Ed. University of California Press, Los Angeles.

Gottlieb, A.G., D.M. Fram, S.F. Whitehead, G.M. Rubin, C.H. Russell, P.A. Castleman, and F.N. Webb. 1979. CLINFO: A friendly computer system for clinical research. XII International Conference on Medical and Biomedical Engineering. Jerusalem, Israel.

Knott, G.D. 1979. MLAB--A mathematical modeling tool. Comp. Prog. in Biomed. 10(3): 271-280.

SAS Institute Inc. 1982. SAS User's Guide: Basics. Cary, NC. 494 pp.

DESIGNING A SCIENTIFIC
PROBLEM SOLVING ENVIRONMENT:
THE NMSU SCIENCE WORKBENCH

Walt Conley, Brian M. Slator,
Matthew P. Anderson, and
Richard A. Sitze

ABSTRACT

This paper presents a plan of action for the development of an interactive computing environment for scientific
researchers. The project's main objective is to provide a
system that enhances productivity of moderate size (3-5)
research teams. The environment contains fully developed
applications packages that can be broadly categorized into
four sub-environments: knowledge-base system, statistics
and representation, text handling, and project management.
Each sub-environment contains a fully integrated suite of
utilities that are accessed in a similar manner. A major
emphasis is being placed on the design of a natural language
interface between the users and background applications
software. Intelligent "Expert Systems," each with a restricted domain, form the communication interface to the
user. This paper describes features of the NMSU Science
Workbench, an approach toward interface development, and our
view of the interface as a suite of expert systems containing knowledge about the application subsystems, the users,
and the various interaction strategies. The Science Workbench is being designed to provide support for the working
scientist--it is intended to save the scientist time and enhance the ability to produce informed decisions.

INTRODUCTION

Integrated Scientific Research

An integrated, intelligent computational environment is being designed to assist scientists in all phases of their research and in project administration. Once the data have been collected, the major tasks are data entry, analysis, and manuscript preparation. In interdisciplinary projects, where large amounts of data are generated and teams of researchers must work together, it is quite difficult to conduct an adequate data analysis.

Complexity is inherent to ecosystem research. An effective approach to complex problems is one the computing industry refers to as "divide and conquer" (Alagić & Arbib, 1978; Zelkowitz et al., 1979); in effect, modularizing the problem and continuing the decomposition until each small piece can be expressed in a few lines of code. For an ecosystem-level project, effective modularization requires discipline specialists. The problem of getting a group of specialists to work together is not trivial; the decomposition implies a high-level project organization, a priori knowledge of the problem, and a compatible spirit that will allow resynthesis of the modular solutions into a full solution of the problem at hand.

One approach to final synthesis is having frequent "workshops." In general, such sessions discuss concepts and the research to be conducted, assignments are apportioned, people go off to their laboratories to do the work, and return to talk about it sometime later.

Interdisciplinary teams should approach data analysis and synthesis as a team, rather than as a confederation of individuals. However, research environments are typically not configured to support the needs of team research. Current computing environments are generally directed either toward individual workers or to massive mainframe computers that may serve several hundred simultaneous users. What is needed is a research computing environment that is specifically designed to meet the needs of research teams of moderate size; the goal being to facilitate the interaction of researchers as a team rather than individuals.

The guiding thesis of this project is that a properly designed computing environment, serving small teams of researchers, will increase productivity and promote total data synthesis. This suggests an exploration of the feasibility and methods of operation of "real-time workshops." In such an environment, data appropriate to the workshop questions have been previously assembled and organized by a research team. The workshop can proceed directly to specific analyses, exploratory investigations, or final graphics for poster, paper, or presentation. Text entry and review are similarly supported. All work proceeds on-line and in real-time. Analyses can be brought to a graphics screen for discussion, review, and rejection or acceptance. Effective use of time is a key ingredient.

A sample of principal investigators from various ecological projects indicates realms of responsibilities and time commitments that can be rearranged to increase productivity. The average principal investigator devotes about 10% of available time to data collections, 20% to data analysis, 30-40% to text manipulation, and 30-40% to project management. Our intent is to reduce the project management requirement by 10-20% (a minimal estimate), and to increase the time available to the researcher for analysis and write-up (a factor of 10 is within reach). Also, various aspects of group dynamics will be facilitated, and the system will enhance _total_ data synthesis--the bringing together of multiple data sets across various disciplines.

Currently available computing "workstations," deemed adequate for many purposes, prove incapable when faced with the need for the elaborate coding required to facilitate communications. Mainframe environments, on the other hand, are geared for an average user and cannot be easily customized for the individual.

What is needed is a medium-size computer, with considerable power, that can be designed to serve the needs of a group of users that, by themselves, do not provide sufficient economic incentive to attract commercial attention. An intelligent environment, supporting perhaps 3-5 concurrent users, and providing peripherals that support plotting, good quality printing, communications, etc., could reasonably be customized for small research teams comprised of people who would prefer to concentrate on the task at hand, rather than on how to get the machine to cooperate.

A reasonable computing environment will support both planned and casual workshops. It will serve groups of researchers that range in size from 1 to perhaps 6 or 7 (our experience shows that more than 6 generates a shift from "doing the work" to "talking about doing the work").

Limitations of Previous Approaches

Integration and synthesis of information at the ecosystem level was a major objective of data collections conducted under the auspices of the International Biological Programme. While such syntheses did in fact occur, it is probably safe to say that the time lag between data collection and final synthesis reduced the excitement as well as the potential impact of the final results. Further, such syntheses usually appear in a book, either written by one or two senior investigators, or a collection of chapters, written by many participants. Such works, which represent monumental effort both in the data collections and in the final write-up, have somehow fallen short of the total blending of data and concepts that promises to provide genuine understanding of ecosystem processes.

The reasons for this shortfall in understanding ecosystem processes can be ascribed to several broad categories of problems with doing ecosystem-level science. First, ecosystem studies are simply too large in both breadth and depth for full comprehension by even the most experienced scientists. The vast amounts of data involved and the wide array of disciplines that need to be covered make the task intimidating at best, and impossible at the worst.

Second, it is seemingly impossible to gather relevant data on all aspects of the ecology of a study site. Because researchers must choose what to do with limited time and resources, specific questions are generally posed and pertinent data are gathered. This, of course, leaves out other questions that are important.

Third, the general pattern in ecological studies is one of high variances in sample data sets. While this problem is serious when dealing with a "single hypothesis" study that works with a limited domain of ecosystem data, it becomes critical when multiple data sets must be compared. This problem is compounded when dealing with data sets that were collected for purposes other than the question at hand.

Finally, the sheer effort required to maintain long-term data sets in a manner so that they remain accessible, and the effort required to conduct a thorough analysis of a widely divergent group of data sets, is simply too high a price to pay. One result is that too many analyses are conducted in "one-shot" fashion, that is, a question is posed, and a statistical hypothesis is tested (often by a programmer). The results are typically obtained days or weeks later. This frequently requires searching through various tape files, merging data sets, and manipulating data so that they exist in the right form for this analysis. If any special programming needs to be done, the delay from question to answer will be even longer. The general level of interest and the time lag to a final answer are inversely related.

Other reasons for the inherent difficulty of doing large-scale ecosystem science could of course be provided. One additional problem is that of doing cooperative team research. Existing support services in the usual university setting are geared toward the individual not the research team. Expensive resources such as libraries and computer centers should be shared; however, a research team requires unique considerations and facilities (which are generally lacking). Facilitation of group effort must include a conscious attempt to understand the psychology and dynamics of small groups of research scientists, and it ought to include the means by which such groups can be encouraged to be productive (see Burgoon et al., 1974; Davis & McCallon, 1975; Bradford, 1976; Trecker & Trecker, 1979). The Science Workbench project is not going to solve all of the problems of team research; however, there are spots where it will contribute to group productivity.

The current Long-Term Ecological Research program of the National Science Foundation has directed attention to such problems of ecosystem research. Recognition that long-term data often provide answers that are different from those obtained from analysis of short-term studies has led to the philosophy of collecting and keeping data over extended periods of time. The promise of the LTER program is that these data will be available in a manner that allows recall and use in future times (implicit in this is use by investigators that did not collect the original data). The level of sophistication in data documentation, management, and analysis required is unprecedented in the ecological sciences. If the LTER promise is to be kept, there must be a significant effort expended on the question of data management (in the broadest sense). Current practices in the LTER program support these requirements, and indeed, significant effort and funds are being expended. Even more effort will be required in the future, as existing data sets continue to grow, and as new data sets are added in response to new understanding of what needs to be done. Long-term success will require a high level of sophistication in data handling.   The philosophy of <u>using</u> the data must guide the keeping of data. A roomful of historic magnetic tapes that no one knows much about is not a desirable consequence of any long-term research program.

DISCUSSION

It would be silly, if not impossible, to attempt the recreation of such programs as operating systems, algorithm packages, a raw database management system, writer's support capabilities, and general statistical routines. Elegant and suitable programs already exist in a form compatible with the Science Workbench project. We do, however, impose the requirement of extensibility; research projects are dynamic, therefore the computing environment must be capable of growing.   Extensibility generally implies that the package should be a language and environment unto itself. Such a package, consisting of a suite of primitive command sequences that can be collected into higher and yet higher levels of capability, is programmable in its own right. This type of package will allow a development team to customize and enhance programs at any level, from incorporating raw code in Assembler or any higher level language to programming shells and interfaces in either a high-level general language or in the supported environment of the system itself. Packages that allow such interactions are typically thought of as "programmers packages" and are not particularly easy to learn how to use. This requirement thus guides the three levels of organization that make up the Science Workbench project.

1.   To design and prepare a suitable interface to each of several computing sub-environments that broadly provide: a) database management; b) statistical analysis and representation; c) writer's support; and d) budgeting and record keeping. These interfaces are viewed as intermediate-level "Expert Systems" (Hayes-Roth <u>et al</u>., 1983).

2.   To design and prepare a suitable "object-oriented" background such that any result from any subset of the overall environment is available to any other subset without further manipulation. Such products must be compatible across the entire computing environment. This implies self-describing data sets and file typing.

3.   To design and prepare a very high-level user-
     interface that provides access and expert help to
     the computing environment within.   This shell will
     emphasize a natural language approach to user com-
     munication, and will provide extensive support and
     help so that biological researchers can concen-
     trate on "doing the science," and not be dis-
     tracted by the operation of the system.

In short, the philosophy is to make the computing power
now enjoyed only by expert computer systems programmers
available to a team of biological researchers.

To describe the entire computing environment under
development would require more space than is available.   In
the sections that follow, additional detail is provided for
the four main sub-environments (database management, statistical analysis and representation, writer's support, and
budgeting and record keeping), computer-user interactions,
and some discussion of the design and implementation.

Database Management

Database management implies the organized storage and
subsequent retrieval of raw data.   A proper Database Management System (DBMS) will allow incorporation of any level of
organized information, as well as information that occurs in
diverse forms ($\underline{e}.\underline{g}.$, precipitation data $\underline{vs}$. keyworded bibliographic citations).   What must be accomplished if the
Science Workbench is to be a success is to provide a Knowledge Base that supports archiving and reporting of complex
subsets of the total stored data.   Such subsets of data must
be available to the analysis and reporting environments in
such a manner that extensive further preparation is not required.   Typed data sets that are self-describing with respect to internal structure and contents are planned (see
Becker & Chambers, 1984b for an example).   A long history of
problems related to data banking biological and environmental data are described in Oppenheimer et al. (1976), and
current efforts on the part of state and federal agencies
are discussed in Cross (1983).

Research environments are dynamic, and the questions
that will be posed of existing data cannot always be predicted.   Further, as understanding increases, some data
being collected will be recognized as unproductive, while
new data sets will be initiated.   The background database
system for the Science Workbench must be capable of dealing

with frequent modification, and must be capable of handling a full data complement in the 300-500 megabyte size range (i.e., thousands and hundreds of thousands of observations for a given research project). Long-term studies present special problems of their own in database considerations, and must carefully consider volume and stability requirements.

Three basic structural models of database systems exist. Hierarchical models are tree structures that provide a "one to many" mapping of data entities. Hierarchical systems, an older technology, are used extensively in banks and many other business environments where the job is more definable and less dynamic than is the case with research programs. Network data structures provide a "many to many" mapping of data entities, and thus are powerful searching structures. Both hierarchical and network data structures lack full flexibility, and generally require that the entire database be unloaded and reloaded when expansion or modification of either the logical or the physical schema is required. This resetting of the pointers is time consuming and makes the database unavailable for use during modification periods.

Relational database structure incorporates two-dimensional tables of rows and columns, with rows as observations and columns as attribute fields. Successive levels of two-dimensional tables are linked (i.e., related) via pointer fields, thus allowing considerable flexibility in overall design. A "relational" database structure (Neely & Stewart 1981; Ross, 1981; Date, 1982) serves the requirements for the Science Workbench project.

Existing database systems are being tested for suitability. Specific problems with the current state of DBMS development include too few data types available, difficulty in managing transitions between micro data (at the level of raw observations) and macro data (at the level of summary analyses), instructions regarding missing values, flags for data quality, minimal compression, and generally poor interface support for human users.

Considerable effort must be expended to code the shell around the chosen DBMS; however, the final result will be a fully functional and powerful background archiving and reporting (i.e., retrieval) package that can be updated interactively, expanded or reduced in size or configuration, and queried with reasonable effort. The Knowledge Base sub-environment will service other sub-environments by producing subsets of data for their use, including numeric information for statistical analysis, backup writing support (e.g., bibliographics, "boilerplate"), and budgeting and other management records.

Statistical Analysis and Representation

A supportive statistical analysis environment should provide both simplicity and power in a manner that reinforces rather than conflicts (i.e., an easy-to-use system ought not constrain its users; a powerful system need not be useful only to experts [Becker & Chambers, 1978a]). The statistical analysis environment must be interactive, with built-in operations and functions that include the usual arithmetic and logical operations, and in addition, a large number of high-level statistical and graphical operations. The whole environment should treat all such functions within a uniform syntactic structure. Further, the environment must be capable of extensive enhancement and expansion. Ecological scientists have developed a large body of theoretical and esoteric analyses that will need to be incorporated along with the more standard analysis capabilities.

The capability to conduct formal statistical analyses that are derived from distinct a priori hypotheses is only a part of the support environment envisioned. Interactive capability allows not only the usual hypothesis-directed statistical analyses to be done, but also provides closer linkage between analyst and data. Extensive exploratory capability with the data are required, including

- support for perusing large and diverse sets of data in a manner that facilitates workshop sessions and data review;
- merging, tabulating, subsetting, transforming, outlier detection, and any required file manipulation;
- graphical review of any configuration of the data of interest, with the ability to move among visual and analytical treatments;
- creating data structures that are self-describing, that is, organized in such a manner that extensive argument lists describing the data structure for each desired analysis are not needed, and assuring that such file structures can be transferred to any other location in the environment without requiring further description;

- creating the results of an analysis as a legal
  data structure and producing a table of the results as a side-effect. Consider for example doing a regression analysis where the residuals are
  fully available, without manipulation, for further
  analysis and/or graphing, and a table of the results is also available, or producing the results
  of an analysis in graphical form for a visual review and making available. any result for further
  analysis (thus catering to a researcher's intuition
  first--this latter case implies a capability to
  input a data value directly from a graphics display);
- the ability to produce the results of a high-level
  operation (e.g., a regression, discriminant functions, or cluster analysis) from the application
  of a single function call, obtainable with minimal
  reference to a configuring argument list;
- provision of an interface language that enables
  new algorithms to be made into system functions,
  whether or not they were originally intended to be
  included;
- ability to incorporate new algorithms, coded in
  any higher level language and wherever they were
  developed, thus encouraging development and testing of new ideas.

A growing body of literature supports what has come to
be called Exploratory Data Analysis Management (Chamber,
1977; Tukey, 1977; Velleman & Hoaglin, 1981; Hoaglin et al.,
1983). The general concept is one of putting researchers
quite close to their data and encouraging total data analysis. "Exploratory Data Analysis" refers to an iterative,
clue-driven trial and error process, while "Management" implies a tracking of the process of moving from raw data to
summary data to final conclusions, thus "managing" the analyst's observations as the analysis proceeds.

At AT&T Bell Laboratories, a new concept for a statistical analysis system which has been evolving since about
1976. Arguing the notion that total data analysis combines
ease of use, power, graphical review, and extensibility, a
series of papers (Becker & Chambers, 1976, 1978a & b; Becker
1978; Chambers, 1978, 1979, 1980; Chambers & Kleiner, 1982;
Chambers et al., 1983) lead to a finalized version of what
is now called S: An Interactive Environment for Data Analysis and Graphics (Becker & Chambers, 1984a & b).

Concurrently, statistical languages (Wexelblat, 1978) and communication of information through proper graphics were being examined (Cleveland, 1979, 1985, Schmid, 1983; Wainer, 1984). These developments built upon the mature algorithm development that had emerged through the 1960's and 1970's (e.g., Rice, 1983). The concepts that evolved were different than what had been developed in statistics packages that were essentially batch-oriented. The BMD (Dixon, 1974) and BMDP (Dixon, 1983) packages have formed a backbone for statistical analyses for over 15 years. Similarly, the Statistical Analysis System (SAS) begun at North Carolina State University, and now a full fledged private institute (see SAS Institute 1981, 1982a & b), has become a standard package of choice at many computer installations. Even though the SAS system is available in "interactive" versions (SAS Institute, 1983), the structural concept in the programming remains as basically a batch environment wherein the user essentially assembles and submits a "card deck" from a terminal.

It is now likely that even a major package such as SAS can be fully supplemented by S; this process will, however, be a necessarily long one. In the interim, access to mainframe batch environments will provide needed support. Although not discussed in detail in this paper, the Science Workbench project has designed full communication capabilities into the working environment. In this manner, services better provided on a mainframe system (they are becoming fewer and fewer) can be accessed and utilized. Similarly, as fully developed distributed systems become available, the Science Workbench will realize it's full power. In this context, we envision easy access to remote computer facilities, and full networking communications at both the local and global levels.

The background thus exists that will allow a customized analysis and representation system to be developed for ecological scientists. This project is implementing the S system as designed by the Becker and Chambers team, and will extend it into such ecological specificity as is required to accomplish our goal of reasonable and complete data analysis for ecologists. Substantial amounts of this ecological specificity are already completed, and remain only to be implemented in a workbench environment. Population models (Conley, 1978; Lenarz & Conley 1980, 1982; Watts & Conley, 1981, 1984; Sterling et al., 1983) have already been incorporated. New generations of such models will make use of such developments as intelligent expert systems (Conley, 1983). Other programs for demographic analyses, community analyses, ordination and classification procedures, and association analyses are already in place. The full implementation of a substantial set of ecological utilities is a long-term project; one that is unlikely to ever be completed. The overall capability should, however, grow in much the same manner as the utilities complement in the LISP machines in the artificial intelligence disciplines, or indeed, in the UNIX operating system, which now offers hundreds of separate utilities.

<u>Writer's Support</u>

Research scientists spend considerable time writing about the results of their research and ideas. Full writer's support is designed into the Science Workbench, in a manner that will encourage documents produced to be more correct grammatically and typographically. More importantly, the writer's environment within the Science Workbench will provide links to the Knowledge Base system where bibliographic and numerical reporting is available, and links with the statistical analysis and representation system so that results produced (including graphical representations) can be freely incorporated into the textual discussion of those results. The environment will also contain an extensive suite of text development and formatting macros that will support style guidelines for a specific set of ecological publication outlets, provide outline structures for proposal formats, and support conversational interaction in letter and memo writing.

Extensive bibliographic support will be available. A research team that maintains a key-word bibliography of relevant citations (annotated to any extent) will have the capability to subset the major citations list in any manner. Bibliographic citations in the master set contain tag fields that define the various components (author, date, title, etc.), allowing automatic rearrangement to accommodate journalistic style requirements. The writer will need only to reference the citation in text; the writer's support environment will extract the citation from the correct files, and will place the citation into a properly structured literature-cited section. Further bibliographic support includes full citation reporting, organized by key words and formatted to provide hardcopy of literature search results (see Conley & Conley, 1984, for example). Similar support is designed for "boilerplate" sections, chunks of text that are used again and again (campus descriptions, curriculum vitae, etc.).

It will be possible to simultaneously prepare a manuscript with tables, figures, footnotes, and any other stylistic form required in the scientific literature. One window on the display screen will contain such material. The writer may stop writing for a moment, and in a second window, conduct an analysis or review of a chosen data set, produce the analysis results, and incorporate the results into the text. Yet a third window might carry system messages and commands. A background window provides a preview of the text formatting that is ongoing as the text is entered. Available correctness checking will provide spelling and other error-fixing support (repeated words are a common problem, and syntax, grammatical construction, and document design checking are well within the realm of possibility). Dictionaries of up to 100,000 (or more) words are feasible, and discipline-specific dictionaries with esoteric argot, scientific names, and any other unusual terminology a research environment might require will be automatically developed as the system is "trained" through the production of manuscripts, proposals, etc. that a team is likely to produce. Automatic creation of tables of contents, lists of figures and tables, and referencing of equations or other items of interest is available. Specialized support for construction of complex tables and any level of mathematical notation are included.

The goal here is to support the production of single researcher or team write-ups in the form of editor-ready manuscripts, ready-to-send proposals, and similar intermediate-state writing. Access to photo-typesetting devices is available, and thus, the production of a fully proofed and ready to print book is also supported. In this case the write-up and all corrections can be handled by the author(s), and the final version transmitted electronically to the publishing outlet.

A proper writer's working environment (Reid, 1983; Lamport, 1984) will contain tools that support powerful and easy editing (Stallman, 1981): "batch" handling of an existing document (Cherry, 1981), interactive handling of an emerging document (Walker, 1981), a combination of these two approaches (Reid, 1980), and utilities (couched in a similar functional form as was described in the section on statistical analysis) that support the management of organization and repair of a document. Implicit in this is support for a team write-up that will require collation and organization after the initial draft is created. A requirement here is the availability of "easy-entry" text editors that require no system knowledge at all, thus allowing visitors and participants in a short-term workshop to contribute without having to learn very much about the computing environment they need to use. It is feasible to provide version tracking of the various drafts a document goes through on its way to publication. Adaptation of the concepts from one of the programming support packages (e.g., Feldman, 1979) is a straight-forward approach to this problem in the short-term, and a full implementation of system tracking and logging will provide the ultimate backtracking capabilities.

Text formatting services are currently being obtained from implementation of T_EX82 (Knuth, 1984). This program meets all of the extensibility criteria described above, and it allows a fully developed writing support environment to be designed and implemented. (TROFF from UNIX, or Brian Reid's SCRIBE could also be utilized, and may be offered in parallel.) Two high-level macro systems have been developed around T_EX82, one of which is a general document-formatting system (Lamport, 1984), and the other is a system that specifically deals with technical mathematical writing and is supported by the American Mathematical Society (Spivak, 1982, 1983). Hewlett-Packard has recently provided an ambitious suite of macros for T_EX (Daniels, 1984). Additional macro development in T_EX that is addressed specifically toward aspects of writing that are unique to ecological scientists will be done as part of the Science Workbench. Again, such specificity is being modularized, and the incorporation of similar capabilties for any other discipline is not a difficult task.

Budgeting and Record Keeping

This sub-environment is currently under review, and specifics are not available. What is intended is a support environment that will handle record keeping, budget development and maintenance, budget projections and review, inventory management, personnel records, audit trail records, correspondence and project documents, and any material that is properly deemed office management. Among the recent developments in computing, office automation is perhaps the most polished and sophisticated. A number of available packages will suffice for the rather mild requirements (relative to a business with a hundred employees and thousands of customers) of the typical research team environment.

The Operating System

Among the many available operating systems, one is clearly superior for the tasks required of the Science Workbench. The UNIX[1] operating system, developed at AT&T Bell Laboratories is fully extensible, and possesses sufficient power to completely serve the needs of the Science workbench. If there is an emerging standard for operating systems, UNIX comes closer than most (considering medium-size computers) to meeting the criteria. UNIX is a programmers environment and is not friendly in any usual sense of the word (e.g., Norman, 1981). It is sufficient to indicate here that UNIX is fully extensible, and that it is the intent of this project to simply hide the operating system from the average user. The full power of the programmer support utilities (etc.) within UNIX will remain intact and available. An entrance to the literature on UNIX can be obtained from Kernighan and Ritchie (1978), Ritchie (1978), Bourne (1983), McGilton and Morgan (1983), Kernighan and Pike (1984), and Sobell (1984).

It is also important that the operating system support an adequate spectrum of high-level programming languages. FORTRAN is still a good language for certain kinds of algorithm development, but more importantly, it is required for historical purposes. Other languages ultimately required for system development include LISP and PROLOG for symbolic manipulation (for coding the user interfaces), and a selection of the recent structured languages such as Pascal and Ada. C comes with the UNIX operating system. Some of these languages (especially the symbolic manipulators) are not available on all operating systems or for all hardware configurations. Some planning is thus required.

The User Interface

In the long-term, a comfortable and reasonable user-interface for the Science Workbench is the most difficult

---

[1]UNIX is a trademark of AT&T Bell Laboratories.

task of all. A fully developed interface will allow natural language (conversational) commands to be entered, will parse them into proper parts, and will "understand" (in some sense) vagaries of spelling and syntax applied by different users. This implies a level of artificial intelligence which is not available at the present time. Natural language systems do work, however, when the domain of interest is restricted (e.g., Woodyard & Hamel, 1980). For an excellent theoretical review of natural language communication and computers see Wilks (1983).

Additionally, considerable effort is being devoted to determining how to readily communicate maximum amounts of information to the user. This involves display structures (Foley & Wallace, 1974; Engel & Granda, 1975; Ball & Hayes, 1981; Knapp et al., 1982; Simpson, 1982) and the manner in which information is transferred from the computer display to the user. The Science Workbench project is currently serving as a test case for various methods of user-computer communication.

Interface Design. The interface to the workbench will benefit from the existing body of knowledge on what constitutes a good interface (Badre & Schneiderman, 1982; Thomas & Schneider, 1984). Research on natural language interfaces, expert knowledge systems, and intelligent error-handling capabilities are all current areas of interest to interface designers who wish to support user requirements. At the least, the interface for the Science Workbench will include consistent rules of interaction across the boundaries of the workbench subsystems, and on-line help texts for each application program that is pertinent to current work context. The Science Workbench will also investigate the utilization of the following more advanced interface concepts:
1. Natural language translation of user input allowing for conversational interactions.
2. Cooperative error resolution with the aid of effective error messages and possible corrections suggested (Hayes & Szekely, 1983).
3. Personalization of interactions to individual users based upon models of past interactions.
4. Expert consultation and tutoring facilities containing knowledge of the subsystems (Duda & Shortliffe, 1983; Stoddard, 1983).
The incorporation of these interface features into the Science Workbench will be brought about as a series of

separate projects, while maintaining a focus on the larger problem of an integrated, functional environment that serves users who are not expert programmers. The method for deriving a <u>user-defined</u> (Good <u>et al</u>., 1984) interface that supports conversational interactions is discussed in the following section.

<u>Restricted Natural Language Dialogues</u>.   The initial step in the interface design process is the creation of independent routines and lexicons that act as separate natural language processors for each of the applications provided. One goal is to create a systematic process for the construction of application-dependent natural language interfaces. While these independent interfaces are unique to the applications, the design process should be general enough such that only the content--and not the form--of the interface construction process changes as new applications are introduced into the system.

It is possible to avoid the difficult problem of natural language understanding by computers through the restriction of the dialogue content and its syntactic form. Computer applications that accept natural (non-formal) language input have been shown to be successful within a variety of task domains when these domains contain a relatively small and well-defined world of objects and actions. Winograd (1973) has demonstrated the dramatic effect of constraining the dialogue for a natural language system; however, the objects and actions of the Science Workbench applications are not as clearly defined or predictable as those of Winograd's blocks world. For this reason, actual user interactions with the application software will provide the framework for the interface design. Successful examples of applications that accept natural language input include an appointment calendar (Kelly, 1983), a computerized mail system (Wixon <u>et al</u>., 1983), and a clinical database system (Woodyard & Hamel, 1980). The key to success appears to be a reliance upon the language restrictions enforced by these systems.

Kelly (1983) has reported a method for empirically obtaining a restricted vocabulary and grammar through the analysis of protocols from actual users. This methodology involves the simulation of a natural language interface via the translation of user directives by a human interpreter into the required application-dependent grammar. This "hidden" interpreter (a person who is well trained in the application program's formal language) intercepts and re-codes input before it is relayed to the application program. Log file records of these interactions provide protocols from which a lexicon may be created. The analysis of proto-cols also allows the classification of users' characteristic behavior patterns which may be used to create rules for analyzing and constraining command syntax.

Kelly's strategy is an iterative procedure for the de-sign and testing of an expanding collection of lexical items and syntax rules for natural language processing. It is as-sumed that both the size of the lexicon and the number of parsing rules will reach an asymptotic level where subse-quent enhancements will provide fewer and fewer benefits relative to the amount of additional programming effort (see Kelly, 1983; Wixon et al., 1983).

The intent of using restricted natural language as an interface feature is not to restrict the communicability or creativity of users, but rather to provide a viable means of reducing the effort of forming understandable commands. While these systems appear to be well received by their users, how do the dialogue restrictions affect system and user performance? Speed of response and system performance in general should improve as the number of processing rules applied to user input is reduced. Constructing a lexicon limited in size should have a similar effect.

With regard to user performance, restricted natural language systems may also be at least as efficient as more unruly and verbose systems. Ogden and Brooks (1983) have shown the ability of users to restrict the syntactic form of their commands without increasing the time to complete data-base queries. Kelly and Chapanis (1977) have demonstrated that for human-to-human communication, limited-vocabulary dialogues are as effective as unrestricted vocabularies in that they do not increase problem solution times. These re-sults indicate that natural language processing within well defined domains is feasible when the syntax and vocabulary are constrained appropriately.

The simulation of interface software as described above will also be attempted for the implementation of other interface components such as error resolution and problem consultation. Designing the interface will also consist of attempts to model a user's knowledge of the system or appli-cation program, as well as providing the user with a de-scriptive model of the application software.

<u>The Interface as an Expert System</u>.   Since the Science Workbench is designed as a scientist's assistant to speed and ease the work of a research team, it seems useful to cast the "assistant" properties of the Science Workbench in the context of an expert system. Much of the applications software for the Science Workbench exists; the advantage of the Workbench approach lies in the ease with which a non-programmer (and the expert as well) can use the software. The utility of the Science Workbench then, will stand or fall on the ultimate capability of the human-computer inter-face.

An expert system is a knowledge base of domain-specific rules and facts, an inference mechanism, and a short-term memory, all organized to provide expertise to a user. In the context of the Science Workbench, the domain of expert-ise is the details and workings of a suite of applications packages that serve scientists with large amounts of data to comprehend and summarize.

While some expert systems have a natural language interface to their expertise, our expert systems will <u>be</u> natural language interfaces to their expertise. The domain of knowledge for a software package includes information about the capabilities and commands for the package, knowl-edge about the English language and its analogs in the com-mand language of the software package, and advice about how to get the most out of the package. Each of the Science Workbench components (database, statistical analysis, etc.) will have an expert system built around it: an interface that captures knowledge of the package, of natural language, and (eventually) of the needs and expectations of the user.

Each interface will be a natural language front-end to an applications software package and will translate English into the mnemonics of the software package. This will in-clude the mnemonics for commands as well as requests for assistance. Additional expert knowledge of the software will go into an interface <u>help</u> file to supplement the pack-age help functions. In addition, the interface will have knowledge of the workings of the package and will be able to suggest ways to better use the software.

The outer layer of the computing environment will be a high-level expert system that understands, and controls, each of the individual expert systems. This superordinate interface will be the <u>real</u> scientist's assistant--the one that determines what specific software expert to make avail-able. Rather than trying to construct a monolithic Science

Workbench expert interface to satisfy all needs, this modular approach allows for incremental development and extensibility. Software applications and their expert interfaces can be added or removed.

Interface Intelligence. Without learning an interface cannot really be called intelligent. The Science Workbench is designed to operate so that the flow of learning across the human-computer interface occurs in two directions.

The Science Workbench will attempt, in the course of a series of sessions, to implement learning in two ways:

- the system will attempt to "model the user" and use that information to derive the user's experience level and expectations in order to provide better service and support; and
- the system will attempt to draw the user toward more efficient system use by maintaining a transparent window through the interface to the software applications operating in the background.

Modeling the user of the Science Workbench entails learning about the kinds of things the user can and will want to do--establishing a user-history of the use of tools, data sets, and user-defined procedures that are needed. There is also the notion of measuring user expertise and involving this knowledge in a decision of how much detail to provide in response to a "help" query or in providing advice.

Modeling the software of the Science Workbench entails showing the user the mnemonic commands that were translated from the natural language input. Slator and Anderson (1985) argue that users will acquire the use of these mnemonics since they save keystrokes and will speed response time--and that the process of acquiring these mnemonics will be akin to the way native speakers acquire the pidgin-like speech of intelligent but inarticulate foreigners. This "foreign-speak" linguistic register will then act, on the Science Workbench, like a tutor to give the user some feedback on what the system is doing with each request. Thus, the user learns about the software being used, and can tailor requests for more efficient use of system and user time.

PROGNOSIS FOR THE FUTURE

A working natural language interface on a technical graphics package is in place as of this writing. We now believe that by restricting the domain that must be considered, natural language systems can be made to work. The Science Workbench is operative now; however, it lacks the extensive interfacing described herein. It is a non-trivial task to assure proper integration across all of the sub-environments required. Further, details of system design are being constantly redesigned as continued use dictates new approaches. In this manner, we hope to assure that the final product serves the intended audience (see Gould & Lewis, 1985, for a discussion of this process). Projecting current industry trends several years into the future leads to the conclusion that this environment can be made available for about half the cost and with about ten times the power of existing machines. Long-term research programs, with their promise for synthesis, perpetual maintenance, and continued use of the data being collected, provide an ideal test case for the computing environment we have described.

ACKNOWLEDGMENTS

This research is supported by the New Mexico State University Computing Research Laboratory, and The National Science Foundation, Grants BSR-8114466-02 and BSR-8419790.

LITERATURE CITED

Alagić, S. and M.A. Arbib. 1978. The Design of Well-Structured and Correct Programs. Springer-Verlag, NY. 292 pp.

Badre, A. and B. Schneiderman. 1982. Directions in Human-Computer Interaction. Ablex Publishing Corporation, Norwood, NJ. 225 pp.

Ball, E. and P. Hayes. 1981. A test-bed for user interface designs. Association for Computing Machinery 24: 85-88.

Becker, R.A. 1978. Portable graphical software for data analysis. pp. 92-95. In: Proceedings of the 11th Annual Symposium on the Interface; Computer Science and Statistics. A.R. Gallant and T.M. Gerig (eds.). Institute of Statistics, North Carolina State University, Raleigh, NC.

Becker, R.A. and J.M. Chambers. 1976. On structure and portability in graphics for data analysis. pp. 14-21. In: Proceedings of the Ninth Annual Symposium on the Interface, Computer Science and Statistics. D.C.

Hoaglin and R.E. Welsch (eds.). Prindle, Weber & Schmidt, Inc., Cambridge, MA.

Becker, R.A. and J.M. Chambers. 1978a. Design and implementation of the S System for interactive data analysis. Proceedings IEEE Compsac 78: 626-629.

Becker, R.A. and J.M. Chambers. 1978b. GR-Z: A System of Graphical Subroutines for Data Analysis. pp. 409-415. In: Proceedings of Computer Science and Statistics: Tenth Annual Symposium on the Interface. National Bureau of Standards Special Publication 503. National Bureau of Standards, Gaithersburg, MD.

Becker, R.A. and J.M. Chambers. 1984a. S: An Interactive Environment For Data Analysis and Graphics. Wadsworth Advanced Book Program, Belmont, CA. 550 pp.

Becker, R.A. and J.M. Chambers. 1984b. Design of the S System for data analysis. Communications of the ACM 27(5): 486-495.

Bourne, S.R. 1983. The UNIX System. Addison-Wesley Publishing Company. Reading, MA. 351 pp.

Bradford, L.P. 1976. Making Meetings Work. University Associates, La Jolla, CA. 121 pp.

Burgoon, M., J.K. Heston, and J. McCroskey. 1974. Small Group Communication: A Functional Approach. Holt, Rinehart, and Winston, Inc., NY. 214 pp.

Chambers, J.M. 1977. Computational Methods For Data Analysis. John Wiley and Sons, NY. 365 pp.

Chambers, J.M. 1978. Language design and statistical systems. pp. 21-24. In: Proceedings of the 11th Annual Symposium on the Interface. A.R. Gallant and T.M. Gerig (eds.). Institute of Statistics, North Carolina State University, Raleigh, NC.

Chambers, J.M. 1979. Designing statistical software for the new computers. pp. 99-103. In: Computer Science and Statistics: 12th Annual Symposium on the Interface. J.F. Gentleman (ed.). University of Waterloo, Waterloo, Canada.

Chambers, J.M. 1980. Statistical computing: History and trends. The American Statistician 34(4): 238-243.

Chambers, J.M., W.S. Cleveland, B. Kleiner, and P.A. Tukey. 1983. Graphical Methods for Data Analysis. Wadsworth International Group. Belmont, CA. 395 pp.

Chambers, J.M. and B. Kleiner. 1982. Graphical techniques for multivariate data and for clustering. pp. 209-244. In: Handbook of Statistics, Vol. 2. P.R. Krishnaiah and L.N. Kanal (eds.). North-Holland Publishing Company, Amsterdam.

Cherry, L.L. 1981. Comptuer aids for writers. pp 62-67. In: Proceedings of the ACM SIGPLAN SIGOA. Symposium on Text Manipulation. ACM SIGPLAN, Portland, OR.

Cleveland, W.S. 1979. Robust Locally Weighted Regression and Smoothing Scatterplots. Journal of the American Statistical Association, Theory and Methods Section 74(368): 829-836.

Cleveland, W.S. 1985. Elements of Graphing Data. Wadsworth Statistics/Probability Series, Belmont, CA. 285 pp.

Conley, M. and W. Conley. 1984. New Mexico State University College Ranch and Jornada Experimental Range: Summary of Research 1900-1983. New Mexico State University Agriculture Experiment Station. Special Report No. 56. 83 pp.

Conley, W. 1978. Population modeling. pp. 305-320. In: Big Game of North America Ecology and Management. J. Schmidt and D.L. Gilbert (eds.). Stackpole Books, Harrisburg, PA.

Conley, W. 1983. Artificial intelligence and expert systems: Perspective and intent of a computing environment for population analysis. In: Proceedings of National Workshop on Computer Uses in Fish and Wildlife Programs: A State of the Art Review. G. Cross (ed.). VPI & SU, Blacksburg, VA.

Cross, G. (ed.). 1983. Proceedings of National Workshop on Computer Uses in Fish and Wildlife Programs: A State of the Art Review. VPI & SU, Blacksburg, VA.

Daniels, S. 1984. User's Guide to the HP T_EX Macros. Hewlett-Packard Co., Cupertino, CA. Online Documentation.

Date, C.J. 1982. Data structure and corresponding operators. pp. 63-80. In: An Introduction To Database Systems, 3rd Ed. Addison-Wesley, Reading, MA.

Davis, L.N. and E. McCallon. 1974. Planning, Conducting, and Evaluating Workshops. Learning Concepts, Austin, TX. 310 pp.

Dixon, W.J. (ed.). 1974. Biomedical Computer Programs. University of California Press, Los Angeles. 773 pp.

Dixon, W.J. (ed.). 1983. BMDP Statistical Software: 1983 Printing with Additions. University of California Press, Berkeley. 735 pp.

Duda, R.O. and E.H. Shortliffe. 1983. Expert systems research. Science 220(4594): 261-268.

Engel, S.E. and R.E. Granda. 1975. Guidelines for Man/Display Interfaces. IBM No. TR 00.2720. Poughkeepsie Laboratory. 42 pp.

Feldman, S.I. 1979. Make--A program for maintaining computer programs. Bell Laboratories, Murray Hill, NJ. Software--Practice and Experience 9: 225-265.

Foley, J.D. and V.L. Wallace. 1974. The art of natural graphic man-machine conversation. Proceedings of the IEEE 62(4): 462-471.

Good, M.D., J.A. Whiteside, D.R. Wixon, and S.J. Jones. 1984. Building a user-derived system. Communications of the ACM 27(10): 1032-1043.

Gould, J.D. and C. Lewis. 1985. Designing for usability: Key principles and what designers think. Communications of the ACM 28(3): 300-311.

Hayes, P.J. and P.A. Szekely. 1983. Graceful Interaction Through the COUSIN command interface. Carnegie-Mellon University Computer Science Department, Pittsburgh, PA.

Hayes-Roth, F., D.A. Waterman, and D.G. Lenat (eds.). 1983. Building Expert Systems. Addison-Wesley Publishing Company, Reading, MA. 444 pp.

Hoaglin, D.C., F. Mosteller, and J.W. Tukey. 1983. Understanding Robust and Exploratory Data Analysis. John Wiley and Sons, NY. 447 pp.

Kelly, J.F. 1983. Natural Language and Computers: Six Empirical Steps for Writing an Easy-To-Use Computer Application. Unpublished doctoral dissertation. The Johns Hopkins University.

Kelly, M.J. and A. Chapanis. 1977. Limited vocabulary natural language dialogue. International Journal of Man-Machine Studies 9: 479-501.

Kernighan, B.W. and R. Pike. 1984. The UNIX Programming Environment. Prentice-Hall, Inc., Englewood Cliffs. 357 pp.

Kernighan, B.W. and D.M. Ritchie. 1978. The C Programming Language. Prentice-Hall, Inc., Englewood Cliffs. 228 pp.

Knapp, B.G., F.L. Moses, and L.H. Gellman. 1982. Information highlighting on complex displays. pp. 195-215. In: Directions in Human-Computer Interactions. A. Badre and B. Schneiderman (eds.). Ablex Publ., Norwood, NJ. 1982.

Knuth, D.E.   1984.   The Texbook.   Addison-Wesley Publishing Company, Reading, MA.   483 pp.

Lamport, L.   1984.   The LaTex Document Preparation System. Second Preliminary Edition. Online Documentation. Stanford Research Institute, Stanford University, Stanford, CT.   170 pp.

Lenars, M. and W. Conley.   1980.   Demographic considerations in reintroduction programs of bighorn sheep (Ovis). Acta Theriologica 25(7): 71-80.

Lenars, M. and W. Conley.   1982.   Reproductive gambling in bighorn sheep (Ovis): A simulation.   Journal of Theoretical Biology 98: 1-7.

McGilton, H. and R. Morgan.   1983.   Introducing the UNIX System. McGraw-Hill Book Company, NY.   556 pp.

Neely, J. and S. Stewart.   1981.   Fundamentals of relational data organization.   BYTE 6(11): 48-60.

Norman, D.   1981.   The trouble with UNIX.   Datamation Nov. 1981: 138-150.

Ogden, W.C., and S.R. Brooks.   1983.   Query languages for the casual user: Exploring the middle ground between formal and natural languages.   pp. 161-165.   In: Proc. CHI'83 Human Factors in Computing Systems Conference. ACM SIGCHI, Boston.

Oppenheimer, C.H., D. Oppenheimer, and W.B. Brogden.   1976. Environmental Data Management.   Plenum Press, NY.   244 pp.

Reid, B.K.   1980.   A Document Specification Language and its Compiler.   Ph.D. dissertation.   Carnegie-Mellon University, Pittsburgh.

Reid, B.K.   1983.   A writer's work station: Support systems for preparing documentation.   IEEE 83: 522-524.

Rice, J.R.   1983.   Numerical Methods, Software, and Analysis: IMSL Reference Edition.   McGraw-Hill Inc., NY. 661 pp.

Ritchie, D.M.   1978.   UNIX Time-sharing System: A Retrospective.   American Telephone and Telegraph Company. The Bell System Technical Journal 57(6): 1947-1969.

Ross, R.G.   1981.   Data Dictionaries and Data Administration. American Management Association, NY.   454 pp.

SAS Institute Inc.   1981.   SAS/Graph User's Guide.   Cary, NC.   126 pp.

SAS Institute Inc.   1982a.   SAS User's Guide: Basics. Cary, NC.   923 pp.

SAS Institute Inc.   1982b.   SAS User's Guide: Statistics. Cary, NC.   584 pp.

SAS Institute Inc. 1983. SAS Companion for the VM/CMS Operating System. Cary, NC. 161 pp.

Schmid, C.R. 1983. Statistical Graphics. John Wiley and Sons, NY. 212 pp.

Simpson, H. 1982. A human-factors style guide for program design. BYTE April: 108-132.

Slator, B. and M. Anderson. 1985. Pygmalion at the Interface: Impatient Users and Foreign-speak. Computing Research Laboratory Memoranda # 26, April. New Mexico State University.

Sobell, M.G. 1984. A Practical Guide to the UNIX System. The Benjamin/Cummings Publishing Company, Inc., Menlo Park, CA. 428 pp.

Spivak, M. 1982. The Joy of Tex. Tex User's Group American Mathematics Society, Providence, RI.

Spivak, M. 1983. Summary of AMSTEX. Online documentation. Stanford Research Institute, Stanford University, Palo Alto.

Stallman, R.M. 1981. EMACS: The extensible, customizable, self-documenting display editor. pp. 147-160. In: Proceedings of the ACM SIGPLAN SIGOA. A Symposium on Text Manipulation. ACM SIGPLAN, Portland, OR.

Sterling, B., W. Conley, and M. Conley. 1983. Simulations of demographic compensation in coyote populations. Journal of Wildlife Management 47(4): 1177-1181.

Stoddard, M. 1983. Modelling the user in intelligent user interfaces: The UNIX computer consultant. A colloquium presented at New Mexico State University. Las Cruces, NM.

Thomas, J.C. and M.L. Schneider. 1984. Human Factors in Computer Systems. Ablex Publishing Corporation, Norwood, NJ. 276 pp.

Trecker, H.B. and A.R. Trecker. 1979. Working With Groups, Committees and Communities. Follett Publishing Company, Chicago. 225 pp.

Tukey, J.W. 1977. Exploratory Data Analysis. Addison-Wesley. Reading, MA. 688 pp.

Velleman, P.F. and D.C. Hoaglin. 1981. Applications, Basics, and Computing of Exploratory Data Analysis. Duxbury Press, Boston.

Wainer, Howard. 1984. How to display data badly. The American Statistician 38(2): 137-147.

Walker, J.H. 1981. The document editor: A support environment for preparing technical documents. pp. 62-67.

In: Proceedings of the ACM SIGPLAN SIGOA. Symposium on Text Manipulation. ACM SIGPLAN, Portland.

Watts, T. and W. Conley. 1981. Extinction probabilities in a remnant population of Ovis canadensis mexicana. Acta Theriologica 26: 393-405.

Watts, T. and W. Conley. 1984. Reproductive potential and theoretical rates of increase for feral goat (Capra hircus) populations. Journal of Wildlife Management 48(3): 814-822.

Wexelblat, R.L. 1978. What is a language for statistical computing? pp. 25-37. In: Proceedings of the Computer Science and Statistics: Eleventh Annual Symposium on the Interface. A. Ronald Gallant and Thomas M. Gerig (eds.). North Carolina State University, Raleigh.

Wilks, Y. 1983. Machine translation and the artificial intelligence paradigm of language processes. pp. 61-111. In: Computer in Research 2. W.A. Sedelow and S.Y. Sedelow (eds.). Walter de Gruyter and Co., Berlin, NY.

Winograd, T. 1973. A process model of natural language understanding. pp. 152-186. In: Computer Models of Thought and Language. R.C. Schank and K.M. Colby (eds.). W.H. Freeman, San Francisco.

Wixon, D., J. Whiteside, M. Good, and S. Jones. 1983. Building a user-defined interface. In: Proc. CHI'83 Human Factors in Computing Systems Conference. Boston.

Woodyard, M. and B. Hamel. 1980. A natural language interface to a clinical data base management system. pp. 41-62. In: Computers and Biomedical Research. R.W. Stacy and Bruce D. Wasman (eds.). Academic Press, Inc., New York.

Zelkowitz, M.V., A.C. Shaw, and J.D. Gannon. 1979. Principles of Software Engineering and Design. Prentice Hall, Inc., Englewood Cliffs, NJ. 338 pp.

SUMMARY AND FUTURE DIRECTIONS
OF RESEARCH DATA MANAGEMENT
IN ECOLOGY

G. Richard Marzolf and Melvin I. Dyer

THE METHOD OF SCIENCE

Philosophers of science identify the basis of science as "that which our senses tell us is the truth" or "reality is what our senses perceive" (i.e., a keen eye is essential and often sufficient). The validation of this statement takes various forms, but the most powerful is the ability to predict. If what our senses have revealed is true, then, armed with that truth or understanding, we should be able to predict the course of natural events. In some cases, prediction is possible with few formalized statements of understanding. Furthermore, the definition of prediction has been stretched to include spatial aspects, for example, statements about the state of nature in other places, and "post-dictions" about events in the pre-historic past given that a defined set of circumstances is met (Toulmin, 1961). This construct forms the basis for the use of experiment as a means of learning.

Technology has increased the extent and acuity of the human senses, sometimes to an extraordinary degree. Microscopy, photography, thermometry, sonar, gas spectroscopy and isoptopic tracers are a few common examples of technical advances that have influenced ecology. Figure 1 illustrates the sequence of physical and mental processes that is called science; asterisks identify where modern research data management (RDM) has the potential to have its greatest impact by further extending human capability.

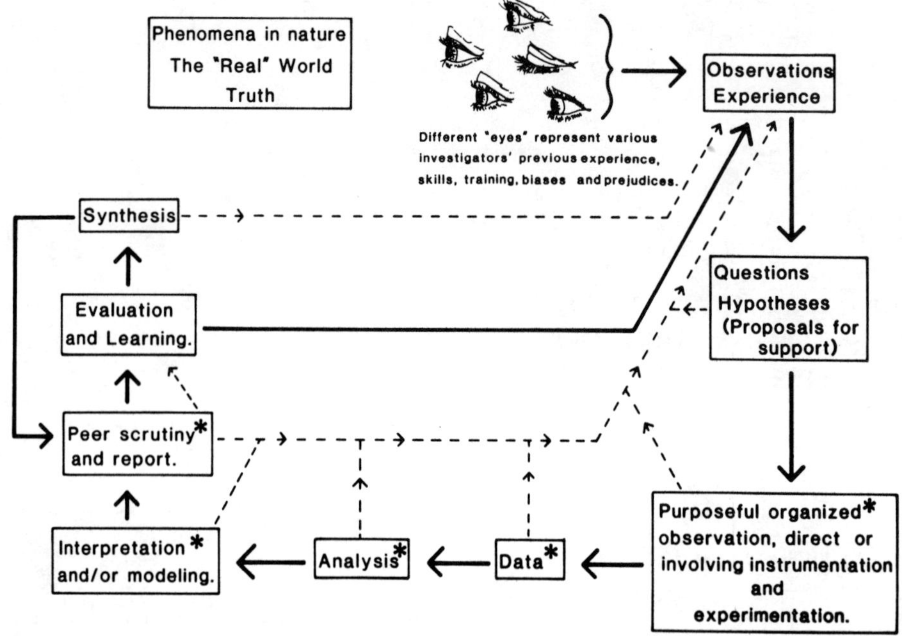

Fig. 1. An illustration of the scientific process. The compartments marked with asterisks are places where research data management is most likely to influence the efficiency and productivity of scientists.

OBSERVATIONS AND EXPERIENCE

Initial observations of nature are casual, driven by awareness or curiosity. Rarely do two observers see nature the same way since they have had different interests, experiences, and talents. These led them to different educational experiences and training disciplines (development of different skills) and so to different biases and prejudices. When complex phenomena in nature are the objective of inquiry, there is a great opportunity for these differences to persist, thus there is more progress to be made by overcoming impediments to useful communication. There has not been much opportunity, at this early level of scientific inquiry, for RDM to make substantive changes. There is, however, great opportunity to expand our powers of observation because data required to drive complex models can be collected, managed, and used. Some ideas being addressed today simply could not have been developed or would not be considered without the ability to handle large data sets.

QUESTIONS, HYPOTHESES, AND SUPPORT FOR SCIENCE

Observation and scholarship lead to questions and planning for formal investigations. Plans for formal investigations of ecosystems are being presented more often in hierarchical sequences (Allen & Starr, 1982). The formation of research teams to conduct ecosystem investigation have not often been organized--at least not in the context of broad hierarchical approaches.

Execution of investigation is usually preceded by a request for financial support. Such proposals are one of the earliest points in the scientific process where written feedback from peers can influence an investigator's experience and, therefore, his approach. If new questions can now be approached, then this phase of investigation is where modern RDM techniques might now substantially alter how ecology can progress.

PURPOSEFUL OBSERVATION AND EXPERIMENT

The degree of organization at this stage of an investigation is often critical to the success of subsequent analyses and interpretation. Here, again, research data management provides significant contributions (e.g., interfaces with analytical instruments, field data loggers, optical scanners for digitizing maps, etc.). Introducing RDM techniques at this point in the planning process substantially improves the efficiency with which an investigation might be conducted. New cartographic techniques and uses of remote sensing (Klemas, this volume; Seilkop, this volume) will lend themselves well to ecological analyses of spatial patterns and to corrective decisions on environmental issues.

ANALYSIS

The astonishing power of modern analytical software is a potential danger because it allows conceptual errors to be made precisely and rapidly. In the past, continual reevaluation was common since we performed many calculations by hand (see Hurlbert, 1984, for examples of recent misapplication of analytical inference). Analytical procedures should not be confused with RDM, though RDM has the potential to alter mental processes in a substantial way because it forces closer attention to data formats and encourages error-checking procedures, thus increasing reliability of large databases.

The challenges of handling ecological data are being addressed creatively, with important developments in collaborative "workbench" ideas (Conley et al., this volume) that utilize the skills and talents of several people to probe complex problems. Attempts to develop software that is more easily usable by scientists (Russell, this volume) promise to heighten the potential for application of these rapidly developing techniques. A request for sharing specific software, especially software developed for microcomputers, was also heard at this symposium (see also Inouye, 1985).

INTERPRETATION

The ability to rapidly manipulate, merge, compare, overlay, or otherwise arrange data for evaluation increases the likelihood that new insights will occur. Interpretation is served further by graphic representations of information since it is easier to inspect relationships and search for pattern(s). Once conclusions have been drawn, imaginative graphics are useful for communication of the new knowledge to the extent that RDM practices make this technically possible; this is an important aid to interpretation and communication.

PEER SCRUTINY AND REPORT

Peer scrutiny of written reports submitted for publication is not likely to be influenced by RDM, though electronic technology is rapidly changing the rate of publication and the way that scientists are communicating published information.

CONCLUSIONS

Long-term ecological data from well designed research projects deserve careful handling. If managed properly, the value of such ecological data will increase. Emerging techniques for better handling and storage of data also should increase its accessibility, making it more understandable and more easily analyzed by the original collectors and by other investigators with novel ideas about ecological phenomena. Future needs are being anticipated as scientists and their parent institutions accept this archival burden. This burden can be minimized with careful and attentive application of data management principles, setting the stage for new questions.

Techniques and protocols for research data management in ecology have been discussed by Zinnel and Marozas (this volume). Examples of how data management problems are identified and approached at large installations and small biological field stations were given by Gurtz (this volume). The applications of these techniques and of the technology that supports them promises to make this facet of ecology robust and dynamic.

The separateness of data managers and scientists is a recurrent theme. This separation may be a temporary condition; disappearing as those who now call themselves data managers become more involved in research and as newly trained ecologists with RDM skills enter the research community. This separation was created originally by the increased demands on ecology in the late 1960's and rapid and diverse technological developments in the 1970's; many scientists, already in the middle and late stages of their careers, were not trained to use RDM tools. These scientists, too busy to sort through the available technology or to keep up with the rapid changes, have made the expedient choice: they have simply hired younger scientists with the appropriate skills as the data managers.

Formal research data management is only beginning to make substantive alterations in the way that ecology is conducted, with the possible exception of increased use of the enormous databases assembled from the monitoring programs of government agencies.

CHALLENGES

Libraries constitute the conventional mechanism for storage and maintenance of reported new knowledge, the primary literature. The purchasing and collection policies of libraries affect how data are managed, at least in the traditional end product form, the scientific publication. Scientific societies, supporting the publication of journals, play a critical role as part of the library's infrastructure. The _quid_ _pro_ _quo_ for the more costly library subscription rates is the assurance of quality scholarship resulting from the peer reviews performed by the society's editorial board and it's members serving as _ad_ _hoc_ reviewers.

Scientific databases (raw data, data summaries, etc.) associated with research sites representing sources of information potentially accessible by researchers are not submitted to the same level of peer scrutiny. This is potentially one of the weakest points in the development of a robust system. Currently, there is no mechanism for the review of data, or its documentation, before traditional publication is sought. Quality assurance techniques may be applied unevenly or not at all, reasons for collecting the data may not be carefully articulated, the methods may not be well documented, collectors of the original data may be unknown, or important corollary information may be unavailable. It is a substantial challenge to those taking responsibility for archiving long-term data sets to develop mechanisms for review of these data, especially if they are to be made generally available to the research community.

In most cases, peer reviewers require careful statements of the objectives of the investigations. Without knowing the questions to be answered with a given data set, it will be nearly impossible to judge their value (quality for that purpose). This may serve to make peer review of databases a moot issue, but achieving the goal of greater accessibility of ecological data carries with it a caution about how those data might be used. This may become a major difficulty, and overcoming it a major challenge, for science and data managers in the near term.

A challenge that is more remote, but not less significant, is that of finding ways to use the large environmentally oriented databases that have been developed over approximately two decades in government agencies and at national laboratories (e.g., the Geoecology Data Base at Oak Ridge [Olson et al., 1980], National Weather Service data, NOAA's NEDRES [see Freeman, this volume], NASA's LANDSAT images, EPA's water quality databases). These data serve various legislative objectives or agency mandates (i.e., weather prediction, policy setting, warning of environmental dangers). These data represent massive amounts of information about the Earth's environmental states, all of which represent large-scale conditions mediated by ecological processes that are usually studied at a smaller scale.

RDM problems encountered by these two different ecologically oriented approaches (site specific and agency level) differ in their requirements for solution (Krummel & Dyer, 1984). For instance, processes under investigation at ecological field research sites (e.g., Long-Term Ecological Research sites) do not receive the same level of capital investment and need not have the same degree of accessibility as the larger, well endowed programs, but the intensive ecological research at LTER sites precludes many uncertainties about the accuracy of data. Nevertheless, the utility

of the results of site-related LTER for application to the
larger spatial scale remains uncertain. On the other hand,
large databases, managed by government to address environmental problems, contain a great many uncertainties about
their accuracy, but their utility is clearer. The intel-
lectual distance between these scales of inquiry is ill
defined, but it is large, and the theoretical mechanisms for
melding them for useful interpretation are only beginning to
emerge.

RATIONALE

Ecology is at an important juncture. An opportunity to
solve a significant number of problems lies with the creative use of ecosystem preserves dedicated to ecological investigation. The LTER sites are examples of places where
this opportunity is most clear, but the opportunity is not
limited to this subset of biological field stations. Scientists responsible for the development of research at such
sites are sensitive to the needs for long-term data sets
(Likens, 1983a & b). The need to archive data and to document them carefully is now understood and possible; that is,
the role of modern research data management is clear.
There are many ways to view ecosystem investigation
(contrast the viewpoints of MacMahon et al., 1978, and Allen
et al., 1984), but unless coordinated attention is devoted
to all aspects that contain spatial and temporal elements
(i.e., some hierarchical association), progress will be
slow.
As RDM use expands we are confident that it will make
ecologists more efficient, more responsible in terms of the
archival mandate that accompanies long-term data sets, and
perhaps more productive, especially in terms of achieving
syntheses of data from disparate disciplines working at different scales. Maximizing the improvement in efficiency,
reliability and archival certainty with RDM is the goal, and
in doing so we want to improve the capacity for understanding ecosystem function and for dealing with complex environmental problems.

ACKNOWLEDGMENTS

Accepting the assignment to comment on the future of
data management at this meeting was audacious. In a field
moving so rapidly, the risk of falling short made the task
uncomfortable. Paul Risser and Jerry Franklin made useful

comments on the original manuscript. Marsha Conley, Susan
Stafford and Bill Michener helped with the organizational
suggestions for later drafts. They all reduced the risk and
ultimately made the assignment much more comfortable for us.
We acknowledge the support of the National Science Foundation grant No. 8012166 to Kansas State University and of Oak
Ridge National Laboratory, operated by Martin Marietta
Energy Systems, Inc. under contract No. DE-AC05-840R21400
with the Department of Energy.

LITERATURE CITED

Allen, T.F.H. and B. Starr. 1982. Hierarchy: Perspectives
    for Ecological Complexity. University of Chicago
    Press, Chicago.

Allen, T.F.H., R.V. O'Neill, and T.W. Hoekstra. 1984. In-
    terlevel Relationships in Ecological Research and Man-
    agement: Some Working Principles from Hierarchy The-
    ory. US Forest Service General Technical Report
    RM-110. Rocky Mountain Forest and Range Experiment
    Station, Ft. Collins, CO. 11 pp.

Conley, W., B.M. Slator, M.P. Anderson, and R.A. Sitze.
    1985. Designing a Scientific Problem-Solving En-
    vironment: The NMSU Science Workbench. pp. 383-409.
    In: Research Data Managment in the Ecological Sci-
    ences. W. Michener (ed.). Belle W. Baruch Library in
    Marine Science, No. 16. University of South Carolina
    Press, Columbia.

Freeman, R.R. 1985. The National Environmental Data Refer-
    ral Service: A publicly available on-line data cata-
    log. pp. 143-154. In: Research Data Management in
    the Ecological Sciences. W. Michener (ed.). Belle W.
    Baruch Library in Marine Science, No. 16. University
    of South Carolina Press, Columbia.

Gurtz, M.E. 1985. Development of a research data manage-
    ment system: Factors to consider. pp. 23-38. In:
    Research Data Management in the Ecological Sciences.
    W. Michener (ed.). Belle W. Baruch Library in Marine
    Science, No. 16. University of South Carolina Press,
    Columbia.

Hurlbert, S.H. 1984. Pseudoreplication and the design of
    ecological field experiments. Ecol. Monogr. 54(2):
    187-211.

Inouye, D. 1985. Technological tools of interest to ecolo-
    gists. Bull. Ecol. Soc. Amer. 66(1): 31-32.

Klemas, V. 1985. Remote sensing of coastal and marine resources: Data analysis and management. pp. 279-300. In: Research Data Management in the Ecological Sciences. W. Michener (ed.). Belle W. Baruch Library in Marine Science, No. 16. University of South Carolina Press, Columbia.

Krummel, J.R. and M.I. Dyer. 1984. Consumers in agroecosystems: A landscape perspective. pp. 55-72. In: Agricultural Ecosystems: Unifying Concepts. R. Lowrance, B.R. Stinner, and G.J. House (eds.). John Wiley and Sons, NY.

Likens, G.E. 1983a. A priority for ecological research, address of the past president. Bull. Ecol. Soc. of Amer. 64(4): 234-243.

Likens, G.E. 1983b. Beyond the shoreline: A watershed-ecosystem approach. Verh. Internat. Verein. Limnol. 22: 1-22.

MacMahon, J.A., D.L. Phillips, J.V. Robinson, and D.J. Schimpf. 1978. Levels of biological organization: An organism centered approach. Bioscience 28: 700-704.

Olson, R.J., C.J. Emerson, and M.K. Nungesser. 1980. Geoecology, a county level environmental data base for the conterminous United States. Environmental Sciences Division Publ. No. 1537. Oak Ridge National Laboratory, Oak Ridge, TN.

Russell, C.H. 1985. Advances in scientific software. pp. 373-381. In: Research Data Management in the Ecological Sciences. W. Michener (ed.). Belle W. Baruch Library in Marine Science, No. 16. University of South Carolina Press, Columbia.

Seilkop, S.K. 1984. Statistical interpolation procedures for mapping ecological data. pp. 227-246. In: Research Data Management in the Ecological Sciences. W. Michener (ed.). Belle W. Baruch Library in Marine Science, No. 16. University of South Carolina Press, Columbia.

Toulmin, S.E. 1961. Foresight and Understanding: An inquiry into the aims of science. Indiana University Press.

Zinnel, K.C. and M.F. Marozas. 1985. Computer data entry techniques used in scientific applications. pp. 61-72. In: Research Data Management in the Ecological Sciences. W. Michener (ed.). Belle W. Baruch Library in Marine Science, No. 16. University of South Carolina Press, Columbia.

INDEX